# Precision Medicine in Clinical Practice

Mandana Hasanzad
Editor

# Precision Medicine
# in Clinical Practice

 Springer

*Editor*
Mandana Hasanzad
Medical Genomics Research Center
Tehran Medical Sciences
Islamic Azad University
Tehran, Iran

ISBN 978-981-19-5084-1      ISBN 978-981-19-5082-7   (eBook)
https://doi.org/10.1007/978-981-19-5082-7

This Springer imprint is published by the registered company Springer Nature Singapore Pte Ltd.
The registered company address is: 152 Beach Road, #21-01/04 Gateway East, Singapore
189721, Singapore

*To my parents,*

*Source of encouragement, support, and love*

# Acknowledgments

The production of this book is truly a collaborative effort. I am greatly indebted to a variety of people. I am truly grateful for their time and efforts.

I would also like to extend my special thanks to my colleagues in the Personalized Medicine Research Center, Endocrinology and Metabolism Clinical Sciences Institute, Tehran University of Medical Sciences, Tehran, Iran.

There are many people at Springer Nature whom I deeply respect and whose efforts are amazing.

# Contents

# Principles of Precision Medicine

Bagher Larijani ⓘ, Hamid Reza Aghaei Meybodi ⓘ,
Negar Sarhangi ⓘ, and Mandana Hasanzad ⓘ

**What Will You Learn in This Chapter?**
Healthcare is quickly moving toward precision medicine, which appears to offer a better understanding of human physiology through genetic knowledge and insight and technological advancements. Precision medicine is necessary to alleviate unnecessary adverse reactions to medical care which can result from the current one-size-fits-all approach, technologies that encourage the healthcare ecosystem to develop and deliver genetic-based care and manage customizations. Accordingly, this chapter has a special focus on the introducing of precision (personalized) medicine. At the beginning of this chapter, the general definition of synonymous terms of personalized medicine will be discussed.

B. Larijani · N. Sarhangi
Personalized Medicine Research Center,
Endocrinology and Metabolism Clinical Sciences
Institute, Tehran University of Medical Sciences,
Tehran, Iran

H. R. Aghaei Meybodi
Endocrinology and Metabolism Research Center,
Endocrinology and Metabolism Clinical Sciences
Institute, Tehran University of Medical Sciences,
Tehran, Iran

M. Hasanzad (✉)
Personalized Medicine Research Center,
Endocrinology and Metabolism Clinical Sciences
Institute, Tehran University of Medical Sciences,
Tehran, Iran

Medical Genomics Research Center, Tehran Medical
Sciences, Islamic Azad University, Tehran, Iran

At the end of this chapter, precision medicine and evidence-based medicine will be addressed to provide the future of the medical practice.

**Rationale and Importance**
Precision medicine is an emerging medical practice that utilizes an individual's genetic profile to direct decisions taken in the field of disease prevention, prediction, and personalized treatment.

Precision medicine, because it is concentrated on the unique genetic makeup of each patient, is beginning to overcome the limitations of conventional medicine. It is increasingly enabling healthcare providers to shift the emphasis on medicine from response to the prevention and also the prediction of the disease susceptibility, especially in common diseases.

Precision medicine improves the health impact of existing treatments by enhancing the matching process between patients and treatments and by improving patient understanding of the risk of serious side effects.

The rationality behind precision medicine is understanding the different genetic backgrounds which have impacts on the response of individuals to therapeutic interventions.

The concept of "one medicine for all patients with the same disease" does not hold, and a more individualized approach is needed because of significant individual variation; some individuals show no response, while others show a strong response.

## 1.1    An Introduction to Precision Medicine in Clinical Practice

By the mid-twentieth century, health professionals had developed a certain kind of individualized approach to the treatment of patients. The rise of genetics came in the twentieth century. The huge amount of scientific discoveries made in the field of genomics has been supposed to allow the personalized/precision medicine approach to move from a previously hopeful dream to an effective truth. Precision medicine became more meaningful at the beginning of the twenty-first century with the integration of the Human Genome Project, which leads to the transformation of personalized medicine from an idea to a reality. The project adopted a new approach linking the genetic makeup of individuals and their health [1–4].

Genomics many claims revolutionize the medical practice and healthcare by enabling early diagnosis and disease management to be more precisely targeted at each patient.

In the evidence-based medicine approach, many existing drugs are authorized and developed based on their efficiency in a large population of individuals, but future medicines are developed as personalized solutions to the needs of a particular patient.

Each individual has a highly specific genomic, transcriptomic, proteomic, and metabolic profile which can contribute to specific pathological symptoms of disease, response to treatment, and disease severity.

Clinical practice encounters major challenges, including emerging rapidly spreading new infectious diseases, fast-growing common diseases such as type 2 diabetes (T2D) and cancers, changes in the clinical manifestations of some diseases in the treatment process (e.g., drug-resistant or adverse drug reactions), and population shifts (i.e., aging).

In the context of common disorders, the conventional "one-drug-fits-all" approach involves trial and error before effective treatment is established. And clinical trial data for a new drug shows only the average response of the study group.

The theory and statement for personalized medicine have attracted the greatest attention among many exciting fields. The individualized approach appears to be a critical feature of healthcare in the coming decades. Personalized/precision medicine aims to improve treatment outcomes through new molecular taxonomy for disease and reduce adverse drug reactions (ADRs) that affect both the clinicians and the patients.

The potential of precision medicine applies to all clinical disciplines including oncology, cardiology, and all stages of the disease development that many benefits for patient care have been mentioned. Some evidence-based examples are briefly explained in each chapter of this book.

## 1.2    Definition of Precision Medicine

The personalized medicine idea is not new and is traced back at least to the time of Hippocrates. He believes in the individuality concept of disease. He said, "It is even more essential to remember what kind of individual the disease has than what kind of disease the individual has" [5].

An early example of personalized medicine was the first known blood compatibility test for transfusion using blood typing methods, the genetic basis in favism, and cytochrome P450 2D6 function determination [6, 7].

The term personalized medicine appeared in a publication that discussed the change in the role of family physicians in the modern world of medicine and technology. The personalized medicine approach is considered as an art of the medicine [8].

The role of pharmacogenetics (as a part of personalized medicine) in clinical practice has been published by Gupeta et al. [5]. No officially recognized consensus on the definition of personalized medicine exists. The term "personalized medicine" appeared in the literature in MEDLINE in 1999 by an article entitled "New Era of Personalized Medicine: Targeting Drugs for Each Unique Genetic Profile" [9].

Various terms, including personalized medicine, precision medicine, p4 medicine, individualized medicine, and stratified medicine, have been used interchangeably, to describe the concept of personalized medicine.

In particular, the concept of targeted therapy has had a significant impact in one area of discipline, called oncology. The National Cancer Institute (NCI) provides personalized medicine definition as follows: "A form of medicine that uses information about a person's genes or proteins to prevent, diagnose, or treat disease." In cancer, personalized medicine uses specific information about a individual's tumor to help make a diagnosis, plan treatment, find out how well the treatment is working, or make a prognosis.

Examples of personalized medicine include using targeted therapies to treat specific types of cancer cells, such as HER2-positive breast cancer cells, or using tumor marker testing to help diagnose cancer, also called precision medicine [10].

The Personalized Medicine Coalition has defined personalized medicine as "the application of genomic and molecular data to better target the delivery of health care, facilitate the discovery and clinical testing of new products, and help determine a person's predisposition to a particular disease or condition" [11].

In several reports, personalized medicine is defined by emphasizing the signature of genetics for getting more effective therapies as well as disease predisposing and early interventions that might prevent disease or delay disease progression [12].

In a brief definition, personalized medicine is the selection of the right drug at the right dose for the right patient at the right time [13].

Several stages were described for personalized medicine in different studies: patient's risk analysis to allow early detection and/or prevention; increased diagnostic accuracy by better definition of diseases and phenotype description, targeted treatment pharmacogenomics advancement, integration of genomics data and its derivatives, including transcriptomics, proteomics, metabolomics with clinical health records, evaluation of clinical outcomes, and the infectious environment and its various properties [14–17].

Individualized medicine described individual drug metabolism in the context of pharmacogenomics some contexts, including gene therapy, stem cell therapies, and cancer vaccines are considered in the individualized medicine definition [18–22].

Precision medicine is defined for the first time by having three main characteristics: the ability to recognize the presence of these causal elements of disease, a knowledge of what causes a disease, and the ability to treat the origin or causes efficiently [23].

In 2011, the National Research Council of the US National Academies in their report "Toward Precision Medicine" was provided a particular definition for precision medicine as precise disease taxonomy based on molecular data. The potential of genomics as an emerging technology was introduced for the investigation of the molecular features of the disease [24].

Precision medicine is defined as a state-of-the-art molecular profiling, which helps establish accurate diagnostic, prognostic, and therapeutic approaches tailored to a patient's needs [25].

Trusheim and colleagues defined stratified medicine as "where therapies are matched with specific patient population characteristics using clinical biomarkers" [26].

The clinical biomarker which could be used in diagnosis and targeted therapy as an example of stratified medicine was seen in BCR-ABL-positive tyrosine kinase genotype in chronic myeloid leukemia patients who are likely to respond to imatinib (Gleevec®), an inhibitor of this kinase [26].

P4 medicine was introduced by the development of the systems biology approach. The term p4 medicine as an alternate term of personalized medicine stands for its personalized, predictive, preventive, and participatory features by applying the "omics" approach, i.e. genomics, transcriptomics, and proteomics [27].

Pharmacogenomics (PGx) is one of the most important components of personalized medicine which focuses on the association between genetic variations and drug response [28].

The role of genetic variants in the modulation of variability in drug actions was proposed by the physician-scientist Sir Archibald Garrod around the year 1900. He presented the term "chemical individuality" [28].

## 1.3 Pharmacogenomics

The history of the genetic basis for drug response phenotypes dates back to the early 1950s. The word "pharmacogenetics" was first coined by Friedrich Vogel in Heidelberg, Germany, in 1959 [28]. The antimalarial drug primaquine causes acute hemolytic crises in individuals with the glucose 6-phosphate dehydrogenase (G6PD) deficiency [28]. Another adverse drug reaction was reported for succinylcholine that is administered as anesthesia. A genetic variant in gene encoding pseudocholinesterase causes adverse drug reactions like apnea [28].

The term pharmacogenomics is now used to explain how multiple genetic variants across the genome (DNA and RNA) can affect drug response, while pharmacogenetics is the study of DNA variations related to drug response [29]. The drug response pathway is performed through pharmacokinetics (PK) which is drug absorption, distribution, metabolizing, and elimination, or pharmacodynamics (PD), which is modifying drug target or by disrupting the biological pathways that shape a patient's pharmacologically sensitive [30].

Pharmacogenomics information can help physicians decide medication selection, dose adjustment, and treatment period and prevent adverse drug reactions. Furthermore, pharmacogenetics can contribute to the development of new therapeutic agents [30–33]. Only 30–60% of prescriptions are clinically successful, and 7% of all hospital admissions are partly related to adverse drug reactions every year.

Every gene contains single nucleotide polymorphisms (SNPs) that occur throughout the human genome in every 1000–3000 base pairs [34]. It is shown that certain genetic polymorphisms are associated with anticancer drugs response.

Variability of drug response is a major concern which is indicated in the era of personalized medicine. Over the last decade, significant progress has been made in our knowledge of the contribution of genetic differences in pharmacokinetics and pharmacodynamics to interindividual variability in drug response.

The human genome consists of nearly 20,000 protein-coding genes. Maybe the most common variations are SNPs, described as single-base differences that exist between individuals. More than 22 million SNPs have been observed in the human genome [35].

SNPs which result in the substitution of amino acids are referred to as non-synonymous. Non-synonymous SNPs that exist in coding regions of the gene (e.g., exons) may have an impact on protein's activity and show a considerable impact on drug responses which can affect protein metabolism and transport.

Synonymous polymorphisms do not result in the substitution of amino acids; however, those happening in the gene regulatory region (e.g., promoter region, intron) may change the gene expression pattern and the amount of protein. Other types of variations that may impact gene expression or protein conformation include insertion-deletion of polymorphisms (indels), copy of number variants (CNVs), and short tandem repeats (STR) [36].

## 1.4 Precision Medicine in Clinical Practice

The precision medicine have emerged that would bring approximately dramatic changes in healthcare systems.

The concept of precision medicine is relatively new but appears to hold promising results. Some of the potential benefits of precision medicine are discussed below [4, 37–39].

### 1.4.1 The Effectiveness of Care

Currently, physicians do not fully understand how certain treatments will affect a particular patient. With precision medicine, medical providers can apply personalized treatment methods for each of their patients, thus increasing the likelihood of recovery.

## 1.4.2   Preventive Medicine

Early diagnosis of genetically caused disease is possible through genetic screening methods, and prevention of such disease is possible by an understanding of individual risk.

## 1.4.3   Cost-Effectiveness

Precision treatment by pharmacogenetics approach can reduce the cost of care, and increasing the chance of more effective treatments and decreasing adverse drug reactions.

## 1.4.4   New Taxonomy

Precision medicine proposes a new classification for diseases and categorizes them by genetic variations rather than symptoms.

## 1.4.5   Population Healthcare

The study of genetic patterns in the population as a whole can help identify and develop the causes of particular diseases and develop specific treatments. Consequently, precision medicine can reduce trial and errors in the clinical practice and take into account preventive measures for common diseases.

Eventually, precision medicine is aimed to be used in prevention and personalized treatment approaches for all health problems. At present, its daily application in many disease states is relatively growing. Precision medicine is already routinely used in certain areas of medicine, such as cancer care.

Precision medicine components in use today usually involve the following:

*Genomic testing (sometimes referred to as molecular or genetic testing)* aims to identify alterations in disease-related chromosomes, genes, or proteins. Genomic and technological methods have enabled the potential to rapidly test biological specimens for mutations of interest at a significantly reduced cost to the patient. Although the first human genome has taken more than 10 years to complete, commercial companies are now offering testing of target genes with turnaround times of days to weeks. These tests are commonly provided as a panel of targeted genes with gene coverage generally ranging from assessment of hotspot regions (well-characterized mutational sites within the gene) to full gene sequencing. Analysis of protein expression by immunohistochemistry or panels could also be used to evaluate molecular aberration. Despite technological developments, genomic testing still has a considerable economic burden on the patient [40, 41]. Targeted therapies affect specific disease-driven molecules, such as cancer molecules that promote angiogenesis or affect cell growth and tumor progression [42, 43].

*Targeted treatments* are drugs that interfere with specific genes (molecular targets) involved in a given disease. Numerous targeted therapies are being established and authorized for cancer treatment management. The majority of drugs currently available fall into two different types of drugs [44, 45]:

- Commercial antibodies aimed to track particular protein targets in cancer cells or other related cells are named monoclonal antibodies.
- Chemicals targeting specific molecules or pathways are called small-molecule drugs.

*Genetic markers* are genetic characteristics that provide information about an individual (such as the risk for disease or likelihood to respond to a particular treatment).

Genomic information has been used in the delivery of healthcare, and now genetic markers for prevention, treatment, and survival have been identified [46]:

- Risk markers help screen patients efficiently.
- Prognosis markers help us understand who is at risk of rapid progression, recurrence, and

outcomes based on their genetic makeup, not on the treatment chosen.

- Predictive markers help guide medical decisions, such as adverse drug reactions (e.g., pharmacogenomics).
- Response markers determine the patient's response to a specific treatment.
- Recurrence and toxicity markers affect the long-term quality of life of a patient once cancer treatment has been completed.

Accordingly, patients will live longer with fewer complications from their treatment.

## 1.5 Precision Medicine and Evidence-Based Medicine

The main difference between evidence-based medicine (EBM) and traditional medicine is that EBM requires better evidence than has traditionally been the case. Evidence-based medicine refers to the integration of clinical experience, patient preferences, and best available evidence in the decision-making process related to patient healthcare [47].

Guidelines are provided for evidence-based medicine from the highest level of evidence derived from multiple randomized controlled clinical trials to solve specific clinical problems [48].

EBM and precision medicine has been developed based on medical evidences and genomics profile, respectively. Precision medicine differs greatly from EBM. EBM seeks to determine the best practice approach for a patient who is showing general knowledge of the population but precision medicine is the individualization of care by focuses on the unique characteristics of a particular patient [47, 49].

Medical professionals encounter several problems for prescriptions in the context of evidence-based medicine: a significant percentage of lack of efficacy in some medicines, EBM promoting the standardized use of therapy that does not address response variations in each patient, high incidence of adverse drug reactions, and clinical trials focusing on taking statistical information on the general populations and applying them to the patient. EBM ignores the outliers, but PM focuses on the outliers [50, 51].

Predictions based on mechanical knowledge of the genetic or environmental effects of drug reactions may be incorporated into the design of randomized clinical trials (RCTs). EBM guidelines may be more PM-friendly by incorporating a more patient-centered approach. EBM and PM are both complementary in their approaches, such that there is a need for collaboration between experts in both fields in the advancement of science in clinical practice and its applications in the treatment of each patient [51].

## 1.6 Genomics Precision Medicine

Recent technological advancements in genomics and many other OMIC sciences (transcriptomics, proteomics,...) have revolutionized the conventional and future practice of clinical medicine.

Diagnosis of most challenging rare diseases is now feasible with a high degree of accuracy with the new technologies. It is now possible to diagnose "gene-specific" and "genome-driven" genetic disorders.

Genomic precision medicine is a revolution that will empower patients to take control of their healthcare. There is a growing awareness of pharmacogenomics (PGx) as a key part of personalized medicine. Since June 2018, over 250 FDA-approved drugs are labeled for prescribing based on the patient's genomic profile, a number that has tripled since 2014 [52, 53].

Genomic medicine is a promising medical discipline that applies genomic information about an individual as part of their clinical care in prediction, prevention, and tailored treatment in precision medicine approach; this definition is presented by the National Human Genome Research Institute (NHGRI) [54]. Genomic medicine is capable of revolutionizing the healthcare of patients with rare or common diseases to implement precise diagnosis, improved disease risk assessment, prevention through screening programs, and personalized treatment. By understanding the genetic architecture of many diseases, the gap between basic and clinical research has been quickly filled. Therefore, we are entering a new era in clinical medicine.

With the new concept of genomic medicine, primary care achieves its goal in maximizing health benefits and minimizing unnecessary harms to patients [55]. The applications of precision medicine are numerous, but it truly requires genomic medicine potential. The adoption of genomic medicine in patient care got more attention for achieving high-quality evidences for supporting the clinical decisions in common diseases [56]. A key milestone in genomic medicine was the human genome project in 2003, but this success is the only one of many milestones in the journey of genomic medicine from Mendel to next-generation sequencing (NGS) [57].

The achievement arising from the Human Genome Project was the beginning of the postgenomic era, rather than the end of one [58]. Today, genetic testing using high-throughput approaches is pursued by a growing number of physicians. Rapid development in high-throughput technologies such as "next-generation" DNA sequencing and genome-wide association study (GWAS) has facilitated the use of genomic medicine to perform better management in several diseases from Mendelian to complex disease. Moreover, whole exome sequencing (WES) has been used in the workup of patients with undiagnosed conditions [59, 60]. Hence, genomic medicine achievements lead to a major

clinical advance in the management of common diseases including different types of cancer and cardiovascular disease in the context of precision medicine. Additionally, remarkable advancement has been obtained in pharmacogenetics and pharmacogenomics through analyzing genetic variants [61].

## 1.7 "Omics" and Precision Medicine

The suffix "omics" in science and technology such as genomics, transcriptomics, and proteomics refers to such technologies which have been applied in the development of personalized (precision) medicine. Some of the important "omics" with impact in clinical practice of medical disciplines are described in various chapters of this book.

During the last decade, omics science has revolutionized translational medicine [62] Omics (X-omics) describes high-throughput experimental technologies, providing the tools for widely monitoring disease processes at a molecular level that focuses on big data. The publication of the full human genome sequence was a breakthrough in the history of omics research [63, 64].

The suffix "ome" derives from "chromosome" and includes a complete set of biological fields such as genomics, transcriptomics, proteomics, metabolomics, and other omics. The "omics" approach implies a comprehensive, or global, evaluation of a set of molecules.

Traditional molecular methods are time-consuming and not adequately efficient, while omics sciences which are based on high-throughput analytical methods have proven to be accurate and more efficient, enabling scientists to better understand the molecular architecture of common diseases [65, 66]. Multi-omics (X-omics) is a neologism that provides a tremendous opportunity for improvement for precision medicine (Fig. 1.1).

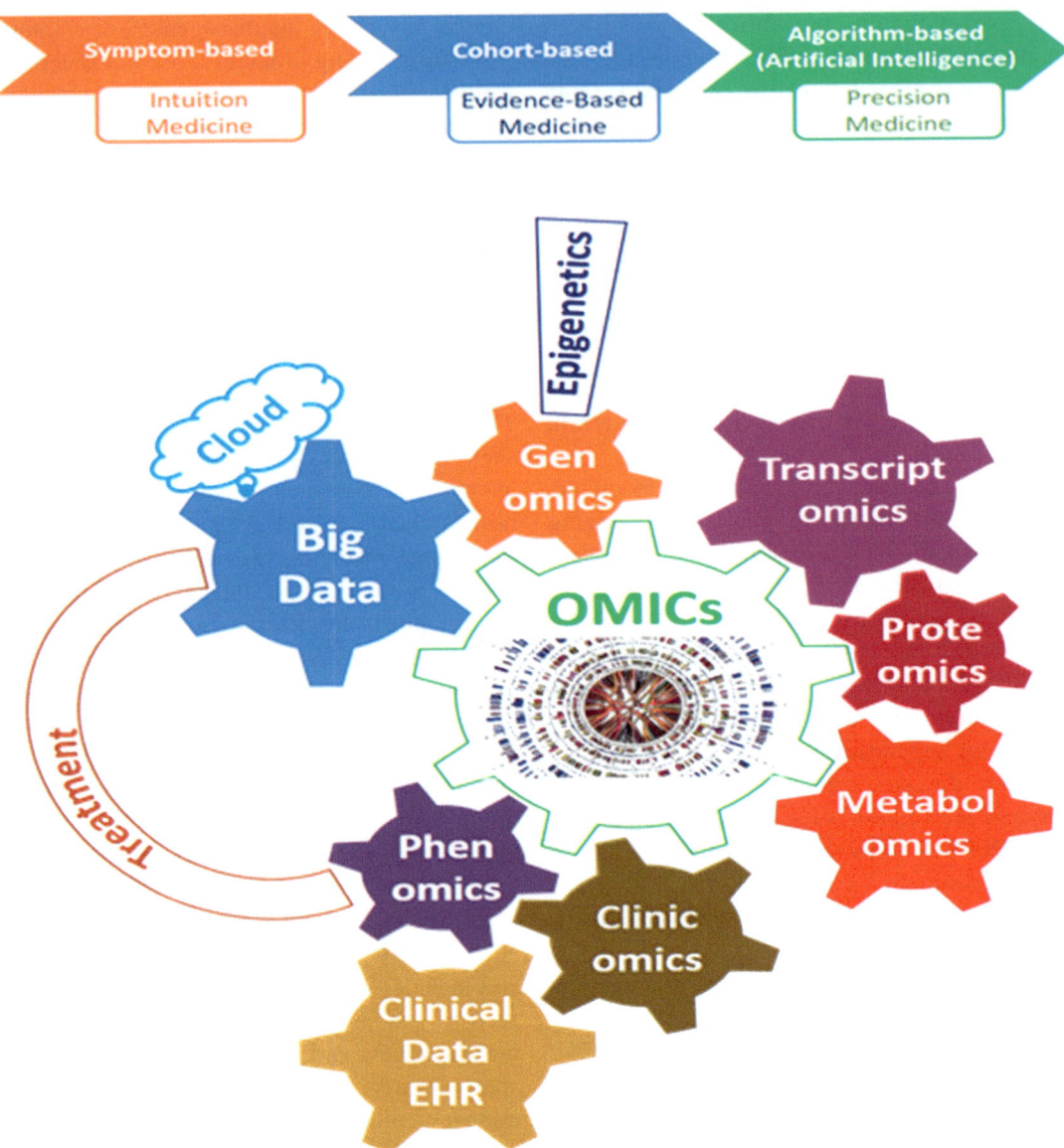

**Fig. 1.1** Precision medicine approach is a journey which has been passed through intuition medicine and evidenced-based medicine. This new approach should apply "omics" technologies. Multi-omics technologies including genomics, transcriptomics, proteomics, metobolomics, etc. by producing a large amounts of data and in combination with patient phenotypes integrate a great scientific revolution in the practice of medicine

Precision medicine offers a way to change the clinical approaches which provide precise prevention, diagnosis, and treatment options. With the development of next-generation sequencing (NGS) and RNA sequencing (RNA-Seq) technologies, precision medicine is becoming attractive and practical that holds great promise for future success.

## 1.8 Personalized Medicine, Artificial Intelligence, and Digital Twin

Noncommunicable diseases (NCDs) mainly type 2 diabetes mellitus (T2DM), cancers, cardiovascular diseases (CVDs), and chronic respiratory diseases (CRDs) are the leading cause of death worldwide [67]. The complexity of common diseases is related to the involvement of thousands of genes that have different patterns among patients with a similar diagnosis. It shows poor diagnostics which only relies on a small number of biomarkers with limited specificity or sensitivity.

Digital and genomic medicine may be able to overcome this problem by processing and integrating massive data from digital devices, imaging, electronic health records, and omics [68].

Digital equipment and services play an essential role in supporting both physicians and patients in today's medical and healthcare practices. Digital twins are an engineering idea that has been applied to complicated systems. The developing digital twin technology is being acclaimed as an intriguing and promising method for advancing medical research and improving clinical and public health outcomes. Digital twins can help care systems to be more personalized and proactive. A virtual model of a physical object having dynamic, bi-directional links between the physical thing and its corresponding twin in the digital domain is known as a digital twin [69, 70].

The digital twin parts include the physical component in the physical space, the digital representation of the physical component in virtual space, and the links between the two, that is, information moving between the physical and digital components [71].

Another significant modern aspect of a digital twin is its capacity to predict how the process will perform. Prediction accuracy gradually is increasing through novel technologies such as artificial intelligence (AI). Medicine and public health will be revolutionized by intelligent AI, which combine data, knowledge by different algorithms.

Personalized medicine needs the collection and analysis of massive amounts of data (big data from a mix and health records), and new approaches like digital twins and AI accelerates this process. Digital twins are high-resolution models of patients that computationally medicate with hundreds of drugs to determine the best drug for the patient [71]. Digital twins will be important in offering highly personalized treatments and interventions, and their main features, such as digital thread tracing and monitoring, will allow them to do so. The banks of human digital twins may one day be crucial for very successful clinical trial matching, among other applications.

## References

1. Gameiro GR, Sinkunas V, Liguori GR, Auler-Júnior JOC. Precision Medicine: changing the way we think about healthcare. Clinics. 2018;73:e723.
2. Seyhan AA, Carini C. Are innovation and new technologies in precision medicine paving a new era in patients centric care? J Transl Med. 2019;17(1):114.
3. FS C, Varmus H. A new initiative on precision medicine. N Engl J Med. 2015;372:793–5.
4. Horne R, Bell JI, Montgomery JR, Ravn MO, Tooke JE. A new social contract for medical innovation. Lancet. 2015;385(9974):1153–4.
5. Gupta R, Kim J, Spiegel J, Ferguson SM. Developing products for personalized medicine: NIH research tools policy applications. Per Med. 2004;1(1):115–24.
6. Carson PE, Flanagan CL, Ickes C, Alving AS. Enzymatic deficiency in primaquine-sensitive erythrocytes. Science. 1956;124(3220):484–5.
7. Mahgoub A, Dring L, Idle J, Lancaster R, Smith R. Polymorphic hydroxylation of debrisoquine in man. Lancet. 1977;310(8038):584–6.
8. Gibson WM. Can personalized medicine survive? Can Fam Physician. 1971;17(8):29.
9. Langreth BR, Waldholz M. New era of personalized medicine: targeting drugs for each unique genetic profile. Oncologist. 1999;4(5):426.
10. National Cancer Institute. NCI dictionary of cancer terms–personalized medicine. https://www.cancer.gov/publications/dictionaries/cancer-terms/def/personalized-medicine
11. Abrahams E, Ginsburg GS, Silver M. The personalized medicine coalition. Am J Pharmacogenomics. 2005;5(6):345–55.
12. Moon H, Ahn H, Kodell RL, Baek S, Lin C-J, Chen JJ. Ensemble methods for classification of patients for personalized medicine with high-dimensional data. Artif Intell Med. 2007;41(3):197–207.

13. Meaney E, Sierra-Vargas P, Meaney A, Guzmán-Grenfell M, Ramírez-Sánchez I, Hicks JJ, et al. Erratum to "Does Metformin Increase Paraoxonase Activity in Patients with the Metabolic Syndrome? Additional Data from the MEFISTO Study". Clin Transl Sci. 2015;8(6):873.

14. Hong K-W, Oh B. Overview of personalized medicine in the disease genomic era. BMB Rep. 2010;43(10):643–8.

15. Brand A. Public health genomics and personalized healthcare: a pipeline from cell to society. Drug Metabol Drug Interact. 2012;27(3):121–3.

16. Møldrup C. Beyond personalized medicine. Pers Med. 2009;6(3):231–3.

17. Fierz W. Challenge of personalized health care: to what extent is medicine already individualized and what are the future trends? Med Sci Monit. 2004;10(5):RA111–RA23.

18. Srivastava P. Drug metabolism and individualized medicine. Curr Drug Metab. 2003;4(1):33–44.

19. Gravitz L. The story "A Fight for Life that United a Field". Nature. 2011;478:163–4.

20. Baker M. Reprogramming Rx. London: Nature Publishing Group; 2011.

21. Hall JG. Individualized medicine. What the genetic revolution will bring to health care in the 21st century. Can Fam Physician. 2003;49:12.

22. Graham-Rowe D. Overview: multiple lines of attack. Nature. 2011;480(7377):S34–S5.

23. Boguski MS, Arnaout R, Hill C. Customized care 2020: how medical sequencing and network biology will enable personalized medicine. F1000 Biol Rep. 2009;1:73.

24. Cancer Diagnosis Progran (CDP). Director's Challenge: Toward a Molecular Classification of Cancer. https://cdpcancergov/scientific_programs/specs/1/challenge/directors_challenge_initi ativehtm#:~:text=Classification%20of%20Cancer-,Overview,molecular%20alterations%20in%20 human%20tumors (1999).

25. Mirnezami R, Nicholson J, Darzi A. Preparing for precision medicine. N Engl J Med. 2012;366(6):489–91.

26. Trusheim MR, Berndt ER, Douglas FL. Stratified medicine: strategic and economic implications of combining drugs and clinical biomarkers. Nat Rev Drug Discov. 2007;6(4):287–93.

27. Hood L. A personal journey of discovery: developing technology and changing biology. Annu Rev Anal Chem. 2008;1:1–43.

28. Meyer UA. Pharmacogenetics–five decades of therapeutic lessons from genetic diversity. Nat Rev Genet. 2004;5(9):669–76.

29. Food and Drug Administration, HHS. International Conference on Harmonisation; Guidance on E15 Pharmacogenomics Definitions and Sample Coding; Availability. Notice. Fed Reg. 2008;73(68):19074.

30. Relling MV, Evans WE. Pharmacogenomics in the clinic. Nature. 2015;526(7573):343–50.

31. Relling MV, Dervieux T. Pharmacogenetics and cancer therapy. Nat Rev Cancer. 2001;1(2):99–108.

32. Weinshilboum R, Wang L. Pharmacogenomics: bench to bedside. Discov Med. 2009;5(25):30–6.

33. Wheeler HE, Maitland ML, Dolan ME, Cox NJ, Ratain MJ. Cancer pharmacogenomics: strategies and challenges. Nat Rev Genet. 2013;14(1):23–34.

34. Sachidanandam R, Weissman D, Schmidt SC, Kakol JM, Stein LD, Marth G, et al. A map of human genome sequence variation containing 1.42 million single nucleotide polymorphisms. Nature. 2001;409(6822):928–34.

35. National Center for Biotechnology Information, National Library of Medicine. Database of single nucleotide polymorphisms (dbSNP). National Center for Biotechnology Information, National Library of Medicine… 2015.

36. Robert F, Pelletier J. Exploring the impact of single-nucleotide polymorphisms on translation. Front Genet. 2018;9:507.

37. Kasztura M, Richard A, Bempong N-E, Loncar D, Flahault A. Cost-effectiveness of precision medicine: a scoping review. Int J Public Health. 2019;64(9):1261–71.

38. Ginsburg GS, Phillips KA. Precision medicine: from science to value. Health Aff (Millwood). 2018;37(5):694–701.

39. Mathur S, Sutton J. Personalized medicine could transform healthcare. Biomed Rep. 2017;7(1):3–5.

40. Shendure J, Findlay GM, Snyder MW. Genomic medicine—progress, pitfalls, and promise. Cell. 2019;177(1):45–57.

41. Horton RH, Lucassen AM. Recent developments in genetic/genomic medicine. Clin Sci (Lond). 2019;133(5):697–708.

42. Ke X, Shen L. Molecular targeted therapy of cancer: the progress and future prospect. Front Lab Med. 2017;1(2):69–75.

43. Seebacher N, Stacy A, Porter G, Merlot A. Clinical development of targeted and immune based anti-cancer therapies. J Exp Clin Cancer Res. 2019;38(1):156.

44. National Cancer Institute. Targeted Cancer Therapies. https://www.cancergov/about-cancer/treatment/types/targeted-therapies/targeted-therapies-fact-sheet.

45. Slastnikova TA, Ulasov AV, Rosenkranz AA, Sobolev AS. Targeted intracellular delivery of antibodies: the state of the art. Front Pharmacol. 2018;9:1208.

46. Krzyszczyk P, Acevedo A, Davidoff EJ, Timmins LM, Marrero-Berrios I, Patel M, et al. The growing role of precision and personalized medicine for cancer treatment. Technology. 2018;6(03n04):79–100.

47. Masic I, Miokovic M, Muhamedagic B. Evidence based medicine–new approaches and challenges. Acta Inform Med. 2008;16(4):219.

48. Goldberger JJ, Buxton AE. Personalized medicine vs guideline-based medicine. JAMA. 2013;309(24):2559–60.

49. Tonelli MR, Shirts BH. Knowledge for precision medicine: mechanistic reasoning and methodological pluralism. JAMA. 2017;318(17):1649–50.

50. Kumar D. The personalised medicine: a paradigm of evidence-based medicine. Ann Ist Super Sanita. 2011;47:31–40.

51. de Leon J. Evidence-based medicine versus personalized medicine: are they enemies? J Clin Psychopharmacol. 2012;32(2):153–64.

52. Drozda K, Pacanowski MA, Grimstein C, Zineh I. Pharmacogenetic labeling of FDA-approved drugs: a regulatory retrospective. JACC Basic Transl Sci. 2018;3(4):545–9.

53. Food and Drug Administration (ADA). Table of pharmacogenomic biomarkers in drug labeling (2018). https://www.fda.gov/drugs/science-and-research-drugs/table-pharmacogenomic-biomarkers-drug-labeling

54. National Human Genome Research Institute (NHGRI). Genomics and Medicine. https://www.genomegov/health/Genomics-and-Medicine

55. Davies K. The era of genomic medicine. Clin Med. 2013;13(6):594.

56. Rubanovich CK, Cheung C, Mandel J, Bloss CS. Physician preparedness for big genomic data: a review of genomic medicine education initiatives in the United States. Hum Mol Genet. 2018;27(R2):R250–R8.

57. Roden DM, Tyndale R. Genomic medicine, precision medicine, personalized medicine: what's in a name? Clin Pharmacol Ther. 2013;94(2):169–72.

58. Sykiotis GP, Kalliolias GD, Papavassiliou AG. Hippocrates and genomic medicine. Arch Med Res. 2006;37(1):181–3.

59. Raza S, Hall A. Genomic medicine and data sharing. Br Med Bull. 2017;123(1):1–11.

60. Manolio TA, Green ED. Leading the way to genomic medicine. Am J Med Genet C Semin Med Genet. 2014 Mar;166C(1):1–7.

61. Semiz S, Dujic T, Causevic A. Pharmacogenetics and personalized treatment of type 2 diabetes. Biochem Med (Zagreb). 2013;23(2):154–71.

62. Manzoni C, Kia DA, Vandrovcova J, Hardy J, Wood NW, Lewis PA, et al. Genome, transcriptome and proteome: the rise of omics data and their integration in biomedical sciences. Brief Bioinformatics. 2018;19(2):286–302.

63. Yan S-K, Liu R-H, Jin H-Z, Liu X-R, Ye J, Shan L, et al. 'Omics' in pharmaceutical research: overview, applications, challenges, and future perspectives. Chin J Nat Med. 2015;13(1):3–21.

64. Venter JC, Adams MD, Myers EW, Li PW, Mural RJ, Sutton GG, et al. The sequence of the human genome. Science (New York, NY). 2001;291(5507):1304–51.

65. Bluett J, Barton A. Precision medicine in rheumatoid arthritis. Rheum Dis Clin. 2017;43(3):377–87.

66. Au TH, Wang K, Stenehjem D, Garrido-Laguna I. Personalized and precision medicine: integrating genomics into treatment decisions in gastrointestinal malignancies. J Gastrointest Oncol. 2017;8(3):387.

67. Hunter DJ, Reddy KS. Noncommunicable diseases. N Engl J Med. 2013;369(14):1336–43.

68. Topol EJ. A decade of digital medicine innovation. Sci Transl Med. 2019;11(498):eaaw7610.

69. Kamel Boulos MN, Zhang P. Digital twins: from personalised medicine to precision public health. J Pers Med. 2021;11(8):745.

70. Björnsson B, Borrebaeck C, Elander N, Gasslander T, Gawel DR, Gustafsson M, et al. Digital twins to personalize medicine. Genome Med. 2020;12(1):1–4.

71. Grieves M. Digital twin: manufacturing excellence through virtual factory replication. White Paper. 2014;1:1–7.

# Principles of Pharmacogenomics and Pharmacogenetics

**2**

Mandana Hasanzad ⬤, Negar Sarhangi ⬤,
Leila Hashemian, and Behnaz Sarrami

**What Will You Learn in This Chapter?**
The reader will learn about the fundamentals of pharmacogenomics and its applications in different clinical settings. There will be sections that include each disease state individually, such as mental health, cardiology, and oncology, which will be explained in detail along with their relevant genes that affect the way a medication would work for that individual. There will also be sections on variations in drug transporter genes and drug target genes along with resources and clinical guidelines that are available for them. Case studies included will give more context on the relevance of gene variations and how it plays a role in patients' dosing recommendations and

change in medication regimen. Toward the end of this chapter, there will a section about various resources available for looking up a drug-gene interaction, phenoconversion, and guidelines that are out there. The goal is a solid understanding of the concept of pharmacogenomics and its clinical value and utility in any clinical setting.

**Rationale and Importance**
Pharmacogenomics offers an interesting possibility to improve patient care by optimizing pharmaceutical selection and dose, lowering the risk of adverse effects, and so incorporating personalized medicine concepts.

It's important to understand how genetic variations can play a role in how the body responds to medication by learning the pharmacokinetics and what the drug does to the body by learning about pharmacodynamics. In addition, it's important to know how medications interact with each other as many individuals are on multiple medications. Lastly, knowing how environmental factors and our social behavior impact the way our genes work which in turn affect the medications we are taking is so crucial. It is vital to learn all the dynamics and the in-depth knowledge at a molecular level as that plays a huge role in truly personalizing a patient's medication right from the start of a treatment plan. Remember that pharmacogenomics is a part of precision medicine. Other factors including drug-drug interaction, age, kidney function, the timing of medication, phenoconversion, and epigenetics,

M. Hasanzad
Personalized Medicine Research Center, Endocrinology and Metabolism Clinical Sciences Institute, Tehran University of Medical Sciences, Tehran, Iran

Medical Genomics Research Center, Tehran Medical Sciences, Islamic Azad University, Tehran, Iran

N. Sarhangi
Personalized Medicine Research Center, Endocrinology and Metabolism Clinical Sciences Institute, Tehran University of Medical Sciences, Tehran, Iran

L. Hashemian
Medical Genomics Research Center, Tehran Medical Sciences, Islamic Azad University, Tehran, Iran

B. Sarrami (✉)
Creighton University School of Pharmacy and Health Professions, Omaha, NE, USA

© The Author(s), under exclusive license to Springer Nature Singapore Pte Ltd. 2022
M. Hasanzad (ed.), *Precision Medicine in Clinical Practice*,
https://doi.org/10.1007/978-981-19-5082-7_2

all, play a key part in the bigger picture that needs to be discussed about.

## 2.1 Introduction to Pharmacogenomics and Pharmacogenetics

The origins of pharmacogenomics are unknown; however, Pythagoras, a Greek philosopher and mathematician, observed in 510 BC that a subset of people who ate broad beans (*Vicia faba*) suffered from potentially fatal hemolytic anemia. This reaction was eventually related to a hereditary deficit in the enzyme glucose-6-phosphate dehydrogenase (G6PD), which also makes people susceptible to hemolysis from drugs like rasburicase and the antimalarial primaquine. The antimalarial drug primaquine causes acute hemolytic crises in individuals with glucose 6-phosphate dehydrogenase (G6PD) deficiency [1]. The Clinical Pharmacogenetics Implementation Consortium (CPIC) has provided a guideline as to which G6PD genotypes are associated with G6PD deficiency in males and females [2]. Another adverse drug reaction was reported for succinylcholine that is administered as anesthesia. A genetic variant in gene encoding pseudocholinesterase causes adverse drug reactions like apnea [1]. The role of genetic variants in the modulation of variability in drug actions was proposed by the physician-scientist Sir Archibald Garrod around the year 1900. He presented the term "chemical individuality" [1, 3].

The history of the genetic basis for drug response phenotypes dates back to the early 1950s. The word "pharmacogenetics" was first coined by Friedrich Vogel in Heidelberg, Germany, in 1959 [1]. Several family studies undertaken in the 1960s and 1980s identified patterns of inheritance for many medication effects, leading to molecular research that revealed many of the features' inherited determinants. The first polymorphic human drug-metabolizing gene, CYP2D6, was cloned and described in 1987 [4]. Like other areas of genetics, the speed of pharmacogenetic discoveries has been accelerated after the completion of the Human Genome Project. The term pharmacogenomics is now used to explain how multiple genetic variants across the genome (DNA and RNA) can affect drug response, while pharmacogenetics is the study of a DNA variation related to drug response [5].

The International Council for Harmonisation of Technical Requirements for Pharmaceuticals for Human Use (ICH), a worldwide consortium of regulatory agencies, has defined "pharmacogenomics" as the study of variations of DNA and RNA characteristics as related to drug response and "pharmacogenetics" as the study of variation in DNA sequence as related to drug response [5].

Pharmacogenomics information can help physicians decide medication selection, dose adjustment, and treatment period and prevent adverse drug reactions. Furthermore, pharmacogenetics can contribute to the development of new therapeutic agents [6–8].

## 2.2 Pharmacodynamics and Pharmacokinetics

To effectively understand how medications work in each disease state, a solid understanding of pharmacokinetics (PK) and pharmacodynamics (PD) is vital [9].

Pharmacokinetics is understanding what the body does to a drug as it enters the body. The drug goes through a series of steps called ADME for short: absorption, distribution, metabolism, and excretion [10, 11].

Absorption reflects the bioavailability of a medication which varies depending on the route it ends the body. The distribution looks at the rate a drug accumulates at the site of its action. Metabolism is the most important process of how a medication is eliminated but also activates medications that are in an inactive state or prodrugs.

Finally, excretion is the last step of the ADME process which helps elimination from the body [12].

Pharmacodynamics provides an insight into how a drug affects the body [13]. This describes the relationship between the concentration of a drug and its response. This includes presynaptic binding, postsynaptic receptors, and the interaction between the two [14].

Pharmacokinetics and pharmacodynamics have to be considered together for treatment optimization, rather than looking at each in isolation [15]. The main branches of clinical pharmacology are linked by both the PK and PD which are presented by the concentration-time profile and intensity of the response, respectively. Therefore, a deep understanding and knowledge of the two branches are vital before moving forward in PGX [15]. Both PK and PD, although powerful tools, do not represent a complete picture if looked at in isolation [16].

The process of drug response is performed through pharmacokinetics that is drug absorption, distribution, metabolizing, and elimination or pharmacodynamics which is modifying drug target or by disrupting the biological pathways that shape a patient's pharmacologically sensitive [6].

## 2.3 Phenoconversion

Cytochrome P450 2D6 (CYP2D6) is responsible for the metabolism of about 25% of the most commonly prescribed medications and is also highly polymorphic. Many medications are CYP2D6 inhibitors, and the US Food and Drug Administration (FDA) has recognized some of those medications on their "Top 300 Drug" list [17]. Some of the strongest CYP2D6 inhibitors are bupropion, fluoxetine, paroxetine, and quinidine. The inhibition can increase the levels of medications that are being metabolized by the same enzyme. Some medications are prodrugs (inactive), and the metabolism activates them.

So, inhibition of those can decrease or eliminate their therapeutic effect. Adding an inhibitor can alter the phenotype of an individual.

There is a well-known phenomenon in PGX called phenoconversion where an individual's phenotype is converted from an extensive metabolizer (EM) or intermediate metabolizer (IM) to a poor metabolizer (PM). This is usually due to extrinsic factors such as the interaction between medications that inhibit the metabolism of the other as discussed earlier [18, 19]. We can also consider it as a mismatch between a genotype and phenotype due to some non-genetic factors that occur, for example, multiple medications or extrinsic factors such as smoking. It can also be caused by age and diseases [20].

For example, an individual taking codeine (prodrug) is an NM (normal metabolizer) of CYP2D6 and can convert to morphine and have its therapeutic effect of pain relief. This patient then adds fluoxetine (a CYP2D6 inhibitor) to her regimen. The combination caused a phenoconversion, from NM to PM, and is not able to convert codeine to its active metabolite, therefore no longer achieving pain relief.

Another example is the combination of clozapine and the most commonly prescribed antibiotic ciprofloxacin. Ciprofloxacin inhibits the metabolism of clozapine, causing the phenotype of a PM.

Phenoconversion has been one of the major challenges of public health as it is not well understood and not integrated into standard practice [17]. Recently, the University of Florida (UF) Health has created a web-based calculator tool for clinicians to aid in the integration [17, 21].

According to a retrospective analysis done from an acute care psychiatric inpatient clinic, at admission, there was nine phenoconversion detected and eight at discharge. The medications involved were esomeprazole, sertraline, and duloxetine which accounted for 71% of the phenoconversion at admission and 76% at discharge. This tells us a lot about understanding not just the genotype and its matching phenotype but also how extrinsic factors and multiple medications affect each other [22].

## 2.4 Pharmacogenomics Nomenclature

The Clinical Pharmacogenetics Implementation Consortium (CPIC) was established in 2009 as a joint effort of PharmGKB (https://www.pharmgkb.org) and the Pharmacogenomics Research Network (PGRN). The CPIC guidelines are aimed to help physicians in understanding how available genetic test results should be used to optimize treatment strategies, which to date have produced 26 clinical guidelines (https://cpicpgx.org/guidelines/) [7].

Allele function and phenotype terms should be standardized; therefore, CPIC developed consensus terms that can be utilized to define pharmacogenetic allele function and related phenotypes [23].

Many pharmacogenes utilize the star allele nomenclature system. Pharmacogeneticists used a "star" nomenclature (e.g., *CYP2C19\*2*) to characterize gene variants (also known as "pharmacogenes") that underpin drug response variability.

The CYP2C19 gene is considered as an example. Variations in the *CYP2C19* liver enzyme can result in different drug metabolism response and unexpected drug serum levels.

For *CYP2C19*:

- The *1/*1 genotype indicates normal activity of *CYP2C19*.
- The *1/*3 genotype indicates intermediate activity of *CYP2C19*.
- The *2/*2 genotype indicates low activity of *CYP2C19*.
- The *17/*17 genotype indicates high activity of *CYP2C19*.

In most cases, the *1 allele corresponds to the fully functional reference allele or haplotype (wild type) depending on the subpopulation in which the gene was initially studied in. It is

**Table 2.1** Pharmacokinetic and pharmacodynamics nomenclature

| Gene | Genotype | Alleles status |
|---|---|---|
| PK genes | Normal metabolizers (NMs) | Normal alleles |
| | Poor metabolizers (PMs) | Two loss-of-function alleles |
| | Intermediate metabolizers (IMs) | One loss-of-function allele |
| | Ultrarapid metabolizers (UMs) | Carrying gain-of-function alleles or gene duplications |
| PD genes | Positive or negative | Positive or negative for high-risk alleles |

important to understand that this does not necessarily indicate that it is the most common allele in all populations. Other designations (*2, *3, etc.) refer to haplotypes carrying one or more variants.

Some star alleles may contain multiple variants; for example, *CYP2C19\*4* denotes an allele defined by the presence of two single nucleotide polymorphisms (SNPs), and differentiating this allele from individuals bearing only one of the SNPs might be difficult.

To date, the majority of variants investigated have partially or inhibited the function of the encoded protein. Sometimes, variants increase the activity of drug-metabolizing enzymes (Table 2.1).

## 2.5 Genetic Variations

Certain genetic variants are found to be associated with clinical differences in drug disposition between individuals, which include susceptibility risk to adverse drug reactions, and the likelihood of therapeutic response or efficacy [8].

Human DNA sequences are nearly 99.9% identical, yet due to the size of the genome, this leads to 4 to 5 million base-pair differences between any two persons.

There are approximately 20,000 protein-coding genes in the human genome. The single nucleotide polymorphism is by far the most frequent. SNPs are considered to be a common type of genetic variation. SNPs may have a significant impact on the protein target via a variety of mechanisms. Every gene contains SNPs that occur throughout the human genome in every 1000–3000 base pairs [24]. Evidence already exists that connects certain polymorphisms with responses to anticancer drugs. Polymorphisms are commonly seen in genes encoding drug metabolism, drug transporter, and drug-target proteins which are in the PD pathway. Drug availability at the target site can be affected by drug metabolism and transporter genotypes, although a patient's sensitivity to a drug can be affected by the drug target genotype.

According to several studies, genetic factors have been proposed to be responsible for 20% to 80% of drug metabolism and drug reaction differences [25].

Drug metabolism and transporter genotypes can affect drug availability at the target site, whereas drug-target genotype can affect a patient's sensitivity to a drug. Genes encoding proteins involved in drug absorption, distribution, metabolism, and excretion (ADME) play important role in the PK pathway [26].

Variation in the regulatory areas can influence transcription factor binding and, eventually, gene expression. Single gene variants affect PK in two ways [27]:

1. Administration of a prodrug, a pharmacologically inactive molecule that must be bioactivated by drug metabolism to exert therapeutic effects. Genetic variants that result in loss of function of a single drug-metabolizing enzyme might reduce or inhibit drug activity.
2. Single PK variants can have a very large impact during the administration of an active drug with a narrow therapeutic range [27].

Metabolism is an important phase in the PK of drugs. The majority of drugs (90%) are processed by liver enzymes represented by one of six cytochrome P450 gene families (1A2, 2C9, 2C19, 2D6, 3A4, and 3A5) [28]. Identifying how a person's CYP450 genes are expressed is crucial

because most medications must be converted before they can be used therapeutically. Around 80% of the individuals in a population have the common (or "wild type") phenotype and are categorized as "normal or extensive" metabolizers. Individuals who metabolize rapidly are categorized as "rapid/intermediate or ultra-rapid," while those who metabolize slowly or are unable to metabolize the drug are classified as "poor or ultra-slow" [29].

The pharmacogenetics (PGx) machinery incorporates a number of genes that code for enzymes and proteins that are important factors in drug targeting and processing, as well as crucial components of the epigenetic machinery responsible for gene expression control [12].

## 2.6 Pharmacogenomics in Clinical Practice

Despite significant advances in biomedical research over the last century, a significant number of patients do not respond to drug treatment.

According to a US Food and Drug Administration (FDA) report, in 38–75% of patients with common diseases, medication is ineffective which results in increased healthcare costs [30].

There is considerable scientific evidence supporting the value of pharmacogenomics testing for patient care, yet there is no rationale to include this testing as part of the therapeutic engagement.

Individuals respond to drugs in unpredictable ways, and treatment failure or unpleasant or severe drug reaction do occur. These unpredicted reactions could be caused by a variety of factors, including drug-drug interactions, gene-drug interactions, drug-food interactions, gender, pregnancy, age, illness condition, impaired renal or hepatic function, and noncompliance.

As the development of the relevant pharmacogenomics guidelines is critical for clinical practices, the US Clinical Pharmacogenetics Implementation Consortium (CPIC), the Dutch

Pharmacogenetics Working Group (DPWG), the Canadian Pharmacogenomics Network for Drug Safety (CPNDS), and the French National Network (Réseau) of Pharmacogenetics (RNPGx) are among the accessible guidelines [7, 31–33].

While various organizations provide pharmacogenetic clinical practice guidelines, the US Food and Drug Administration (FDA) has recommended against using pharmacogenetic information that has not been reviewed by the FDA to guide treatment decisions [33, 34].

The FDA website published two online tables with FDA-reviewed pharmacogenetic information [35].

The first table provides a listing of biomarkers and sections of individual drug labeling that incorporate the related biomarker [36]. The second table is a pharmacogenetic association table which gives a list of drugs with genetic variations in drug targets, drug transport, drug metabolism, or adverse drug reaction and may have information that is not found in the first table [37].

According to the new categorization, genetic variation that can affect pharmacogenetic information and clinical outcome can classify:

- Genetic variation in the drug-metabolizing enzyme that affects drug metabolism.
- Genetic variation in the drug-protein target led to interindividual differences in dose requirement and altered drug sensitivity.
- Genetic variation in drug transporter that alters drug's biotransformation or excretion.
- Genetic variation associated with adverse drug reaction (where genetic variants increase an individual's risk of an adverse drug reaction by pathways other than drug metabolism or transport).
- Genetic variation associated with therapeutic efficacy (where genetics predicted poor therapeutic response by pathways other than drug metabolism or transport).
- Drug indications specified by biomarkers (in which the drug had at least one FDA-approved indication based on the presence or absence of the biomarker).

The development of pharmacogenomics is happening when several major players rapidly use pharmacogenomics information. The major players are healthcare professionals, academic medical centers, the pharmaceutical industry, drug regulatory agencies, and patients [38].

Over the last 15 years, the cost of pharmacogenomics has dropped dramatically. The testing cost over $300,000 in 2004 and is now less than $300. Insurance companies, healthcare organizations, and physicians are more ready to explore or accept PGx testing, particularly in patients with complex medical conditions or who take many medications (polypharmacy). Ninety percent of Americans have at least one genetic variation that affects drug efficacy and toxicity [39]. According to several reports from different studies and organizations like the Centers for Disease Control and Prevention (CDC), adverse drug events annually cause approximately 1.3 million emergency department visits, 350,000 hospitalizations, 9.7% permanent damage, and 106,000 deaths. ADRs also cause $47 billion hospitalization costs and $76.6 billion in morbidity/mortality [39, 40].

Incorporating patient genotyping into clinical settings can help clinicians make decisions about therapeutic regimens and drug dosages that have the greatest efficacy and minimal risk of toxicity.

## 2.7 Pharmacogenomics in Drug Development

Pharmaceutical companies aim to provide drugs that are both safe and effective. Thousands of new molecular entities are investigated each year, but only a few make it through the drug-development pipeline, get regulatory approval, and hit the market as the success rate is about 16% across different therapeutic areas [41].

The average time required for a drug to progress from the start of clinical trials to achieving regulatory approval is approximately 7.6 years [42].

The high cost of the drug development process is due to the long development time and low success rate. Both the industry and regulatory agencies have recently begun to shift to a new precision medicine approach that promises to deliver the right treatments to the right patients at the right time.

The ability to classify individuals into subpopulations that differ in their susceptibility to a particular disease or response to a particular treatment is critical to the successful implementation of precision medicine. A well-known example in the pharmaceutical industry that shifts the R & D framework toward a precision medicine-based model is seen in AstraZeneca. The company applied the 5R framework, including the right target, right tissue, right safety, right patient, and right commercial potential, to strengthen its capabilities in the drug development process [43].

Genetics discipline plays a major role in accelerating the drug development process. Considering the benefit of the pharmacogenomics approach in drug development, characterization, and detection of drug target, drug-receptor, and drug-metabolizing enzyme or each entity that plays a role in pharmacodynamics, pharmacokinetics, efficacy, and drug safety can help the primary process in the development of new drugs.

The last two decades have had great advancements in genomics and related technology. There are several initiatives, including the Human Genome Project, 1000 Genomes, and HapMap [44].

Determining how genetic variants change protein function and downstream physiologic or pathophysiologic processes can lead to the identification of new therapeutic targets. Pharmacogenomics is being applied to all elements of the drug development process as the pharmaceutical industry progresses toward a full endorsement of precision medicine. Therefore, a thorough understanding of potential relevant variants is essential for developing a clinical development strategy. The use of genetic variants to prescreen patients for preregistration clinical research is likely to limit the indication and, as a result, the commercial potential of the authorized drug.

Collaborative efforts between industry, academia, and regulatory bodies are also critical to the success of pharmacogenomics in drug development. The FDA is encouraging the pharmacogenomics study and has made many steps to establish a regulatory framework for pharmacogenomics in drug development. The European Medicines Agency (EMA) has also taken an interest in pharmacogenomics. Several EU regulatory guidelines have been published as best practices in precision medicine and pharmacogenomics.

In Japan, the Pharmaceuticals and Medical Devices Agency (PMDA) provides a framework for incorporating pharmacogenomics information throughout the drug development process. Therefore, all three organizations, FDA, EMA, and PMDA, recommend investigating the impact of genetic polymorphisms on pharmacodynamic endpoints [45].

## 2.8 Pharmacogenomics in Oncology

Cancer is a common disease, and the global cancer burden is estimated to have risen to 18.1 million new cases and 9.6 million deaths in 2018 [46]. In 2020, there will be an estimated 19.3 million new cancer cases (18.1 million excluding non-melanoma skin cancer) and around 10.0 million cancer deaths (9.9 million excluding non-melanoma skin cancer) [47].

Cancer is a complex disorder caused by genetic changes in the somatic genome of the malignant cell. Many efforts have been made to increase the overall therapeutic efficacies and outcomes of cancer treatments. Personalized cancer medicine is one of them in recent years.

The "genomic transition era" has contributed both to a dramatic increase in our awareness about genomics and the development of technology for rapidly collecting large amounts of genomic data [48].

The principal goal in personalized cancer medicine is the focus on prevention, screening, and treatment based on a person's genetic makeup [49, 50]. Cancer is a complex phenotype that arises from genetic alterations in cancer cells. Cancer cell genomes differ from cancerous cells. Many existing cancer therapeutics eliminate tumor cells but cause collateral damage to normal cells [49, 51].

Pharmacogenetics (PGx) is one of the most important components of personalized medicine which focuses on the association between genetic variations and drug response [1]. As discussed earlier, pharmacogenomics focuses on understanding how genetic variations affect therapeutic efficacy and toxicity. This is especially significant in oncology because cancer is a leading cause of morbidity and mortality in developed countries and failure therapy is frequently fatal. Progress in pharmacogenomics has the potential to revolutionize cancer therapy.

Oncology pharmacogenomics is more complex than pharmacogenomics in other disease states because clinical concerns can be considered for two distinct genomes: the tumor's somatic genome and the patient's germline genome. Although some somatic mutations in a tumor can define a patient's condition and hence therapy options, germline genetic variation studies are also crucial. The somatic genome has genetic diversity that promotes neoplastic transformation, whereas the germline genome is inherited deoxyribonucleic acid (DNA), although both can indeed have treatment consequences [52]. The germline genome can determine cancer predisposition risk information which can be used to detect patients who may benefit from improved cancer screening or cancer prevention programs.

The mutations within the cancer cells may also be heterogeneous. Targeted therapies have been developed for some of the proteins which are activated by somatic mutations (typically tyrosine kinases). Some examples of somatic pharmacogenomics include imatinib which is used to treat Philadelphia chromosome-positive leukemia or the use of B-rapidly accelerated fibrosarcoma (BRAF) inhibitors or epidermal growth factor receptor (EGFR) inhibitors in patients with specific BRAF or EGFR mutations, respectively [53]. When analyzing somatic mutations to find an appropriate targeted therapy, pathway considerations are very critical. Specific somatic mutations are predictive of treatment efficacy for several targeted treatments. These associations are mentioned on drug labels by the FDA [48]. About 42 mutation-targeted medications have been approved for the treatment of various hematologic and oncologic diseases. Oncology indications represent a high rate of all FDA drug label warnings related to pharmacogenomics markers. But only a few numbers of these drugs relate to the germline mutations [54].

Some examples of oncology drugs with significant evidence exist in the gene drug interaction list, and pharmacogenetic testing is recommended for this group (Table 2.2).

The following two examples of germline pharmacogenetics with efficacy and toxicity are to be provided.

**Table 2.2** Oncology genetic association with the highest level of evidence by FDA [48]

| Gene | Protein | Drug | Condition | Outcome |
|---|---|---|---|---|
| TPMT/ NUDT15 | Thiopurine methyltransferase | Thiopurines (mercaptopurine) | Acute lymphoblastic leukemia | Toxicity |
| DPYD | Dihydropyrimidine dehydrogenase | Fluoropyrimidines (5-fluorouracil) (5-FU) | Colorectal cancer, breast cancer, and other gastrointestinal tract cancers | Toxicity |
| G6PD | Glucose-6-phosphate dehydrogenase | Rasburicase | Prophylaxis and hyperuricemia (during chemotherapy) | Toxicity |
| UGT1A1 | Uridine 5′-diphospho-glucuronosyltransferase 1A1 | Irinotecan | Colorectal and small-cell lung cancer | Toxicity |
| CYPD6 | Cytochrome P450 2D6 | Tamoxifen | Breast cancer | Efficacy |

## 2.9 Germline Pharmacogenomics Association with Treatment Efficacy and Toxicity

There are numerous examples of germline pharmacogenetic interactions that potentially predict clinical outcomes including drug efficacy and adverse drug reactions.

### 2.9.1 CYP2D6 and Tamoxifen

Tamoxifen, a selective estrogen receptor modulator, is the most commonly used drug for the treatment of estrogen receptor-positive (ER+) breast cancer (SERMs). Tamoxifen treatment for at least five years is the standard of therapy and is related to a favorable clinical outcome. Tamoxifen impacts cancer cells as an estrogen receptor antagonist, interfering with the estrogenic signaling that stimulates cellular replication and tumor formation [55].

Tamoxifen is not a potent antiestrogen but is metabolically bioactivated to endoxifen by CYP2D6. CYP2D6 constitutes 2–3 percent of total liver CYPs, and because of the genetic polymorphisms, its hepatic protein level varies widely between individuals [56].

CYP2D6 is a highly polymorphic metabolizing enzyme with nearly 100 haplotypes including gene deletions and duplications, as well as polymorphisms in coding and regulatory areas [56].

The CYP2D6 activity predicted by genotype is the key predictor of steady-state endoxifen concentrations in tamoxifen-treated patients. The CYP2D6 activity scoring system is used for genotype-to-phenotype interpretation; each CYP2D6 allele gives an activity score based on its functional activity [57, 58].

The CPIC dosing guideline for tamoxifen recommends the use of alternative hormonal therapy such as an aromatase inhibitor for postmenopausal women or aromatase inhibitor along with ovarian function suppression in premenopausal women for CYP2D6 poor metabo-lizer if aromatase inhibitor use is not contraindicated. For CYP2D6 intermediate metabolizers and CYP2D6 allele combinations resulting in an activity score (AS) of 1, the recommendation is to consider the recommendations stated for the CYP2D6 poor metabolizer. If aromatase inhibitor use is contraindicated, consideration should be given to use a higher, but FDA-approved tamoxifen dose for CYP2D6 intermediate metabolizers and CYP2D6 allele combinations resulting in an AS of 1. For poor metabolizers, higher-dose tamoxifen (40 mg/day) increases, but does not normalize endoxifen concentrations and can be considered if there are contraindications to aromatase inhibitor therapy [57–59].

### 2.9.2 Fluoropyrimidines and DPYD

5-fluorouracil (5-FU) and its oral prodrug capecitabine serve as the basis for first-line combination chemotherapy regimens in a variety of solid tumor malignancies, including pancreatic, breast, and colorectal cancer. 5-FU is metabolized by dihydropyrimidine dehydrogenase (DPYD), which is encoded by the DPYD gene. Reduced DPYD activity results in significantly reduced 5-FU elimination and increased systemic drug concentrations, raising the risk of severe toxicity and mortality [60].

The CPIC dosing guideline for 5-fluorouracil and capecitabine recommends an alternative drug for patients who are DPYD-poor metabolizers with an activity score of 0. In those who are poor metabolizers with an activity score of 0.5, an alternative drug is also recommended, but if this is not considered a suitable therapeutic option, 5-fluorouracil or capecitabine should be administered at a strongly reduced dose with early therapeutic drug monitoring. Patients who are intermediate metabolizers with an activity score of 1 or 1.5 should receive a dose reduction of 50%. Patients with the c.[2846A>T]; [2846A>T] genotype may require a >50% dose reduction [57, 58, 60].

## 2.10 Implementation of Pharmacogenomics in Oncology

The main barrier in pharmacogenomics in oncology is additional in-depth knowledge of tumor biology, as well as the amount of evidence made necessary before proceeding. Any particular biomarker is designated "actionable" and is then used to a large extent. The most serious problems that cancer treatment faces are the emergence of drug resistance and severe adverse effects.

Because most chemotherapy drugs are not tumor-specific, they also cause damage to normal cells. This avoids the use of high doses of medication, which may be required for the eradication of less susceptible tumor cells.

## 2.11 Pharmacogenomics in Psychiatry

Mental illness is very prevalent, and according to the Centers for Disease Control and Prevention (CDC), by February 2021, the percentage of adults suffering from anxiety or depression had risen to 41.5% [61]. It is projected that by 2039, the cost of mental disorders will rise to $6 trillion [62]. In the current practice of psychiatry, the approach has been the trial-and-error process that combines the clinicians' experience and how the patient is clinically presenting. Many psychotropic medications are not effective in every patient or may produce just a partial response [62]. Predicting a response has become a challenge for clinicians. Individual genetic variations have a lot to do with this complex task [62].

In patients diagnosed with major depressive disorder (MDD), the most commonly used medications which are also the first-line therapy are the selective serotonin inhibitors (SSRIs) [63]. According to one of the largest depression trials with a funding of $35 million, the rate of response is only about 50% and about 25% remission rate. The trial highlighted not only the lack of effectiveness of the current therapy but also the need

for a more strategic treatment option [64]. Pharmacogenomics is vital in identifying the genes related to the response in psychopharmacotherapeutic agents to improve treatment outcomes. This also helps reduce the time it takes a patient to achieve mental stability and decrease side effects [63].

Psychiatric diseases (PDs), primarily bipolar disorders (BDs), are chronic recurrent illnesses defined by unexpected mood changes. These modifications resulted in significant mental and social disabilities.

A wide range of drugs, including first- and second-generation antipsychotics, antidepressants, anxiolytics, and, most notably, mood stabilizers, are recommended for PDs treatment. Despite the availability of all of these drugs, psychiatrists continue to encounter challenges in treating PDs, as treatment response is frequently ineffective and the rate of remission is low [65]. This poor clinical outcome is known as a therapeutic failure. Adverse drug reactions occur in 30–50% of psychiatric patients, regardless of the initial choice of psychiatric drug that causes morbidity and mortality in hospital settings.

Pharmacogenomics applications have been most successful in predicting adverse drug reactions than treatment responses. In psychiatry, pharmacogenetic testing is widely available and usually focuses on two cytochrome P450 (CYP) genes, *CYP2D6* and *CYP2C19*, that encode for enzymes involved in the hepatic metabolism of the majority of psychiatric drugs [66]. The presence or absence of functional polymorphisms in these genes is used to determine a person's metabolizer status (poor, intermediate, normal, rapid, or ultrarapid). Based on the existing guidelines, an individual's metabolizer state is then utilized to guide prescription selection and dose [67].

The primary site of action for SSRIs is the serotonin transporter which is responsible for the reuptake. The gene that codes for this transporter is SLC6A4 [68]. There have been many studies looking at the relationship between SCL6A4 and antidepressant response. To recap what happens at the neuron level, let's recap what happens to serotonin when it is produced. In the presynaptic cells, serotonin is stored in the vesicles, so when they are needed, they are transported out of the synaptic cleft where they attach to their receptors on the post-synaptic cleft to cause their therapeutic action. When serotonin is no longer needed, it is taken back up through a transporter called SCL6A4 and stored back in its vesicle for later use. Others are degraded along the way. When serotonin is low, a typical indicator of depression, an SSRI such as fluoxetine can be prescribed. What SSRIs do is they inhibit the uptake of serotonin by the SCL6A4 transport. This can increase the amount of serotonin in the synaptic cleft. A genetic variation of *SCL6A4*, S or Short, is associated with less efficacy of SSRIs and more prone to the side effect profile of that medication class. It is hypothesized that the S variation has lower transporters to be able to recycle serotonin, so the serotonin stays in the synaptic space longer than "normal." If for this patient and SSRI is added, that further increases the serotonin level, which can cause more of the side effects of serotonin and make the medication less efficacious.

Pharmacogenetics testing for antidepressants, antipsychotics, and mood stabilizers can be classified as genes encoding drug-metabolizing enzymes, transporter, and receptors.

## 2.12 Implementation of Psychopharmacogenomics in Psychiatry

There have been significant advancements in pharmacogenomics and expectations of what psychopharmacogenenomics could bring to psychiatric therapy over the years. Nevertheless, there is still concern among many physicians for applying psychopharmacogenomics testing. Some of these concerns are clinical validity and utility of psychopharmacogenomics, cost-effectiveness and reimbursement, regulatory

agencies participatory, training and education of patients and healthcare professionals, ethical considerations, and genetic test availability.

## 2.13 Pharmacogenomics in Pain Management

Pain is one of the most common reasons why patients seek medical treatment. Various types of pain medications are categorized based on their neurophysiological origin and duration. In addition, there are numerous variability in how people respond to the treatment of pain [69].

Pain is a heavy burden to the healthcare system, causing more than $600 billion annually in the management of chronic pain. Lack of pain control, adverse outcomes due to medications, and an increase in the length of hospital stay have all been attributed to the cost [70].

Chronic pain has been considered as a gene x environment interaction, which means having the predisposition combined with environmental factors plays a role in both the development and severity of pain. Therefore, it requires a great deal of knowledge on how the pain signal is transmitted and blocked. This creates a vast variability in the way one responds to various treatments out there for pain [69]. The concept of "one size fits all" is being thought of as ineffective, and the approach of patient-tailored therapy is being used. Precision medicine, therefore, has been looked at when it comes to pain management, and genetic predictors may be partly the cause [69].

Pharmacogenomics has the potential to predict the treatment outcome before a medication is prescribed. Many enzymes are associated with the metabolism of opioids, and depending on the enzyme level activity, the efficacy or the toxicity of the opioids is determined. Variability in opioid metabolism is seen primarily in the enzyme CYP2D6, OPRM1, and their polymorphisms.

### 2.13.1 *CYP2D6*

*CYP2D6* has more than 80 alleles which is the reason behind the phenotype variations. The alleles also vary in ethnic groups [70].

There have been studies looking at ethnic groups that may be prone to one phenotype over the other [70]. About 10% of Caucasians are considered PM. This means inactive medications that require activation by CYP2D6 are not efficacious. Codeine pharmacogenomics study is a well-known example in psychiatry. Codeine is an inactive medication (prodrug) and requires activation of morphine to produce its analgesic effect. On the other side, an individual that is a UM might be converted to morphine too quickly causing adverse effects. There have been well-known documented cases of infant deaths due to opioid toxicity because the breastfeeding mother, given codeine for pain, was an ultrarapid metabolizer [70].

### 2.13.2 *OPRM1*

Genetics review of pain has considered μ-opioid receptor (OPRM1) gene as the major site for analgesics [69]. Studies in mice that had a deletion of the OPRM1 gene highlighted this receptor as the target of commonly used opioids varying the effects of analgesia, dependence, and reward [69].

Pharmacogenomics is a different language that needs to be learned, yet it remains complex to do so. There have been multiple resources that have made efforts to publish evidence-based information which is translated for clinicians to be able to implement in clinical practice. Dutch Pharmacogenomics Working Group (DPWG) in 2005, Clinical Pharmacogenomics Implementation Consortium (CPIC) in 2009, and Pharmacogenomics Knowledge Base (PharmGKB) in 2000 have published evidence-based dosing guidelines based on literature and experts [7, 23, 31, 71–73]. These guidelines are cataloged in the Pharmacogenomics Knowledge Base (PharmGKB) [57, 58].

There is also a PROP™ Pharmacogenetics Calculator developed by the University of Florida that assesses CYP2D6 phenoconversion into practice when a *CYP2D6* genotype is available. The allele types and medications can be entered, and it will show if a phenoconversion happened [21].

## 2.14 Pharmacogenomics in Diabetes

According to the International Diabetes Federation (IDF) Diabetes Atlas 8th Edition (2017), the number of individuals with diabetes worldwide has reached 537 million, and this number with a 46% increase is estimated to be 643 million by 2030 and 783 million by 2045 [74]. Hyperglycemia state of diabetes is due to progressive loss of β-cell mass and/or function which is caused by various genetic and environmental factors [75]. The American Diabetes Association (ADA) Standards of Medical Care in Diabetes recommends individualized treatment to achieve control of hemoglobin A1c (HbA1c) levels quickly after diagnosis of type 2 diabetes mellitus (T2DM) [76]. To date, there are 12 drug classes available for T2DM management and antihyperglycemic therapy:

Biguanides (metformin)

Sulfonylureas: second generation (glyburide, glipizide, glimepiride)

Meglitinides: glinides (repaglinide, nateglinide)

Thiazolidinediones (pioglitazone, rosiglitazone)

Alpha-glucosidase inhibitors (acarbose, miglitol)

Dipeptidyl peptidase-4 (DPP-4) inhibitors (sitagliptin, saxagliptin, linagliptin, alogliptin)

Bile acid sequestrants (colesevelam)

Dopamine-2 agonists (bromocriptine)

SGLT2 inhibitors (canagliflozin, dapagliflozin, empagliflozin)

Glucagon-like peptide-1 (GLP-1) receptor agonists (exenatide, lixisenatide, liraglutide, exenatide, albiglutide, dulaglutide)

Amylin mimetics (pramlintide)

Insulins (lispro, aspart, glulisine, inhaled insulin, human regular, human NPH, glargine, detemir, degludec, and premixed insulin products) [74, 77]

Many patients fail to achieve optimal glycemic control. The degree to which these drugs are efficient or cause adverse drug reactions (ADR) significantly varies within the T2DM population [78, 79]. A major factor is interindividual variability in drug response which is estimated 20–30% of interindividual variability in metabolism, and drug response is related to the genetic factors [80]. To date, several genetic variants have been introduced, the different mechanisms of action on the efficacy and adverse event of antihyperglycemic medications. It gets more and more clear that the treatment outcome of antidiabetic agents depends on variants in a plethora of genes [81–83].

Type 2 diabetes (T2D) is a heterogeneous condition defined by deficiencies in insulin secretion and/or insulin action. This common disease is influenced by genetic and environmental factors. Disease progression is associated with a persistent deterioration in ß cell function as well as an inadequate response to or failure to respond to pharmacologic therapy. Diabetes risk and genetic variations altering drug disposition and/or responsiveness could be associated.

The GWAS method has emerged as a powerful method that has changed the genetic landscapes of T2D and related characteristics, with tremendous success in finding T2D genetic susceptibility loci. More than 400 genomic variants have been introduced associated with T2D [84]. Only about 20% of the overall genetic risk for T2D is explained by GWAS which implies that a considerable percentage of heritability remains unexplained [85]. The main objective of genetic investigations is to translate genetic discoveries into clinically necessary information for improved treatment or cure of T2D.

Despite the advances and challenges, there has been a lot of interest in translating this genetic information into clinical practice, and the T2D pharmacogenomics studies will use the novel information on causal genes/variants at T2D susceptibility loci to help in the molecular taxonomy of T2D and antidiabetic drug selection. It is well understood that there is interindividual variability in pharmacokinetics, pharmacodynamics, and adverse drug reaction to antidiabetic drugs. But compared to oncology or psychiatry, T2D pharmacogenomics is still in its infancy. Diabetic pharmacogenetics can be classified into two categories: (1) drug clinical pharmacology genes that are involved in pharmacokinetics/pharmacodynamics pathways and (2) genomic susceptibility markers underlying T2D pathophysiology [86].

Several gene polymorphisms on the therapeutic response of several antidiabetic medicines have been investigated in recent years. Many independent studies have recently demonstrated genetic heterogeneity in therapy response for major oral antidiabetic drugs (OADs) such as biguanides (metformin), dipeptidyl peptidase-4 inhibitors (DPP-4i), glucagon-like peptide-1 receptor agonists (GLP-1RA), sodium-glucose cotransporter 2 inhibitors (SGLT2i), and sulfonylureas/meglitinides [87].

Metformin is an oral hypoglycemic medication from biguanide drugs used in clinical settings. Metformin's glucose-lowering effect has been demonstrated to have a considerable interindividual variation. Gastrointestinal symptoms were found in 2–63% in different clinical trials [88]. Metformin glycemic response is heritable and can thus be explained in part to genetics. The role of genetics in metformin response variability has been investigated, with an emphasis on pharmacokinetics and pharmacodynamics [89].

Metformin's efficacy is largely dependent on many transport proteins, including members of the OCT family (OCT1-3 (SCL22A1-3), ENT4 (SLC29A4), and MATE1 and MATE2-K (SLC47A1 and SLC47A2)), although is not metabolized in the liver. Polymorphisms in these transporter genes may impact metformin absorption as well as excretion. Consequently, while metformin remains the drug of choice for initiating pharmacological T2D treatment, an increasing number of studies are urging more tailored methods [90, 91].

Sulfonylureas decrease blood glucose levels by inhibiting KATP channels in cells, resulting in increased insulin secretion. Furthermore, SU inhibits hepatic gluconeogenesis (due to higher insulin levels) and insulin clearance in the liver. They are used as a second-line or add-on treatment in the management of T2D [92]. *KCNJ11/ ABCC8*, *IRS1*, *CDKAL1*, *CDKN2A/2B*, *KCNQ1*, and *TCF7L2* are among the various genetic markers that predict sulfonylurea treatment outcomes [93].

GLP1 and gastric inhibitory polypeptides are incretins that are inactivated by DPP4 (GIP). DPP4 inhibitors (DPP-4i) increase the half-life of these incretins, which is associated with increased insulin release and decreased glucagon release. Even while DPP4 inhibitors are usually well tolerated and frequently prescribed as second-line therapy, there is still a substantial interindividual variation in their response to these medications [94]. It seems that genetic alterations of the GLP-1 receptor may change the therapeutic response to *DPP-4i*. *KCNQ1*, *KCNJ11*, *CTRB1/ CTRB2*, *PRKD1*, *CDKAL1*, *IL-6*, *TCF7L2*, *DPP4*, *PNPLA3* are the genes that are associated with response to treatment of T2D with DPP-4i [95].

SGLT2 inhibitors reduce blood glucose by increasing glucose excretion through the urine. SGL2T transporter is encoded by the SCL5A2 gene. *SCL5A2* and *UGT1A9* polymorphisms exhibit some association with response to SGLT2 inhibitors [95].

The GLP-1RA were developed to mimic GLP-1 activity by modifying the structures to resist rapid metabolic degradation. GLP-*1R*, *CNR1*, *SORCS1*, *TCF7L2*, and WFS1 are pharmacogenes associated with GLP-1RA responsiveness [95].

## 2.15 Future Perspective of Pharmacogenetics in Diabetes

It is becoming increasingly clear that the outcome of OAD treatment varies greatly between individuals, implying that a personalized strategy would be preferred. Gene polymorphisms are among the other factors that influence OAD effectiveness. However, recent pharmacogenetic investigations of treatment efficacy in T2D patients discovered the presence of certain limitations that may prevent the implementation of new

knowledge in clinical practice. Furthermore, current guidelines do not consider individual variability in treatment response. Larger studies would bring us closer to establishing tailored treatment for T2D patients and substantially improve clinical outcomes. Pharmacogenomics testing would lead to individual therapeutic benefits for patients, as well as decreased costs for drugs and hospitalization. Furthermore, challenges such as a lack of clinical relevance and implementation knowledge, a lack of established guidelines, and ethical, social, technological, administrative, and economic issues continue to be a concern. Despite the challenges, there are many interesting opportunities for diabetes pharmacogenomics. Technological advances like genome-wide association studies (GWAS) can help pharmacogenomics approach in diabetes management lead to identifying genetic variants that influence drug efficacy and toxicity, finding the mechanism of action of antidiabetic drugs, providing new classification for diabetes, and finding novel targets for drug development based on the molecular basis of diabetes.

## 2.16 Pharmacogenomics Resources

The application of pharmacogenomics in clinical practice needs the understanding of a pharmacogenetics test result by a translation into clinical decision. In recent years, the number of pharmacogenomics publications has increased.

The Clinical Pharmacogenetics Implementation Consortium (CPIC) and the Dutch Pharmacogenetics Working Group (DPWG) have developed pharmacogenomics clinical practice guidelines to help physicians in pharmacogenomics-informed treatment decision-making. These guidelines are used worldwide to interpret pharmacogenomics information into clinical practice [7, 31].

The CPIC is an international consortium of individual volunteers who are interested in facilitating use and providing pharmacogenetic recommended for patient care. As of May 2018, CPIC has published 26 clinical practice guidelines, including recommendations for 26 genes, including over 90 drugs which are freely available on the CPIC website (https://cpicpgx.org/guidelines/) and in PubMed Central (https://www.ncbi.nlm.nih.gov/pmc).

The DPWG guidelines are initially written in Dutch and distributed in The Netherlands. But, the English versions of the DPWG guidelines have been published in 2008 and are currently available through the Pharmacogenomics Knowledge Base (PharmGKB) (https://www.pharmgkb.org).

Pharmacogene Variation (PharmVar) Consortium is a repository for pharmacogene variation that provides a systemic nomenclature system for allelic variations of genes that affect the metabolism of drugs which facilitates pharmacogenetic research (https://www.pharmvar.org/).

Effective pharmacogenomics implementation strategies require clinician/patient education and evidence-based clinical practice guidelines. CPIC and DPWG guidelines are simplifying the implementation of pharmacogenomics in clinical practice by providing specific prescribing recommendations for clinically actionable gene-drug pairs.

## 2.17 Pharmacogenomics Cases in Practice

### 2.17.1 Case Study 1: Cardiology

A 62-year-old female patient has been admitted to the hospital with complaints of chest pain. After numerous tests, it was determined that she has blood clots in her left leg and had a mild stroke. She recovered well and was started on clopidogrel. She had been taking clopidogrel for about a month when she ends up in the hospital again for the same complaints. This time she had a massive stroke that left her partially paralyzed on the right side of her face. A PGx test ordered by a physician revealed the following results:

*Conditions*: diabetes, hypertension, hyperlipidemia

*Medications*: metformin, lisinopril, atorvastatin, and clopidogrel

*PGx results:*

Gene: CYP2C19
Genotype: *2/*3
Phenotype: poor metabolizer (PM)

Clopidogrel is a prodrug and *CYP2C19* is required for its conversion into an active metabolite that has its therapeutic effect. Based on the guidelines and FDA drug labels, patients who are *CYP2C19* PM have decreased efficacy, and alternative antiplatelet medication is recommended. The patient was eventually switched over to apixaban (Eliquis).

### 2.17.2 Case Study 2: Depression

A 29-year-old male patient was admitted to the hospital by a friend who noticed restlessness and uncontrollable muscle movements. He had a history of migraine and depression and had started a new medication a few months ago for his bipolar condition at the usual recommended dose of 400 mg. A PGx test recommended by a pharmacist at the hospital revealed the following results:

*Conditions*: bipolar, depression, and migraine

*Medications*: aripiprazole, rimegepant ODT, and alprazolam

*PGx results:*

Gene: CYP2D6
Genotype: *4/*10
Phenotype: poor metabolizer (PM)

According to the recommendations from the Abilify label, patients with a CYP2D6 PM should be administered half the usual dose. This is due to the risk of increased adverse events as the serum concentration of the medication can increase due to decreased metabolism. The patient's dose was given at a lower dose of 200 mg to start and sent home.

### 2.17.3 Case Study 3: Pain and Phenoconversion

A 32-year-old admitted to a hospital for a broken arm as he fell off a ladder was given a 3-day supply of Tylenol #3. He got his left arm cast and was sent home. He suffered through the weekend with no pain relief and ended up in the same hospital a few days later. However, thinking he might have been a drug seeker, he was denied a refill and given 2 more days' supply of Tylenol #3. While picking up his prescription at his local pharmacy, the PGx-certified pharmacist recommended a PGx test, and the results are as follows:

*Conditions*: depression
*Medications*: fluoxetine and Tylenol #3
*PGx results*:

Gene: CYP2D6
Genotype: *1/*1
Phenotype: extensive metabolizer (EM)

Tylenol #3 is a combination of acetaminophen and codeine. Codeine is an inactive (prodrug) medication. To produce its analgesia, it needs to convert to morphine by using the enzyme CYP2D6. Fluoxetine is also metabolized by the same enzyme, but it also inhibits CYP2D6, therefore inhibiting its own metabolism in the process. Fluoxetine has a higher affinity to CYP2D6 than codeine and therefore does not allow codeine to convert to its active metabolite morphine to create pain relief. This patient was phenoconverted to a PM after codeine was added to his regimen. This patient was eventually switched to morphine for a 5 days' supply and kept on fluoxetine as prescribed.

### 2.17.4 Case Study 4: Schizophrenia and Phenoconversion

A 22-year-old patient suffering from schizophrenia had been well controlled on his current regimen of clozapine. He recently picked up smoking cigarettes as a social norm around his friends. Shortly after, he experiences hallucinations and emotional withdrawal. He makes an appointment with his psychiatrist who has pharmacists working with him as part of the Chronic Care Management (CCM) team. The PGx-certified pharmacist recommends PGx testing, although she already suspects what might be causing his new symptoms.

*Conditions*: schizophrenia
*Medications*: clozapine
*PGx results*:

Gene: CYP1A2
Genotype: *1F/*1K
Phenotype: extensive metabolizer (EM)

The patient is an EM or normal metabolizer (NM) of CYP1A2; however, smoking can significantly alter the metabolism of clozapine causing phenoconversion: normal metabolizer to ultrarapid metabolizer (UM). This causes the serum concentration of clozapine to lower by 20–30% versus non-smokers. In this patient's case, the option was given, to stop smoking if feasible or increase the dose of clozapine.

### 2.17.5 Case Study 4: Comprehensive Case Study

A 73-year-old male presents with symptoms of tremors, dizziness, chronic constipation, and worsening of his depression. He seeks medical attention, and a PGx test is taken. His results are listed below:

*Conditions*: T2D, hypercholesterolemia, restless leg syndrome (RLS), major depressive disorder (MDD), persistent pain, benign prostate hyperplasia (BPH), and gout
*Medications*: Lantus, simvastatin, tamsulosin, citalopram, codeine, allopurinol, and pantoprazole
*PGx results*:

– Gene: *SLCO1B1*
   Genotype: *1/*1
   Phenotype: normal tisk
   Relevant medication: simvastatin
– Gene: *CYP3A4*
   Genotype: *1/*1
   Phenotype: extensive metabolizer
   Relevant medication: tamsulosin
– Gene: *CYP2D6*
   Genotype: *1/*3
   Phenotype: intermediate metabolizer
   Relevant medication: codeine

- Gene: *CYP2C19*
    Genotype: *17/*17
    Phenotype: ultrarapid metabolizer
    Relevant medication: citalopram
- Gene: *HLA-B*
    Genotype: *58:01
    Phenotype: positive
    Relevant medication: allopurinol

HLA-B: The patient has an increased risk for severe cutaneous adverse reactions such as Stevens-Johnson syndrome (SJS) and toxic epidermal necrolysis (TEN) with allopurinol.

*CYP2C19*: Citalopram is rapidly inactivated. A higher dose may be required to offset the inactivation.

*CYP2D6*: The patient might not be getting adequate pain relief due to the slow activation of codeine to morphine. An alternative analgesic may be recommended.

*CYP3A4* and *SCLO1B1*: Normal genotype for both simvastatin and tamsulosin.

# References

1. Meyer UA. Pharmacogenetics–five decades of therapeutic lessons from genetic diversity. Nat Rev Genet. 2004;5(9):669.
2. Relling MV, McDonagh EM, Chang T, Caudle KE, McLeod HL, Haidar CE, et al. Clinical Pharmacogenetics Implementation Consortium (CPIC) guidelines for rasburicase therapy in the context of G6PD deficiency genotype. Clin Pharmacol Ther. 2014;96(2):169–74.
3. Garrod AE. Inborn errors of metabolism: H. London: Frowde and Hodder & Stoughton; 1909.
4. Gonzalez FJ, Skoda RC, Kimura S, Umeno M, Zanger UM, Nebert DW, et al. Characterization of the common genetic defect in humans deficient in debrisoquine metabolism. Nature. 1988;331(6155):442–6.
5. Food, Drug Administration HHS. International conference on harmonisation; guidance on E15 pharmacogenomics definitions and sample coding; availability. Notice. Federal register. 2008;73(68):19074–6.
6. Relling MV, Evans WE. Pharmacogenomics in the clinic. Nature. 2015;526(7573):343.
7. Caudle KE, Klein TE, Hoffman JM, Muller DJ, Whirl-Carrillo M, Gong L, et al. Incorporation of pharmacogenomics into routine clinical practice: the Clinical Pharmacogenetics Implementation Consortium (CPIC) guideline development process. Curr Drug Metab. 2014;15(2):209–17.
8. Giri J, Moyer AM, Bielinski SJ, Caraballo PJ. Concepts driving pharmacogenomics implementation into everyday healthcare. Pharmgenomics Pers Med. 2019;12:305–18.
9. van den Anker J, Reed MD, Allegaert K, Kearns GL. Developmental changes in pharmacokinetics and pharmacodynamics. J Clin Pharmacol. 2018;58(Suppl 10):S10–s25.
10. van den Anker JN, Schwab M, Kearns GL. Developmental pharmacokinetics. Handb Exp Pharmacol. 2011;205:51–75.
11. DiPiro J, Spruill W, Wade W, Blouin R, Pruemer J. Introduction to pharmacokinetics and pharmacodynamics. In: Concepts in clinical pharmacokinetics, vol. 5. Bethesda, MD: American Society of Health-System Pharmacists; 2010.
12. Cacabelos R, Cacabelos N, Carril JC. The role of pharmacogenomics in adverse drug reactions. Expert Rev Clin Pharmacol. 2019;12(5):407–42.
13. Currie GM. Pharmacology, part 1: introduction to pharmacology and pharmacodynamics. J Nucl Med Technol. 2018;46(2):81–6.
14. Lista AD, Sirimaturos M. Pharmacokinetic and Pharmacodynamic principles for toxicology. Crit Care Clin. 2021;37(3):475–86.
15. Derendorf H, Lesko LJ, Chaikin P, Colburn WA, Lee P, Miller R, et al. Pharmacokinetic/pharmacodynamic modeling in drug research and development. J Clin Pharmacol. 2000;40(12 Pt 2):1399–418.
16. Abdel-Rahman SM, Kauffman RE. The integration of pharmacokinetics and pharmacodynamics: understanding dose-response. Annu Rev Pharmacol Toxicol. 2004;44:111–36.
17. Cicali EJ, Elchynski AL, Cook KJ, Houder JT, Thomas CD, Smith DM, et al. How to integrate CYP2D6 Phenoconversion into clinical Pharmacogenetics: a tutorial. Clin Pharmacol Ther. 2021;110(3):677–87.
18. Shah RR, Smith RL. Addressing phenoconversion: the Achilles' heel of personalized medicine. Br J Clin Pharmacol. 2015;79(2):222–40.
19. Shah RR, Smith RL. Inflammation-induced phenoconversion of polymorphic drug metabolizing enzymes: hypothesis with implications for personalized medicine. Drug Metab Dispos. 2015;43(3):400–10.
20. Klomp SD, Manson ML, Guchelaar HJ, Swen JJ. Phenoconversion of cytochrome P450 metabolism: a systematic review. J Clin Med. 2020;9(9):2890.
21. University of Florida Health (UFHealth). Precision Medicine Program; CYP2D6 Phenoconversion Calculator. https://precisionmedicineufhealthorg/phenoconversion-calculator/.
22. Mostafa S, Polasek TM, Sheffield LJ, Huppert D, Kirkpatrick CM. Quantifying the impact of Phenoconversion on medications with actionable Pharmacogenomic guideline recommendations in an acute aged persons mental health setting. Front Psych. 2021;12:724170.

23. Caudle KE, Dunnenberger HM, Freimuth RR, Peterson JF, Burlison JD, Whirl-Carrillo M, et al. Standardizing terms for clinical pharmacogenetic test results: consensus terms from the Clinical Pharmacogenetics Implementation Consortium (CPIC). Genet Med. 2017;19(2):215–23.

24. Group ISMW. A map of human genome sequence variation containing 1.42 million single nucleotide polymorphisms. Nature. 2001;409(6822):928–33.

25. Crews KR, Hicks JK, Pui CH, Relling MV, Evans WE. Pharmacogenomics and individualized medicine: translating science into practice. Clin Pharmacol Ther. 2012;92(4):467–75.

26. Calvo E, Walko C, Dees EC, Valenzuela B. Pharmacogenomics, pharmacokinetics, and pharmacodynamics in the era of targeted therapies. Am Soc Clin Oncol Educ Book. 2016;36:e175–84.

27. Roden DM, McLeod HL, Relling MV, Williams MS, Mensah GA, Peterson JF, et al. Pharmacogenomics. Lancet (London, England). 2019;394(10197):521–32.

28. Lynch T, Price AL. The effect of cytochrome P450 metabolism on drug response, interactions, and adverse effects. Am Fam Physician. 2007;76(3):391–6.

29. Budd WT, Meyers G, Dilts JR, O'Hanlon K, Woody JR, Bostwick DG, et al. Next generation sequencing reveals disparate population frequencies among cytochrome P450 genes: clinical pharmacogenomics of the CYP2 family. Int J Comput Biol Drug Des. 2016;9(1–2):54–86.

30. US Food and Drug Administration (FDA). Paving the way for personalized medicine: FDA's role in a new era of medical product development. https://www.fdanewscom/ext/resources/files/10/10-28-13-Personalized-Medicinepdf (Accessed 26 Nov 2019).

31. Relling M, Klein T. CPIC: Clinical Pharmacogenetics Implementation Consortium of the pharmacogenomics research network. Clin Pharmacol Ther. 2011;89(3):464–7.

32. Picard N, Boyer J-C, Etienne-Grimaldi M-C, Barin-Le Guellec C, Thomas F, Loriot M-A, et al. Pharmacogenetics-based personalized therapy: levels of evidence and recommendations from the French Network of Pharmacogenetics (RNPGx). Therapies. 2017;72(2):185–92.

33. Guo C, Xie X, Li J, Huang L, Chen S, Li X, et al. Pharmacogenomics guidelines: current status and future development. Clin Exp Pharmacol Physiol. 2019;46(8):689–93.

34. Cheng CM, So TW, Bubp JL. Characterization of Pharmacogenetic information in Food and Drug Administration drug labeling and the table of Pharmacogenetic associations. Ann Pharmacother. 2020;55(10):1185–94.

35. Mehta D, Uber R, Ingle T, Li C, Liu Z, Thakkar S, et al. Study of pharmacogenomic information in FDA-approved drug labeling to facilitate application of precision medicine. Drug Discov Today. 2020;25(5):813–20.

36. US Food and Drug Administration (FDA). Table of Pharmacogenomic Biomarkers in Drug Labeling. https://www.fdagov/drugs/science-and-research-drugs/table-pharmacogenomic-biomarkers-drug-labeling. (Accessed 8 Mar 2020).

37. US Food and Drug Administration (FDA). Table of Pharmacogenetic Associations. https://www.fdagov/medical-devices/precision-medicine/table-pharmacogenetic-associations. (Accessed 3 Mar 2020).

38. Weinshilboum R, Wang L. Pharmacogenomics: bench to bedside. Focus. 2006;3(3):739–441.

39. Wysocki K, Seibert D. Pharmacogenomics in clinical care. J Am Assoc Nurse Pract. 2019;31(8):443–6.

40. Centers for Disease Control and Prevention. Medication Safety Program; Adverse Drug Events in Adults. https://www.cdcgov/medicationsafety/adult_adversedrugeventshtml.

41. Kaitin KI. Deconstructing the drug development process: the new face of innovation. Clin Pharmacol Ther. 2010;87(3):356–61.

42. Kaitin KI, DiMasi JA. Pharmaceutical innovation in the 21st century: new drug approvals in the first decade, 2000–2009. Clin Pharmacol Ther. 2011;89(2):183–8.

43. Morgan P, Brown DG, Lennard S, Anderton MJ, Barrett JC, Eriksson U, et al. Impact of a five-dimensional framework on R & D productivity at AstraZeneca. Nat Rev Drug Discov. 2018;17(3):167–81.

44. Check HE. Technology: the $1,000 genome. Nat News. 2014;507(7492):294.

45. Maliepaard M, Nofziger C, Papaluca M, Zineh I, Uyama Y, Prasad K, et al. Pharmacogenetics in the evaluation of new drugs: a multiregional regulatory perspective. Nat Rev Drug Discov. 2013;12(2):103–15.

46. Bray F, Ferlay J, Soerjomataram I, Siegel RL, Torre LA, Jemal A. Global cancer statistics 2018: GLOBOCAN estimates of incidence and mortality worldwide for 36 cancers in 185 countries. CA Cancer J Clin. 2018;68(6):394–424.

47. Sung H, Ferlay J, Siegel RL, Laversanne M, Soerjomataram I, Jemal A, et al. Global cancer statistics 2020: GLOBOCAN estimates of incidence and mortality worldwide for 36 cancers in 185 countries. CA Cancer J Clin. 2021;71(3):209–49.

48. Venter JC, Adams MD, Myers EW, Li PW, Mural RJ, Sutton GG, et al. The sequence of the human genome. Science. 2001;291(5507):1304–51.

49. Chin L, Andersen JN, Futreal PA. Cancer genomics: from discovery science to personalized medicine. Nat Med. 2011;17(3):297–303.

50. Hasanzad M, Sarhangi N, Aghaei Meybodi HR, Nikfar S, Khatami F, Larijani B. Precision medicine in non communicable diseases. Int J Mol Cell Med. 2019;8(Suppl1):1–18.

51. Olopade O, Pichert G. Cancer genetics in oncology practice. Ann Oncol. 2001;12(7):895–908.

52. Meyerson M, Gabriel S, Getz G. Advances in understanding cancer genomes through second-generation sequencing. Nat Rev Genet. 2010;11(10):685–96.

53. National Comprehensive Cancer Network (NCCN). https://www.nccnorg/.
54. Wheeler HE, Maitland ML, Dolan ME, Cox NJ, Ratain MJ. Cancer pharmacogenomics: strategies and challenges. Nat Rev Genet. 2013;14(1):23.
55. Del Re M, Citi V, Crucitta S, Rofi E, Belcari F, van Schaik RH, et al. Pharmacogenetics of CYP2D6 and tamoxifen therapy: light at the end of the tunnel? Pharmacol Res. 2016;107:398–406.
56. Zanger UM, Fischer J, Raimundo S, Stüven T, Evert BO, Schwab M, et al. Comprehensive analysis of the genetic factors determining expression and function of hepatic CYP2D6. Pharmacogenetics. 2001;11(7):573–85.
57. Whirl-Carrillo M, McDonagh EM, Hebert JM, Gong L, Sangkuhl K, Thorn CF, et al. Pharmacogenomics knowledge for personalized medicine. Clin Pharmacol Ther. 2012;92(4):414–7.
58. Whirl-Carrillo M, Huddart R, Gong L, Sangkuhl K, Thorn CF, Whaley R, et al. An evidence-based framework for evaluating pharmacogenomics knowledge for personalized medicine. Clin Pharmacol Ther. 2021;110(3):563–72.
59. Goetz MP, Sangkuhl K, Guchelaar HJ, Schwab M, Province M, Whirl-Carrillo M, et al. Clinical Pharmacogenetics Implementation Consortium (CPIC) guideline for CYP2D6 and tamoxifen therapy. Clin Pharmacol Ther. 2018;103(5):770–7.
60. Amstutz U, Henricks LM, Offer SM, Barbarino J, Schellens JHM, Swen JJ, et al. Clinical Pharmacogenetics Implementation Consortium (CPIC) guideline for dihydropyrimidine dehydrogenase genotype and fluoropyrimidine dosing: 2017 update. Clin Pharmacol Ther. 2018;103(2):210–6.
61. Vahratian A, Blumberg SJ, Terlizzi EP, Schiller JS. Symptoms of anxiety or depressive disorder and use of mental health care among adults during the COVID-19 pandemic—United States, August 2020–February 2021. MMWR Morb Mortal Wkly Rep. 2021;70(13):490–4.
62. Corponi F, Fabbri C, Serretti A. Pharmacogenetics in psychiatry. Adv Pharmacol. 2018;83:297–331.
63. Amare AT, Schubert KO, Baune BT. Pharmacogenomics in the treatment of mood disorders: strategies and opportunities for personalized psychiatry. EPMA J. 2017;8(3):211–27.
64. Sinyor M, Schaffer A, Levitt A. The sequenced treatment alternatives to relieve depression (STAR*D) trial: a review. Can J Psychiatr. 2010;55(3):126–35.
65. Panza F, Lozupone M, Stella E, Lofano L, Gravina C, Urbano M, et al. Psychiatry meets pharmacogenetics for the treatment of revolving door patients with psychiatric disorders. Expert Rev Neurother. 2016;16(12):1357–69.
66. Müller DJ, Rizhanovsky Z. From the origins of pharmacogenetics to first applications in psychiatry. Pharmacopsychiatry. 2020;53(4):155–61.
67. Murphy LE, Fonseka TM, Bousman CA, Muller DJ. Gene-drug pairings for antidepressants and antipsychotics: level of evidence and clinical application. Mol Psychiatry. 2022;27:593–605.
68. Murphy DL, Moya PR. Human serotonin transporter gene (SLC6A4) variants: their contributions to understanding pharmacogenomic and other functional G×G and G×E differences in health and disease. Curr Opin Pharmacol. 2011;11(1):3–10.
69. Peiró AM, Planelles B, Juhasz G, Bagdy G, Libert F, Eschalier A, et al. Pharmacogenomics in pain treatment. Drug Metab Pers Ther. 2016;31(3):131–42.
70. Saba R, Kaye AD, Urman RD. Pharmacogenomics in pain management. Anesthesiol Clin. 2017;35(2):295–304.
71. Relling MV, Klein TE, Gammal RS, Whirl-Carrillo M, Hoffman JM, Caudle KE. The clinical pharmacogenetics implementation consortium: 10 years later. Clin Pharmacol Ther. 2020;107(1):171–5.
72. Hoffman JM, Dunnenberger HM, Kevin Hicks J, Caudle KE, Whirl Carrillo M, Freimuth RR, et al. Developing knowledge resources to support precision medicine: principles from the Clinical Pharmacogenetics Implementation Consortium (CPIC). J Am Med Inform Assoc. 2016;23(4):796–801.
73. Caudle KE, Gammal RS, Whirl-Carrillo M, Hoffman JM, Relling MV, Klein TE. Evidence and resources to implement pharmacogenetic knowledge for precision medicine. Am J Health Syst Pharm. 2016;73(23):1977–85.
74. Sun H, Saeedi P, Karuranga S, Pinkepank M, Ogurtsova K, Duncan BB, et al. IDF diabetes atlas: global, regional and country-level diabetes prevalence estimates for 2021 and projections for 2045. Diabetes Res Clin Pract. 2021;183:109119.
75. Committee ADAPP, Committee: ADAPP. 2. Classification and diagnosis of diabetes: standards of medical care in diabetes—2022. Diabetes Care. 2022;45(Supplement_1):S17–38.
76. Committee ADAPP, Committee: ADAPP. 6. Glycemic targets: standards of medical care in diabetes—2022. Diabetes Care. 2022;45(Supplement_1):S83–96.
77. Committee ADAPP, Committee: ADAPP. 9. Pharmacologic approaches to glycemic treatment: standards of medical care in diabetes—2022. Diabetes Care. 2022;45(Supplement_1):S125–S43.
78. Alomar MJ. Factors affecting the development of adverse drug reactions. Saudi Pharm J. 2014;22(2):83–94.
79. Davies E, O'mahony M. Adverse drug reactions in special populations–the elderly. Br J Clin Pharmacol. 2015;80(4):796–807.
80. Ingelman-Sundberg M, Mkrtchian S, Zhou Y, Lauschke VM. Integrating rare genetic variants into pharmacogenetic drug response predictions. Hum Genomics. 2018;12(1):1–12.
81. Mannino GC, Andreozzi F. Pharmacogenetics of type 2 diabetes mellitus, the route toward tailored medicine. Diabetes Metab Res Rev. 2019;35(3):e3109.
82. Heo CU, Choi CI. Current progress in pharmacogenetics of second-line antidiabetic medications:

towards precision medicine for type 2 diabetes. J Clin Med. 2019;8(3):393.

83. Pearson ER. Pharmacogenetics and target identification in diabetes. Curr Opin Genet Dev. 2018;50:68–73.

84. Cai L, Wheeler E, Kerrison ND, Ja L, Deloukas P, Franks PW, et al. Genome-wide association analysis of type 2 diabetes in the EPIC-InterAct study. Sci Data. 2020;7(1):393.

85. Franks PW, McCarthy MI. Exposing the exposures responsible for type 2 diabetes and obesity. Science. 2016;354(6308):69–73.

86. Huang C, Florez JC. Pharmacogenetics in type 2 diabetes: potential implications for clinical practice. Genome Med. 2011;3(11):76.

87. Mannino GC, Andreozzi F, Sesti G. Pharmacogenetics of type 2 diabetes mellitus, the route toward tailored medicine. Diabetes Metab Res Rev. 2019;35(3):e3109.

88. Nasykhova YA, Tonyan ZN, Mikhailova AA, Danilova MM, Glotov AS. Pharmacogenetics of type 2 diabetes-progress and prospects. Int J Mol Sci. 2020;21(18):6842.

89. Ordelheide AM, Hrabe de Angelis M, Haring HU, Staiger H. Pharmacogenetics of oral antidiabetic therapy. Pharmacogenomics. 2018;19(6):577–87.

90. Florez JC. The pharmacogenetics of metformin. Diabetologia. 2017;60(9):1648–55.

91. Xhakaza L, Abrahams-October Z, Pearce B, Masilela CM, Adeniyi OV, Johnson R, et al. Evaluation of the suitability of 19 pharmacogenomics biomarkers for individualized metformin therapy for type 2 diabetes patients. Drug Metab Pers Ther. 2020;35(2) https://doi.org/10.1515/dmdi-2020-0111.

92. Chaudhury A, Duvoor C, Reddy Dendi VS, Kraleti S, Chada A, Ravilla R, et al. Clinical review of antidiabetic drugs: implications for type 2 diabetes mellitus management. Front Endocrinol. 2017;8:6.

93. Loganadan NK, Huri HZ, Vethakkan SR, Hussein Z. Genetic markers predicting sulphonylurea treatment outcomes in type 2 diabetes patients: current evidence and challenges for clinical implementation. Pharmacogenomics J. 2016;16(3):209–19.

94. Monami M, Cremasco F, Lamanna C, Marchionni N, Mannucci E. Predictors of response to dipeptidyl peptidase-4 inhibitors: evidence from randomized clinical trials. Diabetes Metab Res Rev. 2011;27(4):362–72.

95. Rathmann W, Bongaerts B. Pharmacogenetics of novel glucose-lowering drugs. Diabetologia. 2021;64(6):1201–12.

# Precision Medicine in Oncology and Cancer Therapeutics

**3**

Marius Geanta, Adriana Boata, Angela Brand,
Bianca Cucos, and Hans Lehrach

**What Will You Learn in This Chapter?**
This chapter is dedicated to the scientific milestones which transformed the field of oncology, highlighting the revolution in omic sciences and data science. We identified four distinct time intervals, and for each, we discuss the key innovations that shaped the meaning of the concept "precision oncology."

**Rationale and Importance**
Precision medicine has created many expectations over time, has known many definitions, and has been interpreted in different ways when it comes to implementation. With the increasing number of innovations in the field, there is a need to emphasize what is the true potential of precision medicine at present. The level of complexity and data integration that can be achieved today in relation to cancer control, not only in research but also in clinical practice, should become familiar to all stakeholders. We are all witnessing a rapid transition toward a new reality in which the clinical decisions will become increasingly more complex and marked by nuances, the patient-doctor relationship will change, and healthcare systems will have to adapt. What is clear now is that we have the right tools to understand cancer from multiple angles; oncology best illustrates the concept of precision medicine.

In Europe, the EU Beating Cancer Plan and Cancer Mission are two landmark initiatives in the field of cancer control, which aim to offer a framework for the implementation of all these innovations. Harmonization of actions under

M. Geanta (✉)
Center for Innovation in Medicine,
Bucharest, Romania

United Nations University-Maastricht Economic and Social Research Institute on Innovation and Technology, Maastricht, The Netherlands

KOL Medical Media, Bucharest, Romania
e-mail: marius.geanta@ino-med.ro

A. Boata · B. Cucos
Centre for Innovation in Medicine, Bucharest, Romania

A. Brand
United Nations University-Maastricht Economic and Social Research Institute on Innovation and Technology, Maastricht, The Netherlands

Department of Public Health Genomics, Manipal School of Life Sciences, Manipal Academy of Higher Education, Manipal, India

Faculty of Health, Medicine and Life Sciences, Maastricht University, Maastricht, The Netherlands

Dr. TMA Pai Endowment Chair in Public Health Genomics, Manipal School of Life Sciences, Manipal Academy of Higher Education, Manipal, India

H. Lehrach
Max Planck Institute for Molecular Genetics (MPIMG), Berlin, Germany

Alacris Theranostics GmbH, Berlin, Germany

© The Author(s), under exclusive license to Springer Nature Singapore Pte Ltd. 2022
M. Hasanzad (ed.), *Precision Medicine in Clinical Practice*,
https://doi.org/10.1007/978-981-19-5082-7_3

these programs has the potential to ensure equitable implementation of personalized medicine across the EU.

## 3.1 Introduction

Precision medicine in oncology is defined by the European Society for Clinical Oncology as "the use of therapies that provide benefits to subsets of patients whose tumors have certain molecular or cellular characteristics, most often understood as genomic changes and changes in protein expression" [1]. Since the late 1990s, the potential of genomics in medicine has been recognized, being considered the equivalent of the periodic table in biology and representing the foundation of life [2]. The impact of genomics has changed the entire path of oncology care, from prevention to diagnosis and treatment, and is the basis of the 3P medicine (precise, predictive, and preventive) [3].

Along with genomics, new levels of scientific data can be added for cancer characterization, in addition to traditional clinical data. The development of biotechnology, analytical chemistry, biochemistry, and molecular biology allowed the shaping of the "omic sciences," which are shaping again the meaning of precision medicine. The concept of 3P medicine has evolved into 4P medicine, which adds the "participatory" component, thus recognizing the active role of citizens in health systems. Methods of continuous health monitoring are increasingly accessible to every person (digital applications, wearables) representing an important source of data that informs about medical parameters, but also about behaviors. AI-based diagnostics provide physicians with new tools and skills to make decisions.

Precision medicine is used interchangeably with personalized medicine, the former being used more often by the global community. In Europe, the term personalized medicine is preferred, mostly in the context of EU programs and policies (e.g., Horizon Europe, ICPermed). According to the Council of the EU, *personalized medicine* is defined as a "medical model using characterization of individuals' phenotypes and genotypes (e.g. molecular profiling, medical imaging, lifestyle data) for tailoring the right therapeutic strategy for the right person at the right time, and/or to determine the predisposition to disease and/or to deliver timely and targeted prevention" [4]. The Council acknowledges there is no universally agreed definition of personalized medicine.

In this chapter, we will detail the evolution of scientific innovations in the field of oncology using the term precision medicine, following the ESMO terminology, "as it more accurately reflects the highly precise nature of new technologies that permit dissection of cancer genomes" [1].

The understanding of precision medicine has evolved as each medical discipline becomes a field transformed by new technologies and big data. Precision medicine means data collected at the individual level for better decisions for each person, but it also refers to the integration of this data at the population level for better decisions for specific groups of people (precision public health) [5].

Oncology best illustrates the transformations of precision medicine. On the one hand, "precise" interventions are possible at the individual level (cellular therapies developed for each patient, functional tests that evaluate the individual response to drugs, even Digital Twins). On the other hand, the assessment of individual factors that determine cancer has relevance in public health decisions (screening for hereditary cancers, understanding the contribution of infectious, immune or environmental risk factors involved in cancer mechanisms) [6]. Chronic non-communicable diseases with the greatest burden globally are redefined through the biological revolution that allows deciphering of new mechanisms on multiple biological levels (multiomics) which leads to an abundance of data, relevant both at the individual level and to understanding the diversity of the population [7].

Fighting cancer in the age of precision medicine is a collective mission. The disciplines that help in this fight are longitudinal (genomics, bioinformatics) and require the development of new skills and competences and ultimately will enable the creation of new professions [8].

Since sending the first human to the moon, which required a strategy based on multisectoral collaboration, the moonshot thinking has been applied on Earth to solve the most important challenges of humanity, such as cancer – the Cancer Moonshot (in the USA) and Cancer Mission (in the EU) [9, 10].

## 3.2 Precision Oncology 1.0: New Lenses for an Old Quest

Hippocrates is known for both the first "definition" of *personalized* medicine and *the first description of cancer* based on the observation of its effects in the human body [11]. He believed that the treatment of cancer should be the opposite of the cause of the disease [12]. Over the centuries, the tools to characterize cancer and identify its causes have diversified through the development of pathological anatomy, histology, imaging technologies, and epidemiology. In 1838, at the Institute of Pathology in Berlin, the theory of the cell was established, stating that human tissues are composed of *microscopic* structures, and in the same year, cancer is described as an abnormal clustering of cells and stroma [13].

The development of cancer was understood by the formation of new cells in the affected organs, with the potential to spread to other areas of the body through vascular invasion. It has been shown that there are several forms of cancer even in the same organ and that malignant tumors are different from benign ones. Among the causes of cancer, the following were described: trauma, chronic irritation, smoking, mental illness, alcoholism, and constitutional causes [13].

> Cancers are considered incurable when everybody is ignorant of their microscopic nature, John H. Bennett, pathologist 1840.

Before the Second World War, cancer treatment was mostly surgical, aimed at the complete elimination of the tumor before it can spread in the entire organism. The first attempts to treat cancer by nonsurgical methods appeared at the beginning of the twentieth century, starting from the inventions in physics (the discovery of X-rays) and the observation of the effects of toxic substances (alkylating agents) used during the war period, which were later translated into clinical practice [14]. Compounds derived from these two approaches were the new standards of care for patients with different types of solid tumors and hematological cancer, either alone or in association with surgery [11].

The next wave of innovations was focused on identifying pharmacological agents that could improve systemic cancer treatment and reduce the effects of cytotoxic drugs on healthy cells. By the middle of the twentieth century, it was already known that neoplasms are caused by environmental factors (radiation, chemicals, viruses), and the role of the immune system, the influence of hormones, and even the fact that some cancers are inherited were understood [12]. The theory of aneuploidy was formulated as early as 1914 when it was believed that the abnormal number of chromosomes could be a cause of cancer. The theory is confirmed with the development of cytogenetics, and the "constitutional" causes begin to be elucidated, identifying the mechanisms by which chromosomal changes can lead to the formation of abnormal proteins that alter the cell cycle (e.g., BCR-ABL translocation) and cause the uncontrolled growth of cells [13].

The innovations in precision oncology 1.0 include the transition from observations and symptoms to more accurate diagnosis, new tool tools such as the microscope, the advances in systemic therapy (targeting mechanisms involved in cell cycle and proliferation with chemotherapies), and the increasing understanding of the contribution of modifiable risk factors (environment, lifestyle) [15].

The milestone of the century is the discovery of a new code to decipher cancer. In 1962, Watson and Crick were awarded the Nobel Prize for their *discovery of the molecular structure of DNA*, opening a new phase in cancer research: understanding the genetic basis of the disease [16]. In the following years, many answers were found in the double helix: how common risk factors could act by altering the structure of DNA, how drugs can influence the structure of DNA, or how DNA influences the response to different therapies, even that fragments of DNA can become therapy.

## 3.3 Precision Oncology 2.0: The Double Helix and the Rise of Biotechnology

The development of genetics facilitated the introduction of new tools in the discovery of therapeutic agents. If in the twentieth century the random screening of active substances was the main method to find new drugs, in the 1970s, the discovery of drugs started from the target (*rational drug design*). The list of targeted therapies explored in oncology is beginning to expand from hormonal therapies, tyrosine kinase inhibitors, immunotherapies, to modulators of gene expression.

The discovery of the main gene families had a role in the development of cancer: oncogenes and tumor suppressor genes led to research on therapeutic agents that could antagonize the effects of mutations that affect these genes. In 1979, the TP53 gene, the most frequently mutated in human cancers, was discovered [17]. In the following years, new oncogenes (HER2) and tumor suppressor genes (BRCA1 and 2) are cloned, and the causal relationship with certain types of cancers is understood. For example, mutations in the HER2 gene, which occur in about 25% of women with breast cancer, were associated with reduced survival rates and resistance to chemotherapy [18].

Since 1980, the field of targeted therapies is developing following the discovery of selective kinase inhibitors, which can bind to specific molecules, and, by the mid-1990, monoclonal antibodies, the first types of passive immunotherapy, which can target the proteins expressed on the surface of tumor cells [11].

One of the first success stories that mark the beginning of targeted therapies is the discovery of imatinib, a small molecule that blocks the activity of the BCR-ABL fusion protein, found in chronic myelogenous leukemia (CML). The drug improved the 10-year survival rate from approximately 20% to 80–90% [19].

The transition from biochemistry to biotechnology has allowed the development of therapies that are similar or identical to the complex molecules produced by the human body. In 1975, the first work was published that described the hybridoma technique, a method of generating large amounts of monoclonal antibodies with a high specificity for certain targets. The method has revolutionized medical research and has led to the generation of a significant number of therapies for many types of diseases [20]. In 1997, the FDA approved the first monoclonal antibody used in oncology, rituximab, indicated in a type of non-Hodgkin's lymphoma [21]. One year later, the FDA approved trastuzumab, a monoclonal antibody which targets metastatic breast cancer in which overexpression of HER2 occurs. The therapy was approved simultaneously with an immunohistochemistry test for the selection of patients [22].

Developing drugs along with the diagnostic tests will become standard in the next two decades, with companion diagnostic tests (CDx) defining the disease and the category of patients receiving treatment [23]. Choosing the right treatment for cancer depends on making the correct diagnosis at the molecular level. In stage 2.0 of precision oncology, diseases are reclassified from molecules to symptoms, not the other way around. Recombinant DNA technology, the discovery of PCR, immunohistochemistry, and Sanger sequencing are all new tools to approach cancer.

The first *hallmarks of cancer* were formulated in 2000: unlimited replication potential, high levels of growth factors, loss of the ability to regulate the cell cycle, inhibition of apoptosis, angiogenesis, invasion, and metastasis. In the next decade, two new features will emerge: avoiding the immune response and abnormal metabolic pathways [24].

## 3.4    Precision Oncology 3.0: Cancer, a Diseases of the Genome

The publication of the first human genome after 13 years of research, technological development, and unprecedented collaboration have exceeded expectations, representing a successfully accomplished mission for all humanity, similar to the ambition landing of man on the moon [25]. It was also the beginning of a new era of precision medicine. Massive parallel sequencing became feasible after 2004, allowing for widespread applications in human and tumor genome sequencing. The number of human genomes sequenced increased annually, and it was found that the genomes are in fact 99.9% identical [25]. The mission of the discipline that was born during this period – genomics – will be to decipher what these 0.1% differences mean, understanding the diversity of phenotypes encountered in real life [26].

In the next 10 years, tumors could be tested for larger cohorts of hundreds of patients, mostly by exome sequencing. This stage is marked by the initiation of the international sequencing projects of cancer genomes: The Cancer Genome Atlas (TCGA) initiated in 2005 by NCI followed by the International Cancer Genome Consortium (ICGC), mostly focused on the protein coding regions representing 1–2% of the genome [27]. The development of whole genome sequencing (WGS), RNA sequencing, and new bioinformatic tools contributed to the most complex data set on the entire cancer genome published to date – the Pan-Cancer Analysis of Whole Genomes (PCAWG), a collaboration between TCGA and ICGC [28]. The genomic and transcriptomic profile of 38 types of cancers was detailed, confirming what had become obvious even from the HGP: the human genome is finite, but extremely complex [28]. Many studies were simultaneously published under the Pan-Cancer Atlas, including new data on the role of non-coding regions in the genome, retracing of events leading to malignant cell transformation, and new possibilities for identifying cancers in presymptomatic stages. It

has also become clear that there would be *no magic bullet for curing cancer* [29, 30].

In precision oncology 3.0, all these results from the last two decades of research on cancer genomics is being translated into medical practice, adding another level of complexity to cancer screening, prevention, diagnosis, and treatment. Fifty years ago, there were two main types of hematological cancers, leukemia and lymphoma. Access to molecular data led to the identification of more than 40 types of leukemia and 50 types of lymphoma [31]. Ten years ago, lung cancer was described as two histological subtypes: small cell and non-small cell. Today it can be described by the presence or absence of over 30 genetic mutations. Currently, every new treatment approved for non-small cell lung cancer (NSCLC) requires biomarker testing, and the list of targeted therapies is expanding [32]. In 2020, the FDA approved the three first targeted therapies for NSCLC tumors positive for the MET and RET biomarkers. In 2021, FDA approved the first KRAS targeted therapy, considered an inaccessible target for over 40 years [33]. The EU approval for sotorasib followed a year later.

Until 2010, precision medicine was most often understood as targeted therapies (e.g., anti-VEGF agents or EGFR inhibitors). Over the next decade, the use of biomarkers to guide the therapeutic option is expanding, and multigenic panels are updated as new cancer-relevant genes are identified. A milestone in precision oncology 3.0 is related to the emergence of *tissue-agnostic biomarkers*, announcing another revolution: cancer is no longer defined by the localization [34]. Both FDA and EMA approved pembrolizumab in microsatellite instability-high or mismatch repair-deficient solid tumors, as well as both larotrectinib and entrectinib for NTRK fusion-positive tumors.

The accessibility of high-capacity NGS testing has allowed the development of *comprehensive genomic profiling* (CGP) applications, which involve the simultaneous detection of all classes of genomic changes (SNV, indels, CNV, fusions, genomic markers – TMB, MSI) by evaluating hundreds of genes in one test. Molecular testing in cancer is moving from panels with a limited number of exons to larger panels, including data that can be obtained from germline testing and RNA sequencing. FDA and EMA approved diagnostic panels based on NGS which include 324 or 468 genes and which indicate which patients can be treated with approved oncology drugs or who have indications for a clinical trial [35, 36]. International guidelines (NCCN, ESMO) already recommend comprehensive genomic testing for certain cancers (e.g., *NSCLC*).

MOSCATO 01 is the first study to show that NGS-guided treatment can lead to superior clinical outcomes for patients with solid tumors. Compared to classical treatment strategies, progression-free survival increases if patients are offered targeted therapy according to genomic profiling [37].

More recent evidence shows that all patients with solid metastatic tumors should be tested for germline abnormalities. Moreover, using CGP in patients with cancers of *unknown primary origin* (*CUP*) can identify the tissue of origin in more than half of the cases [38].

*Liquid biopsy* is another approach to cancer diagnosis developed to detect possible molecular abnormalities that work when tissue biopsy is not feasible and later expanded to be relevant for characterizing tumor progression (through repeated testing versus classical biopsy), understanding the mechanisms of resistance, and testing for associated genomic signatures and multiple metastases [39]. The first test using *both liquid biopsy and next-generation sequencing* was approved in August 2020 to identify those patients with non-microcellular lung cancer (NSCLC), the metastatic form, who have certain mutations in the EGFR gene [40]. Three weeks later, a *liquid biopsy pan-tumor test* was approved for solid cancer patients [41].

### 3.4.1 Next-Generation Biopharmaceutical Products

Pharmacological resistance is still a challenge in oncology, even with innovative drugs. Tumor heterogeneity poses a considerable challenge when choosing the right treatment for the right patient at the right time. A single biopsy cannot capture the genomic complexity and heterogeneity of the tumor microenvironment and the scenarios of tumor evolution over time. Multifactorial diseases such as cancer cannot be treated by targeting a single target because they are characterized by complex overlapping molecular pathways. Currently, we are witnessing the fourth stage of biopharmaceutical innovation with *combined therapeutics as standard of care, genetic drugs*, and *hybrid constructs*.

### 3.4.2 Multi-targeted Therapy

Research involving large-scale tumor sequencing and studies that include patients according to their molecular profiles report actionable mutations in driver genes in up to 40% of cases, but the number of patients who are ultimately treated with genomic-guided targeted therapy is much lower (10–15%) [42]. Over time, data have been accumulated showing that genetic factors determine the variability of response to drugs in the proportion of 20–90%.

In the field of small molecules, the most recent approaches that are already explored in clinical practice aim for multiple mechanisms: using multikinase inhibitors (e.g., sorafenib) and combinations of therapies targeting several molecules of the same signaling pathway—*vertical pathway inhibition* (e.g., BRAF and MEK inhibitors in melanoma) [43]. Another strategy explored in clinical trials is the combination of several types of tyrosine kinase inhibitors (BRAF + MEK + PI3K) involved in different pathways [44].

### 3.4.3 New Ways of Reprogramming the Immune System

If in the past the focus was only on the interaction between the target and the ligand, today, the goal is to characterize the effect of a drug on a biological system and how this effect can be optimized. Immuno-oncology has become one of the main pillars in the treatment of cancer, having a key role in modulating the tumor microenvironment. There are several categories of next-generation immuno-oncological drugs already on the market, such as checkpoint inhibitors, CAR-T cell therapies, and vaccines. As in the previous years, in 2020, despite the pandemic, the EMA and the FDA approved more immunotherapies in oncology (35) than chemotherapy (17) or small molecules (31). In 2021, the FDA marks the 100th approval of a monoclonal antibody (dostarlimab, indicated in endometrial cancer) [45].

Monoclonal antibodies have become an important pillar in the treatment of cancer due to the property of binding with specificity to a target molecule. Checkpoint inhibitors have already entered the first line of treatment for several indications, and in certain categories of patients such as those with high-risk melanoma and more recently lung cancer, immunotherapy is also used in the adjuvant stage. In 2021, the first immunotherapy (atezolizumab) was approved for patients with stage II–IIIA lung cancer [46]. In advanced lung cancer, immunotherapy leads to long-term survival. Pembrolizumab doubled overall 5-year survival in PD-L1-positive metastatic NSCLC, compared to chemotherapy [47].

Despite their proven effectiveness, monoclonal antibodies do not always work in monotherapy. Combining technologies already used on a scale for the production of antibodies with new technologies (e.g., AI tools for drug modeling), we are witnessing a modern phase of biotechnology in which new constructs have emerged, based on molecular structures derived from classical antibodies. The new classes of immunotherapies therapies include *bispecific, trispecific, or multispecific antibodies, oligoclonal antibodies, and antibody-drug conjugates (ADC)*. New types of therapeutic constructs are developed with the aim of increasing specificity and facilitating direct interaction between immune cells and tumors while decreasing systemic toxicity.

Advanced biotechnological products such as bispecific antibodies are evaluated in preclinical studies to target neoantigens derived from the RAS gene family [48]. *Trispecific antibodies* allow to bind three different epitopes on cancer cells and represent an important approach that

can address the multifactorial nature of cancer [49]. There are no approved therapies yet, but several agents are in clinical and preclinical trials. *TriKe-s* (trispecific NK cell engagers) are a category of experimental therapies of this class, developed for hematological cancer, which target NK lymphocytes (natural killer) and not T lymphocytes as in previous generations of immunotherapies [50].

### 3.4.4   Advances in Adoptive Cell Therapies

The development of technologies that allow the reprogramming of immune effector cells has opened a new stage in the treatment of cancer, which culminated in the success of CAR-T cell therapies, considered the highest possible degree of personalization of cancer treatment: gene therapy, cell therapy, and immunotherapy at the same time. These involve genetically modifying the patient's T lymphocytes and reintroducing them into the body to specifically recognize and destroy cancer cells. CAR-T was also considered a "living drug," being present in the human body as long as there is a target to be neutralized. However, data about the real long-term persistence of CAR-T cells was unavailable until recently. In February 2022, a study published in *Nature* by *Melenhorst* et al. shows that in two patients with chronic lymphocytic leukemia who achieved a complete remission in 2010, CAR-T cells remained detectable more than 10 years after infusion. CLL is the first cancer in which CAR-T cells were studied [51].

CAR-T cell therapies are a revolution in cancer treatment representing the highest level of personalization and addressing unmet needs, especially for hematological cancers. Current manufacturing of CAR-T therapies involves complex and expensive processes to modify cells ex vivo prior to their administration which represents a barrier to ensuring broader access to such innovative therapies.

One important direction of research is focused on "off-the-shelf" cell therapies, which are moving to early-phase clinical testing, as a promising solution for on-demand, allogenic therapies that could work for broader categories of patients [52]. Another direction explored in trials is targeting multispecific antigens CAR to overcome antigenic heterogeneity, with important applications in solid tumors [53].

New evidence suggests that, at least for some indications, CAR-T therapies could even be generated directly in the body (in vivo *CAR-T therapy*) without the need to remove lymphocytes from the blood, modified using genetic engineering, and re-administered. Different mRNA delivery systems (nanoparticles, viruses) are explored in preclinical trials to enable in situ programming of immune cells [54, 55]. Some approaches are using adeno-associated viruses as a way to make CAR-T therapies on demand and less expensive for human T-cell leukemia [56].

Moreover, novel adoptive cell therapies could even overcome some of the limitations encountered in solid tumor treatment. A recent National Cancer Institute (NCI) pilot study demonstrated the potential *of tumor-infiltrating lymphocytes (TILs)* to activate immune mechanisms against breast tumors. The results suggest that the applications of this therapy could be significant especially in *HR+ breast cancer*, long been considered poorly immunogenic [57].

### 3.4.5 Therapeutic Vaccines

With the COVID-19 pandemic, the potential of RNA-based vaccines has been internationally recognized. Vaccines can also be used with therapeutic intent, being explored today in the quest to treat cancer. The concept is similar to prophylactic vaccines – the immune system is presented with an antigen and then responds when it encounters a cell that expresses that protein [58]. BioNTech has developed a personalized cancer vaccine (BNT111), evaluated in the treatment of metastatic melanoma, which already entered phase II trial [59].

Moderna began working on a highly personalized therapeutic vaccine (mRNA-4157, which can encode up to 34 neoantigens). Sequencing of cancer cells allows the identification of neoanti-gens specific to the patient and subsequently the development of a vaccine tailored to the patient's needs. Another vaccine (mRNA-5671) encodes four of the most common KRAS mutations and is currently being evaluated for the treatment of non-small cell lung cancer, colorectal cancers with microsatellite instability, and pancreatic adenocarcinoma [60].

Oncological vaccines can be combined with already existing immunotherapies to optimize the therapeutic response. It is expected that these types of vaccines may become part of a "cocktail" of oncological therapies (e.g., therapeutic vaccine + checkpoint inhibitors) [61, 62].

Clinical Trials in the Big Data Era: Early-Phase, Biomarker-Driven Trials Change Traditional Models

Traditionally, only patients who remain without therapeutic options are enrolled in clinical trials. Today, the goal of precision medicine is to choose the right treatment for the right patient, so clinical trials have become essential for expanding treatment options for cancer patients. Classical biomedical research uses unique hypotheses in conducting extensive clinical trials (testing the effect of a drug on a large population). In precision oncology 3.0, the criteria for inclusion in clinical trials and the design of these studies changed. Because therapies are accurate, they only work for selected categories of patients who have certain molecular markers. In this data-driven stage, prospective studies can be designed to allow a more precise approach to the study of multifactorial, complex diseases such as cancer. *Basket and umbrella designs* are introduced to optimize the development of drugs that require biomarkers [63]. An umbrella protocol means that the patient's eligibility is defined by the presence of a molecular change and tests several therapies that can address that change. In basket-type trials, patients with different tumors (histologically) but with a certain molecular change receive the same targeted therapy. These have increased the access of patients with rare cancers to innovative cancer therapies [64].

Precision medicine in oncology also means *regulatory changes*. Traditionally, an oncology drug approved for a certain localization had to go through another study to be approved for a different indication. Since 2017, when the first site-agnostic therapy was authorized, there have been changes in the regulation of studies. A drug can be approved today based on data collected from a small group of patients, completely different from the requirements of extensive phase III studies [63, 65]. The study which led to the approval of larotrectinib in 2018 included only 55 patients diagnosed with several types of cancers (according to the traditional localizations) but with a common driver: *NTRK fusions* [66]. Based on the same design master protocol, histology-agnostic studies continued for other therapies, such as *RET inhibitors* [67].

## 3.5 Oncology 4.0: Fighting Cancer from Multiple Fronts in the Fourth Industrial Revolution

In physics, the first dimension (1D) is defined with a single coordinate. We add another coordinate and get the X and Y directions that characterize the 2D world. The world we live in has length, height, and depth. Starting with the fourth dimension, if time is added, the equations get complicated and so is the human cognitive ability. Understanding cancer started from a one-dimensional perspective, moving from location to molecules, to the interaction between molecules, and now the transition is to understand it as a continuum, where it is important to integrate several coordinates, including the temporal dimension.

Most cancers are still diagnosed in symptomatic and terminal stages (e.g., lung cancer), in stages in which many components of a biological system are dysfunctional. Most cancer control interventions have so far been static and focused on isolated components of biological systems. Current technology allows for a characterization of cancer beyond localization, which goes as far as deciphering combinations of genetic mutations, abnormal proteins, and metabolites – even at the level of a single cancer cell. It also allows spatial and temporal understanding of variations between cells and interaction with surrounding tissues. By integrating multiple (multiomic) data sources, *each case of cancer could be approached as if it were a separate disease.*

In precision oncology 4.0, the characterization of cancer will be *high resolution* (single-cell analysis), *dynamic* (testing parameters in multiple points over time), *comprehensive* (multiomic marker testing, transition to molecular signatures and *driver pathway* versus *driver genes*), and *multidimensional* (analysis of complex molecular networks). This phase is dominated by data, and the characterization of cancer becomes a cognitive problem for doctors and researchers. The number of variables on which decision-making processes in oncology practice are based is facing an unprecedented expansion. Studies show that the human being can process up to five factors in the decision-making process [68], but the variables which can influence a decision in oncology can rise up to several thousand if we take into account the omics data.

The *Fourth Industrial Revolution* is defined by blurred lines between the physical, biological, and digital worlds [69]. COVID-19 demonstrates the importance of understanding the individual response to a disease. Now we can see at an unprecedented scale that each person responds differently to COVID-19 infection, vaccines, and therapies. And the intersection of science and technology accurately shows which parts of a biological system are dysfunctional.

### 3.5.1 Single-Cell Multiomics for Precision Medicine

To establish the diagnosis, in the classical approach to cancer, histopathological analysis of the sample is made to identify the tissue and the predominant type of cells, and then the returned results reflect an average. However, cells of the same type behave differently to a variety of stimuli (environmental factors, drugs, interventions, etc.). Understanding complex processes such as metastasis and therapeutic resistance depends on deciphering intratumoral heterogeneity.

General observations, such as the fact that malignant cells next to healthy tissue have the property to invade neighboring tissues, while cells inside the tumor have the maximum capacity for proliferation, with a very active metabolism, were followed by more detailed mechanisms [29]. The development of pathological anatomy and molecular biology allows the analysis of smaller cellular components, from smaller tumor samples. The latest application of genomics is represented by sequencing technologies at individual cell resolution. Also, methods are explored to perform several types of simultaneous analyzes, a strategy called single-cell multiomics (DNA analysis, RNA, proteins, metabolites; e.g., Tea-seq, SNARE-seq), which is proving essential for the characterization of the tumor microenvironment [70, 71].

The goal is to understand how individual cells behave from multiple angles, at various points in time, and to identify the response to physiological or pharmacological stimuli. Tumor heterogeneity is one of the main obstacles in the development of new effective therapeutic options. Non-genetic heterogeneity is a less explored dimension that can predict tumor dynamics and phenotype diversity much better than simply determining genomic profile [72]. During the COVID-19 pandemic, for example, single-cell sequencing technologies were used to study the response of immune cells to infection and to study the mechanisms by which some people experience severe forms of the disease, while others are asymptomatic [73].

Single-cell analysis platforms allow early identification of indicators of *response or relapse to CAR-T therapy*. Despite impressive advances in CAR-T cell therapies for haematological cancers, over 50% of patients report disease progression. Over time, clinical and biochemical markers that can predict the response to CAR-T therapies have been studied, such as disease burden and LDH values. In acute lymphoblastic leukemia (B-ALL), for example, studies report that between 30 and 95% of cases show a loss of CD19 antigen, associated with relapses [74]. Single-cell proteomic platforms can measure the levels of molecules secreted by immune cells and can predict the response to CAR-T cell therapies in hematology cancers and the response to TILs in solid cancers (e.g., lung cancer), according to recent data [75]. In the case of solid cancers, this method can be used to obtain an indicator called PSI (Polyfunctional Strength Index) which shows the number of cells capable of secreting several types of cytokines. This works as a marker that shows the potency of cell therapy but also the effectiveness of therapeutic vaccines [76].

Despite the significant advantages proved with single-cell analysis, the current methods are still facing some limitations. The positional information of various cells is lost when single cells are dissociated from the tissue [77]. This information is important to understand the cell stages involved in carcinogenesis. Spatial genomics and transcriptomics are complementary technologies that are revolutionizing the study of cells and tissues. DNA sequencing can be performed at the level of tumor cells in relation to the spatial location. This method preserves the tumor architecture by analyzing intact tissue sections and allows the heterogeneity of neoplastic cells to be captured [78].

### 3.5.2 Network Medicine and AI

*Network medicine* can optimize the implementation of precision medicine by overcoming several limitations of classical siloed approaches: characterization of intratumoral heterogeneity; genomic, transcriptomic, and epigenomic interactions; patient stratification; and identification of new biomarkers [79].

It has become obvious that the presence of driver mutations is not enough to inform the best answer to therapy or patient evolution. In many cases, no driver mutations are identified. The new dimension in precision medicine involves modeling the interactions between somatic tumor mutations and capturing information about the pathways involved in carcinogenesis. Systems biology approaches based on network analysis can also be applied to genomic data. Biological systems are complex. Patients are different not only by genome but also by other types of interactions both between genes and with other levels of biological organization. There is a transition from the identification of driver mutations to the characterization of driver pathways [79].

Most hematological cancers cannot be treated solely based on the molecular profile. Multiple myeloma (MM) is a type of hematological cancer characterized by great clinical and molecular heterogeneity. Patients initially respond to standard therapy; however, most of them relapse and need to follow multiple lines of treatment. About 15% of cases are in the high-risk category and have a relapse in the first 2 years after diagnosis [80]. Recently, Mount Sinai Medical School proposed a new method for classifying multiple myeloma (MM), based on the MM-PSN (Multiple Myeloma Patient Similarity Network) algorithm, which integrates complex data sets from the genomic and transcriptomic analysis. The data was processed using a computerized system that identified 12 subgroups of the disease, characterized by different risk profiles and for which specific therapies can be identified. The functional characterization of the subgroups within the MM-PSN network provides information with immediate implications for the choice of therapy and clinical trials. Each subgroup may respond differently to targeted therapies and immunotherapies. For example, the groups of patients who do not respond to anti-BCMA CAR-T therapy or the patients who have the best response to targeted venetoclax therapy have been defined [81].

### 3.5.3 Functional Tests and Organoid-Driven Precision Medicine

Testing of molecular biomarkers so far enabled us to evaluate whether a target is present or not, which is no longer the only factor influencing clinical decision-making. In real life, patients may present with the same actionable mutations but have different responses to therapies. Precision oncology 4.0 implies transitioning from measuring a static property of cells and tissues to a greater emphasis on assays that measure dynamic properties such as the effects certain interventions could have in vivo. In the last decade, technologies based on tumor organoids derived from patients have been developed to obtain tissue models of different types of cancers, and their applications have ranged from basic research to drug development [82]. Organoids are three-dimensional cellular structures that recreate the characteristics of the organs to be studied in vitro [83].

These preclinical models allow linking the tumor profile with possible treatments in a systems biology approach. *Larset* et al. recently reported a study in which organoid cultures from over 1000 patients showed genomic and transcriptomic similarities with the original tumor. A high-capacity platform based on *neural networks* can assess the specific heterogeneity of the patient in terms of response to therapy, with applicability in many types of solid cancers [82]. Another important development is related to biobanks (living organoid biobanks) containing organoids representing different types of cancers which facilitate the development of drugs, research, and tools for personalized medicine [84].

The development of ex vivo organoid models for drug screening. The contributions in the field of engineering in biology have made it possible to overcome the barriers associated with these tests. Increasingly complex microphysiological systems (e.g., organ-on-chip) are being developed to simulate the effects of cancer in the tumor microenvironment but also the interaction with the immune system (3D tumor models, tumor-immune co-cultures). In the next 5 years, it is estimated that function precision medicine (FPM) will become a standard tool for clinical evaluation in oncology. FPM is based on the exposure of different tissues from patients to drugs, which can provide data on the contribution of non-genomic mechanisms [85].

A study published in 2021 reports on a technique based on high-resolution microscopy and AI, to determine the effect of cancer drugs on individual cells (both tumor and normal) – pharmacoscopy. This was used to test the efficacy of 139 oncological therapies on tumor cells taken from patients with aggressive hematological cancers. For more than half of them, guiding therapy using the functional test resulted in a clinically significant benefit: survival rates without disease progression increased by at least 30% compared to standard treatment. For the first time, a functional test to guide cancer therapy called single-cell functional precision medicine (scFPM) is feasible [86].

### 3.5.4 A New Era of Biomarkers: Genomic Signatures, Composite Biomarkers, Driver-Pathway Biomarkers, Synthetic Biomarkers

Single-gene biomarkers are accessible from a testing and data analysis perspective, but they offer a one-dimensional view of cancer. As the diversity of tumor cells and interaction with the human body is exposed, new types of biomarkers are emerging: *multigenic biomarkers* (integrating data on multiple genes) and *composite biomarkers* (which include genomic data along with other types of data). Extensive research projects in oncology aimed at sequencing a large number of tumor genomes have identified new challenges related to the definition of complex phenomena that occur in the genome. Certain genomic changes that occur in cancer, such as *chromotripsy*, are so complex that there is no unanimously accepted definition of them or means to quantify them [28].

*Genome instability* and *tumor mutational burden (TMB)* are biomarkers mostly used in research, but which have also been in practice in recent years, for several types of tumors. TMB-H (tumor mutational burden high) is used to identify individuals who may benefit from immunotherapy on the assumption that these solid tumors are immunogenic [87]. TMB is considered an agnostic tumor genomic signature, which describes the overall rate of somatic mutations but does not dif-

ferentiate between mutations. Not all mutations generate neoantigens, and not all of them are recognized by T lymphocytes. Certain tumor types may respond very well to checkpoint inhibitors even with moderate levels of TMB. In the case of Merkel cell carcinoma, for example, good responses with immunotherapy are obtained, even in cases with low TMB [88]. Evidence published last year shows that TMB-H is not a universal biomarker that can predict response to checkpoint inhibitors immunotherapy in patients with solid tumors, as previously assumed. In the study, TMB-H was used successfully as a predictive biomarker in only one subgroup of the population included in the study: cancers in which the number of CD8-positive T lymphocytes (CD8+) correlates with TMB-H [89]. Research is currently focused on the possibility of qualitative evaluation of TMB and more insights into the neoantigenic makeup [42].

Furthermore, recent studies show that *composite biomarkers* allow the assessment of the response to immunotherapy with greater accuracy than individual ones [90]. The immune response, the microbiome, the genomic profiling, and the previous lines of treatment can influence how checkpoint inhibitors work. CheckMate 275 and PURE-01 are studies showing the usefulness of composite biomarkers to guide immunotherapy [91, 92].

The new generations of biomarkers will be developed through "big data" analysis, based on machine learning algorithms, which integrate thousands of genes, transcripts, proteins, and metabolites, which allows better integration of the flow of information between biological molecules.

The integration of various tumor data sets also allows the identification of *immune archetypes*. Tumors contain several types of immune cells, in addition to T lymphocytes which are the main target of immunotherapies. Cancers that occur in different locations in the body may be immunologically similar. Some cases of melanoma are more immunologically close to lung cancer, for example, than other types of melanomas. The classification according to immune archetypes would allow optimizing the selection of patients for clinical trials, indicating which tumors would be more responsive to immunotherapy [93].

2–5% of newly diagnosed cancers are *cancers of unknown primary origin*, a term dating back to the 1940s. 85% of cases present with poorly differentiated tumors with chromosomal instability and an unfavorable prognosis [94]. Randomized studies have shown that guiding treatment according to histological criteria does not work, with chemotherapy being the only treatment available. Recent NGS-based studies have shown that CUP tumors have at least 4–5 genomic abnormalities, and about 90% of cases have a unique molecular profile. Given this molecular heterogeneity, research directions include the use of multiple biomarkers to characterize diseases currently in the CUP category, the identification of molecular signatures, and the exploration of several therapies in clinical trials [95].

Detecting tumors before symptoms appear could maximize the effectiveness of medical and surgical interventions in localized cancers. Biomarkers released from early lesions cannot be easily detected, have a short circulation time, and are not a high enough concentration. A new class of diagnostic tools builds on advances in synthetic biology and chemistry to develop synthetic biomarkers that function as sensors that amplify biological markers released by tumors in the early stages of development. Such biomarkers are currently being evaluated in preclinical studies, and for the translation into the clinic, it will be necessary to characterize the biology of early lesions and tumors in situ. The Human Tumor Atlas Network (HTAN) project, initiated by the NCI Cancer Moonshot, will make comprehensive three-dimensional molecular and cellular maps of a variety of cancers at different stages of evolution. Thus, the mechanisms involved in the transition from precancerous to malignant formations can be captured at high resolution.

### 3.5.5 Driver Networks as Biomarkers

In the classical understanding, a biomarker is an objective characteristic (molecular, anatomical, physiological, or biochemical), indicative of a normal or pathological biological process. Although they are synonymous with precision medicine and promise to revolutionize medical practice, there are still limitations to the use of classical gene-centric biomarkers, which follow a linear relationship and do not capture the mechanisms that occur in the pathways in which these genes are involved. Certain driver genes cannot be targeted directly because of their structure or because they cause a loss of function – *tumor suppressor genes* (this is why so far targeted therapies mostly work against oncogenes). The discovery of biomarkers goes toward mechanism-centric approaches, which take into account gene regulation networks, network-based on protein interactions, and so on.

*Synthetic lethality* is a concept known since the 1920s but which has been translated into the development of oncological drugs only in the last 15 years. This mechanism implies that the loss of function of some genes is tolerated when it occurs in isolation, but not in combination with other genes. For example, in breast cancer, biomarkers that conventionally evaluate a single gene (ERBB2, BRCA1, BRCA2, or more recently PIK3CA) are conventionally tested in clinical practice. In recent years, however, interest has shifted toward the examination of mutations occurring in multiple genes such as the molecular pathway HRR (homologous recombination repair), which includes BRCA1 and 2. Testing of other genes besides BRCA1 and 2 for the assessment of DNA repair deficiency allows the expansion of the categories of patients who can benefit from treatment with a PARP inhibitor (the first class of cancer drugs that rely on synthetic lethality).

### 3.5.6 The Future of Precision Oncology: The Rise of Digital Twins

The goal to personalize medicine is an old theme approached in different ways across centuries. While technology enables doctors and researchers to be more precise, the ambition of medical interventions tailored specifically to the person's needs has yet to reach its full potential. In the previous sections, we have searched through history on what precision means in relation to cancer, from the basic understanding of human anatomy and interaction with the environment to "dissecting" cells and molecules. Isolated systems, states, and mechanisms of the human body are being described with higher precision every day, and this knowledge allows for better research and better care. What is clear now is that we have the best tools to understand cancer from multiple angles. With the advent of network medicine, we are going one step forward, trying to connect the dots that make us unique. Ultimately, the quest is to find better models for choosing the right option for the individual, among many variables that are rapidly accumulating.

Many more answers can be found if we go back to the definition of the Fourth Industrial Revolution: the connectivity between the physical, digital, and biological worlds. Humanity has learned that in complex situations we cannot avoid making mistakes and many industries, disciplines (aerospace, automotive, climate), and even cities are using digital models to reduce the risk of errors and optimize production cycles or products.

The same can be applied in medicine to choose the right intervention for the right patient/person by filtering many mistakes before the intervention is applied to real human beings [96]. The *Digital Twin* is a model which integrates many possible variables about an organism with modern technologies, such as sensors, data analytics, and machine learning [97].

In oncology, digital twins could be used to identify individuals who are predisposed to cancer or to test possible interventions that could stop or delay the onset of disease. For therapy choices, a digital model could be used to predict the effects of a range of drugs and to truly choose *the best treatment for the right patient at the right time* [98]. Studies have shown medications are ineffective for 38–75% of people with common diseases [97]. Digital twins can identify the therapies that work best for each person, reducing both the risk to the individual and the costs to the healthcare systems.

Computer-aided models are shaping the present and the future of medical care and prevention but also the future of well-being. Each individual could have their own model or set of models from the moment of birth to old age to be protected against potential mistakes in medical care, and in treatment, but also against other factors related to behaviors or environment.

Furthermore, the same could work for healthcare institutions, which can also have their own types of digital twins to make their processes more efficient. By connecting different types of data (biological data, health records, patient-generated data, health determinants), the digital twin technology can advance research on personalized medicine, and it can promote the development of a system of continuous learning and improving decision-making [97].

Forbes predicts that digital twins will become one of the biggest healthcare tech trends in 2022 [99]. Digital twins will be key to delivering highly personalized approaches to routine cancer care, as costs of big data technologies decline [100].

Personalized medicine has been a transversal theme of many European and national health policies for several years, with oncology being the most advanced field for implementation. Recently, the European Commission has launched the Implementation Roadmap for the EU Beating Cancer Plan which includes an important action related to the creation of a repository of digital twins in healthcare supporting cancer treatment [101] (Fig. 3.1).

| Programme | Objective | Actions supporting the implementation of precision medicine in oncology | Timing |
|---|---|---|---|
| EU Beating Cancer Plan Roadmap | Supporting cancer prevention and care through new cancer research and and innovative ecosystem | Create a 'Knowledge Centre on Cancer' | 2021-2025 |
| | | Launch 'European Cancer Imaging Initiative' (ECII) | 2021-2025 |
| | | Cancer patients securely accessing and sharing electronic health record through European HealthData Space (EHDS) | 2021-2025 |
| | | Support secondary access to data | 2021-2025 |
| | | Repository of digital twins in healthcare supporting cancer treatment | 2021-2024 |
| | | Expand the European Cancer Information System (ECIS) | 2021-2025 |
| | | Launch Horizon Europe partnerships: Innovative Health Initiative (IHI) & Partnership on Transformin Health and Care | 2021-2025 |
| | Ensuring high standards in cancer care | Creation of 'National Comprehensive Cancer Infrastructures' and EU network* | 2021-2025 |
| | | European Initiative to understand Cancer (UNCAN. eu)* | 2021-2025 |
| | | Set up Partnership on Personalised Medicine | 2021-2023 |
| | | Roadmap to personalised prevention | 2021-2022/2024 |
| | | High Performing Computing to rapidly test existing molecules and new drug combinations | 2021-2023 |
| | | Assist researchers working on personalised cancer treatments through tailored support and new digital platforms | 2021-2023 |
| | | Launch 'Genomic for Public Health' project | 2021-2025 |
| | | 1+ Million Genomes Initiative | 2021-2025 |
| | Improving the quality of life for cancer patients, survivors and carers | Create a tailor-made 'Cancer Survivor Smart-Card' | 2021-2025 |
| | | Create the 'European Cancer Patient Digital Centre' | 2021-2025 |
| Cancer Mission Implementation Plan | Improve the understanding of caner | Establish the 'UNCAN.eu' platform by 2023* | 2021-2023 |
| | | Better understand healthy versus cancer cells at individual and population level | 2021-2028 |
| | | Better understand cancer-patient molecular, cell, organ, organismal interactions (2023-2029): | 2023-2029 |
| | | Determine the role of genetics in cancer | 2024-2030 |
| | Prevent what is preventable | Develop a one-stop cancer information centre on prevention | 2022-2026 |
| | | Boost reseatch and innovation into risk assessment | 2021-2023 |
| | | Conduct implementation reseatch on cancer prevention | 2022-2030 |
| | | Establish synergies on prevention with other missions | 2023-2026 |
| | Screening and early detection | Optimise and improve access to existing screening programmes | 2024-2029 |
| | | Develop new methods and technologies for screening and early detection | 2021-2026 |
| | | Develop early predictors/tests | 2025-2030 |
| | Optimise diagnostics and treatment | Support the creation of a Network of Comprehensive Cancer Infrastructures (CCIs) by 2025* | 2021-2025 |
| | | Develop twinning programmes | 2024-2028 |
| | | Develop a clinical trial programme on diagnostics | 2023-2027 |
| | | Develop a clinical trial programme on treatments (optimising treatments through advanced personalised medicine approaches) | 2022-2023 |
| | Support quality of life | Collect and analyse data on today's unmet needs of cancer patients and survivors | 2021-2025 |
| | | Set up of the European Cancer Patient Digital Centre | 2021-2023 |
| | | Develop early predictors for quality of life | 2024-2030 |
| | | Design monitoring programmes for survivors of childhood cancer | 2021-2027 |

**Fig. 3.1** Actions supporting the implementation of precision medicine in oncology extracted from the EU Beating Cancer Plan Roadmap and Cancer Mission Implementation Plan

## 3.6  Implementing Precision Oncology in Europe

### 3.6.1  How Do We Make P4 Medicine a Reality in Medical Practice by Involving Patients, Doctors, Members of the Medical Community, and Citizens?

Starting from the simplest definition, precision medicine involves choosing the right treatment for the right patient at the right time. We need to standardize the right tools to help doctors make that choice but also standardize what precision medicine means to each stakeholder. Precision medicine has created many expectations over time. From the patient's perspective, in traditional health systems, there is a tendency to consider that more is better when it comes to medical care and to look for alternatives. As the number of diagnostic and therapeutic solutions increases, patients may come to the doctor with many questions about their usefulness. Every patient wants the best option for his care, and medical professionals should be prepared to discuss risks, benefits, and especially alternatives.

Precision medicine will change the traditional *relationship between doctor and patient*. The path toward precision medicine implementation will require formal, informal, and non-formal education programs for both the academia and the wider medical community, as well as the development of educational resources for patients and citizens to optimize the doctor-patient relationship in the context of clinical decisions, which are increasingly more complex and marked by nuances. In other words, we need *personalized communication and educational programs* for personalized medicine.

In Europe, the EU Beating Cancer Plan and Cancer Mission are the two landmark initiatives in the field of cancer, which have been recently launched and are in the early stages of implementation. These will be complemented by new initiatives such as the European Partnership for Personalized Medicine (EP PerMed will be operational in 2023/2024), in which an active role will be played by the already established International Consortium for Personalized Medicine. Harmonization of actions under these programs has the potential to ensure equitable implementation of personalized medicine across the EU.

# References

1. Yates LR, Seoane J, Le Tourneau C, et al. The European Society for Medical Oncology (ESMO) precision medicine glossary. Ann Oncol. 2018;29:30–5.
2. Jørgensen JT. Twenty years with personalized medicine: past, present, and future of individualized pharmacotherapy. Oncologist. 2019;24:e432–40.
3. The evolution of personalized healthcare and the pivotal role of European regions in its implementation | Personalized Medicine. https://www.futuremedicine.com/doi/10.2217/pme-2020-0115. (Accessed 13 Feb 2022).
4. EUR-Lex - C:2015:421:FULL - EN - EUR-Lex. https://eur-lex.europa.eu/legal-content/EN/TXT/?uri=OJ%3AC%3A2015%3A421%3AFULL. (Accessed 12 Feb 2022).
5. Iriart JAB. Medicina de precisão/medicina personalizada: análise crítica dos movimentos de transformação da biomedicina no início do século XXI. Cad Saúde Pública. 2019;35:e00153118.
6. Roberts MC, Fohner AE, Landry L, Olstad DL, Smit AK, Turbitt E, Allen CG. Advancing precision public health using human genomics: examples from the field and future research opportunities. Genome Med. 2021;13:97.
7. Park H-A. Are we ready for the fourth industrial revolution? Yearb Med Inform. 2016:1–3. https://doi.org/10.15265/IY-2016-052.
8. Vicente AM, Ballensiefen W, Jönsson J-I. How personalized medicine will transform healthcare by 2030: the ICPerMed vision. J Transl Med. 2020;18:180.
9. EU Mission: Cancer. In: Eur. Comm. - Eur. Comm. https://ec.europa.eu/info/research-and-innovation/funding/funding-opportunities/funding-programmes-and-open-calls/horizon-europe/eu-missions-horizon-europe/cancer_en. (Accessed 13 Feb 2022).
10. Cancer Moonshot^SM - National Cancer Institute. https://www.cancer.gov/research/key-initiatives/moonshot-cancer-initiative. (Accessed 13 Feb 2022); 2016.
11. Falzone L, Salomone S, Libra M. Evolution of cancer pharmacological treatments at the turn of the third millennium. Front Pharmacol. 2018;9:1300.
12. Frontiers | Milestones in Personalized Medicine: From the Ancient Time to Nowadays—the Provocation of COVID-19 | Genetics. https://www.frontiersin.org/articles/10.3389/fgene.2020.569175/full. (Accessed 12 Feb 2022).
13. Hajdu SI. A note from history: landmarks in history of cancer, part 3. Cancer. 2012;118:1155–68.
14. History of Cancer Treatments: Radiation Therapy. https://www.cancer.org/cancer/cancer-basics/history-of-cancer/cancer-treatment-radiation.html. (Accessed 12 Feb 2022).
15. History of Cancer Epidemiology: 18th Century to Present. https://www.cancer.org/cancer/cancer-basics/history-of-cancer/cancer-epidemiology.html. (Accessed 13 Feb 2022).
16. The Nobel Prize in Physiology or Medicine 1962. In: NobelPrize.org. https://www.nobelprize.org/prizes/medicine/1962/summary/. (Accessed 13 Feb 2022).
17. Soussi T. The history of p53. EMBO Rep. 2010;11:822–6.
18. Lakhtakia R, Burney I. A brief history of breast cancer. Sultan Qaboos Univ Med J. 2015;15:e34–8.
19. Iqbal N, Iqbal N. Imatinib: a breakthrough of targeted therapy in cancer. Chemother Res Pract. 2014;2014:e357027.
20. Mitra S, Tomar PC. Hybridoma technology; advancements, clinical significance, and future aspects. J Genet Eng Biotechnol. 2021;19:159.
21. James JS, Dubs G. FDA approves new kind of lymphoma treatment. Food and Drug Administration. AIDS Treat News. 1997:2–3.
22. Research C for DE and (2021) FDA approves fam-trastuzumab deruxtecan-nxki for HER2-positive gastric adenocarcinomas. FDA.
23. Jørgensen JT. Oncology drug-companion diagnostic combinations. Cancer Treat Res Commun. 2021;29:100492.
24. Hallmarks of Cancer: New Dimensions | Cancer Discovery. https://cancerdiscovery.aacrjournals.org/content/12/1/31. (Accessed 13 Feb 2022).
25. Shendure J, Findlay GM, Snyder MW. Genomic medicine — progress, pitfalls, and promise. Cell. 2019;177:45–57.
26. Genetics vs. Genomics Fact Sheet. In: Genome.gov. https://www.genome.gov/about-genomics/fact-sheets/Genetics-vs-Genomics. (Accessed 14 Feb 2022).
27. Zhang J, Bajari R, Andric D, Gerthoffert F, Lepsa A, Nahal-Bose H, Stein LD, Ferretti V. The international cancer genome consortium data portal. Nat Biotechnol. 2019;37:367–9.
28. Campbell PJ, Getz G, Korbel JO, et al. Pan-cancer analysis of whole genomes. Nature. 2020;578:82–93.
29. The evolutionary history of 2,658 cancers | Nature. https://www.nature.com/articles/s41586-019-1907-7. (Accessed 13 Feb 2022).
30. Analyses of non-coding somatic drivers in 2,658 cancer whole genomes | Nature. https://www.nature.com/articles/s41586-020-1965-x. (Accessed 13 Feb 2022).
31. Three Ways Biotech is Changing How We Treat Cancer. https://www.amgen.com/stories/2019/12/three-ways-biotech-is-changing-how-we-treat-cancer. (Accessed 12 Feb 2022).
32. Kerr KM, Bibeau F, Thunnissen E, Botling J, Ryška A, Wolf J, Öhrling K, Burdon P, Malapelle U, Büttner R. The evolving landscape of biomarker testing for non-small cell lung cancer in Europe. Lung Cancer. 2021;154:161–75.

33. Huang L, Guo Z, Wang F, Fu L. KRAS mutation: from undruggable to druggable in cancer. Signal Transduct Target Ther. 2021;6:1–20.

34. Seligson ND, Knepper TC, Ragg S, Walko CM. Developing drugs for tissue-agnostic indications: a paradigm shift in leveraging cancer biology for precision medicine. Clin Pharmacol Ther. 2021;109:334–42.

35. Colomer R, Mondejar R, Romero-Laorden N, Alfranca A, Sanchez-Madrid F, Quintela-Fandino M. When should we order a next generation sequencing test in a patient with cancer? eClinicalMedicine. 2020; https://doi.org/10.1016/j.eclinm.2020.100487.

36. Takeuchi S, Okuda S. Knowledge base toward understanding actionable alterations and realizing precision oncology. Int J Clin Oncol. 2019;24:123–30.

37. High-Throughput Genomics and Clinical Outcome in Hard-to-Treat Advanced Cancers: Results of the MOSCATO 01 Trial | Cancer Discovery. https://cancerdiscovery.aacrjournals.org/content/7/6/586. (Accessed 14 Feb 2022).

38. Cobain EF, Wu Y-M, Vats P, et al. Assessment of clinical benefit of integrative genomic profiling in advanced solid tumors. JAMA Oncol. 2021;7:525–33.

39. Kwo L, Aronson J. The promise of liquid biopsies for cancer diagnosis; 2021.

40. Commissioner O of the (2020) FDA approves first liquid biopsy next-generation sequencing companion diagnostic test. In: FDA. https://www.fda.gov/news-events/press-announcements/fda-approves-first-liquid-biopsy-next-generation-sequencing-companion-diagnostic-test. (Accessed 13 Feb 2022).

41. ESMO FDA Approves First Comprehensive Pan-Tumour Liquid Biopsy Test for Patients with Advanced Cancer. https://www.esmo.org/oncology-news/fda-approves-first-comprehensive-pan-tumour-liquid-biopsy-test-for-patients-with-advanced-cancer. (Accessed 13 Feb 2022).

42. Appleman LJ. Multifactorial, biomarker-based predictive models for immunotherapy response enter the arena. JNCI J Natl Cancer Inst. 2020;113:7–8.

43. Ryan MB, Cruz FF de la, Phat S, Myers DT, Wong E, Shahzade HA, Hong CB, Corcoran RB (2020) Vertical pathway inhibition overcomes adaptive feedback resistance to KRASG12C inhibition. Clin Cancer Res 26:1633–1643.

44. Rittler D, Molnár E, Baranyi M, Garay T, Hegedűs L, Aigner C, Tóvári J, Tímár J, Hegedűs B. Horizontal combination of MEK and PI3K/mTOR inhibition in BRAF mutant tumor cells with or without concomitant PI3K pathway mutations. Int J Mol Sci. 2020;21:7649.

45. FDA approves 100th monoclonal antibody product. https://www.nature.com/articles/d41573-021-00079-7. (Accessed 13 Feb 2022).

46. (2021) Adjuvant Immunotherapy Approved for Lung Cancer - National Cancer Institute. https://www.cancer.gov/news-events/cancer-currents-blog/2021/fda-adjuvant-atezolizumab-lung-cancer. (Accessed 13 Feb 2022).

47. Five-Year Outcomes With Pembrolizumab Versus Chemotherapy for Metastatic Non–Small-Cell Lung Cancer With PD-L1 Tumor Proportion Score ≥ 50% | Journal of Clinical Oncology. https://ascopubs.org/doi/full/10.1200/JCO.21.00174. (Accessed 13 Feb 2022).

48. Douglass J, Hsiue EH-C, Mog BJ, et al. Bispecific antibodies targeting mutant RAS neoantigens. Sci Immunol. 2021;6:eabd5515.

49. Garfall AL, June CH. Trispecific antibodies offer a third way forward for anticancer immunotherapy. Nature. 2019;575:450–1.

50. Felices M, Kodal B, Hinderlie P, Kaminski MF, Cooley S, Weisdorf DJ, Vallera DA, Miller JS, Bachanova V. Novel CD19-targeted TriKE restores NK cell function and proliferative capacity in CLL. Blood Adv. 2019;3:897–907.

51. Melenhorst JJ, Chen GM, Wang M, et al. Decade-long leukaemia remissions with persistence of CD4+ CAR T cells. Nature. 2022;602(7897):503–9. https://doi.org/10.1038/s41586-021-04390-6. Epub 2022 Feb 2.

52. Depil S, Duchateau P, Grupp SA, Mufti G, Poirot L. 'Off-the-shelf' allogeneic CAR T cells: development and challenges. Nat Rev Drug Discov. 2020;19:185–99.

53. Improving the ability of CAR-T cells to hit solid tumors: Challenges and strategies - ScienceDirect. https://www.sciencedirect.com/science/article/pii/S1043661821006204. (Accessed 14 Feb 2022).

54. Parayath NN, Stephan MT. In situ programming of CAR T cells. Annu Rev Biomed Eng. 2021;23:385–405.

55. Luginbuehl V, Abraham E, Kovar K, Flaaten R, Müller AM. Better by design: what to expect from novel CAR-engineered cell therapies? Biotechnol Adv. 2022;58:107917.

56. Nawaz W, Huang B, Xu S, Li Y, Zhu L, Yiqiao H, Wu Z, Wu X. AAV-mediated in vivo CAR gene therapy for targeting human T-cell leukemia. Blood Cancer J. 2021;11:1–12.

57. Zacharakis N, Huq LM, Seitter SJ, et al. Breast cancers are immunogenic: immunologic analyses and a phase II pilot clinical trial using mutation-reactive autologous lymphocytes. J Clin Oncol. JCO.21.02170. 2022. https://doi.org/10.1200/JCO.21.02170. Epub 2022 Feb 1. PMID: 35104158; PMCID: PMC9148699.

58. Jou J, Harrington KJ, Zocca M-B, Ehrnrooth E, Cohen EEW. The changing landscape of therapeutic cancer vaccines—novel platforms and Neoantigen identification. Clin Cancer Res. 2021;27:689–703.

59. BioNTech Announces First Patient Dosed in Phase 2 Clinical Trial of mRNA-based BNT111 in Patients with Advanced Melanoma | BioNTech. https://investors.biontech.de/news-releases/news-release-details/biontech-announces-first-patient-dosed-phase-2-clinical-trial/. (Accessed 14 Feb 2022).

60. Nierengarten MB. Messenger RNA vaccine advances provide treatment possibilities for cancer. Cancer. 2022;128:213–4.

61. Liao J-Y, Zhang S. Safety and efficacy of personalized cancer vaccines in combination with immune checkpoint inhibitors in cancer treatment. Front Oncol. 2021;11:663264.

62. Miao L, Zhang Y, Huang L. mRNA vaccine for cancer immunotherapy. Mol Cancer. 2021;20:41.

63. Garralda E, Dienstmann R, Piris-Giménez A, Braña I, Rodon J, Tabernero J. New clinical trial designs in the era of precision medicine. Mol Oncol. 2019;13:549–57.

64. Basket and umbrella trials: untapped opportunities in rare disease. Clin. Trials Arena; 2021.

65. Photopoulos J. The future of tissue-agnostic drugs. Nature. 2020;585:S16–8.

66. Drilon A, Laetsch TW, Kummar S, et al. Efficacy of Larotrectinib in TRK fusion–positive cancers in adults and children. N Engl J Med. 2018;378:731–9.

67. FDA approves first RET inhibitor | Nature Biotechnology. https://www.nature.com/articles/s41587-020-0568-2. (Accessed 12 Feb 2022).

68. De Maria MR, Di Sante G, Piro G, et al. Translational research in the era of precision medicine: where we are and where we will go. J Pers Med. 2021;11:216.

69. The Fourth Industrial Revolution: what it means and how to respond. In: World Econ Forum https://www.weforum.org/agenda/2016/01/the-fourth-industrial-revolution-what-it-means-and-how-to-respond/. (Accessed 13 Feb 2022).

70. Yu T, Scolnick J. Complex biological questions being addressed using single cell sequencing technologies. SLAS Technol. 2021; https://doi.org/10.1016/j.slast.2021.10.013.

71. Chen S, Lake BB, Zhang K. High-throughput sequencing of the transcriptome and chromatin accessibility in the same cell. Nat Biotechnol. 2019;37:1452–7.

72. Dagogo-Jack I, Shaw AT. Tumour heterogeneity and resistance to cancer therapies. Nat Rev Clin Oncol. 2018;15:81–94.

73. Single-cell immunology of SARS-CoV-2 infection | Nature Biotechnology. https://www.nature.com/articles/s41587-021-01131-y. (Accessed 13 Feb 2022).

74. Spiegel JY, Patel S, Muffly L, et al. CAR T cells with dual targeting of CD19 and CD22 in adult patients with recurrent or refractory B cell malignancies: a phase 1 trial. Nat Med. 2021;27:1419–31.

75. Creelan BC, Wang C, Teer JK, et al. Tumor-infiltrating lymphocyte treatment for anti-PD-1-resistant metastatic lung cancer: a phase 1 trial. Nat Med. 2021;27:1410–8.

76. Vuksanaj K. The Polyfunctional strength index: from concept to solution. GEN - Genet Eng Biotechnol News; 2019.

77. Nagasawa S, Kashima Y, Suzuki A, Suzuki Y. Single-cell and spatial analyses of cancer cells: toward elucidating the molecular mechanisms of clonal evolution and drug resistance acquisition. Inflamm Regen. 2021;41:22.

78. Kleshchevnikov V, Shmatko A, Dann E, et al. Cell2location maps fine-grained cell types in spatial transcriptomics. Nat Biotechnol. 2022;40:1–11.

79. Ozturk K, Dow M, Carlin DE, Bejar R, Carter H. The emerging potential for network analysis to inform precision cancer medicine. J Mol Biol. 2018;430:2875–99.

80. Rajkumar SV. Multiple myeloma: 2018 update on diagnosis, risk-stratification and management. Am J Hematol. 2018;93:981–1114.

81. Bhalla S, Melnekoff DT, Aleman A, et al. Patient similarity network of newly diagnosed multiple myeloma identifies patient subgroups with distinct genetic features and clinical implications. Sci Adv. 2021;7:eabg9551.

82. Larsen BM, Kannan M, Langer LF, et al. A pan-cancer organoid platform for precision medicine. Cell Rep. 2021;36:109429.

83. Alhaque S, Themis M, Rashidi H. Three-dimensional cell culture: from evolution to revolution. Philos Trans R Soc B Biol Sci. 2018;373:20170216.

84. Liu L, Yu L, Li Z, Li W, Huang W. Patient-derived organoid (PDO) platforms to facilitate clinical decision making. J Transl Med. 2021;19:40.

85. Letai A, Bhola P, Welm AL. Functional precision oncology: testing tumors with drugs to identify vulnerabilities and novel combinations. Cancer Cell. 2022;40:26–35.

86. Kornauth C, Pemovska T, Vladimer GI, et al. Functional precision medicine provides clinical benefit in advanced aggressive hematologic cancers and identifies exceptional responders. Cancer Discov. 2022;12:372–87.

87. Fancello L, Gandini S, Pelicci PG, Mazzarella L. Tumor mutational burden quantification from targeted gene panels: major advancements and challenges. J Immunother Cancer. 2019;7:183.

88. Sha D, Jin Z, Budzcies J, Kluck K, Stenzinger A, Sinicrope FA. Tumor mutational burden (TMB) as a predictive biomarker in solid tumors. Cancer Discov. 2020;10:1808–25.

89. McGrail DJ, Pilié PG, Rashid NU, et al. High tumor mutation burden fails to predict immune checkpoint blockade response across all cancer types. Ann Oncol. 2021;32:661–72.

90. Bedard ELR, Abraham AG, Joy AA, Ghosh S, Wang X, Lim A, Shao D, Loebenberg R, Roa WH. A novel composite biomarker panel for detection of early stage non-small cell lung cancer. Clin Investig Med Med Clin Exp. 2021;44:E15–24.

91. Bandini M, Ross JS, Raggi D, et al. Predicting the pathologic complete response after Neoadjuvant Pembrolizumab in muscle-invasive bladder cancer. J Natl Cancer Inst. 2021;113:48–53.

92. Galsky MD, Saci A, Szabo PM, Han GC, Grossfeld G, Collette S, Siefker-Radtke A, Necchi A, Sharma P. Nivolumab in patients with advanced platinum-resistant Urothelial carcinoma: efficacy, safety, and biomarker analyses with extended follow-up from CheckMate 275. Clin Cancer Res. 2020;26:5120–8.

93. Discovering dominant tumor immune archetypes in a pan-cancer census: Cell. https://www.cell.com/cell/fulltext/S0092-8674(21)01426-4?_returnURL= https%3A%2F%2Flinkinghub.elsevier.com%2Fretr ieve%2Fpii%2FS0092867421014264%3Fshowall% 3Dtrue. (Accessed 12 Feb 2022).

94. Dietrich P-Y, Rodriguez-Bravo V, Baciarello G, Fizazi K, Patrikidou A. Redefining cancer of unknown primary: is precision medicine really shifting the paradigm? Cancer Treat Rev. 2021;97:102204.

95. Kato S, Alsafar A, Walavalkar V, Hainsworth J, Kurzrock R. Cancer of unknown primary in the molecular era. Trends Cancer. 2021;7:465–77.

96. Ogilvie L. The future of health care: deep data, smart sensors, virtual patients and the internet-of-humans. In: Futur. - Eur. Comm; 2016. https://ec.europa.eu/futurium/en/content/future-health-care-deep-data-smart-sensors-virtual-patients-and-internet-humans. (Accessed 15 Feb 2022).

97. Kamel Boulos MN, Zhang P. Digital twins: from personalized medicine to precision public health. J Pers Med. 2021;11:745.

98. Björnsson B, Borrebaeck C, Elander N, et al. Digital twins to personalize medicine. Genome Med. 2019;12:4.

99. Marr B The Five Biggest Healthcare Tech Trends In 2022. In: Forbes. https://www.forbes.com/sites/bernardmarr/2022/01/10/the-five-biggest-healthcare-tech-trends-in-2022/. (Accessed 15 Feb 2022).

100. Hernandez-Boussard T, Macklin P, Greenspan EJ, Gryshuk AL, Stahlberg E, Syeda-Mahmood T, Shmulevich I. Digital twins for predictive oncology will be a paradigm shift for precision cancer care. Nat Med. 2021;27:2065–6.

101. Europe's Beating Cancer Plan. In: Have Your Say. https://ec.europa.eu/info/law/better-regulation/have-your-say/initiatives/12154-Europe%E2%80%99s-Beating-Cancer-Plan_en. (Accessed 15 Feb 2022).

# Precision Medicine in Cardiovascular Disease Practice

**4**

Ali Sheikhy ⓘ, Aida Fallahzadeh ⓘ,
Hamid Reza Aghaei Meybodi ⓘ,
and Kaveh Hosseini ⓘ

**What Will You Learn in This Chapter?**
In this chapter, we will get familiar with personalized medicine in the field of cardiology, the genetic basis of most common cardiovascular diseases, and the role of genetics in pharmacotherapy. We will also discuss the ethical issues in personalized medicine and the perspective of this field in cardiology.

A. Sheikhy · A. Fallahzadeh
Tehran Heart Center, Cardiovascular Diseases Research Institute, Tehran University of Medical Sciences, Tehran, Iran

Cardiac Primary Prevention Research Center, Cardiovascular Diseases Research Institute, Tehran University of Medical Sciences, Tehran, Iran

Non-Communicable Disease Research Center, Endocrinology and Metabolism Population Sciences Institute, Tehran University of Medical Sciences, Tehran, Iran

H. R. Aghaei Meybodi
Personalized Medicine Research Center, Endocrinology and Metabolism Clinical Sciences Institute, Tehran University of Medical Sciences, Tehran, Iran

K. Hosseini (✉)
Tehran Heart Center, Cardiovascular Diseases Research Institute, Tehran University of Medical Sciences, Tehran, Iran

Cardiac Primary Prevention Research Center, Cardiovascular Diseases Research Institute, Tehran University of Medical Sciences, Tehran, Iran

**Rationale and Importance**
Personalized medicine is important in early diagnosis; choosing the best treatment options, including the most suitable pharmacotherapy in familial arrhythmias; and preventing adverse drug reactions in the field of cardiology. Recognizing the best treatment and preventive strategy is individualized. It is crucial for healthcare providers to apply the most appropriate approach to patients.

## 4.1 Introduction

Personalized medicine (PM) is a concept that modifies therapeutic strategies according to each individual's genomic, epigenomic, and proteomic profiles [1]. The major concept of PM is the treatment and care of patients with a particular condition while considering individual alterations in genetics, exposures, and lifestyle [2]. Cardiovascular diseases (CVD), the most common cause of death all over the world [3], have genetic risk factors, and the pharmacokinetics of cardiology drugs have a broad spectrum of different genotypes [4]. Moreover, genome-wide associated (GWA) studies have revealed several genetic variants that are associated with cardiology conditions such as cardiomyopathies, arrhythmias, and coronary artery diseases. Thus, determining genetic information and applying PM strategies are useful in the effective preven-

**Fig. 4.1** Personalized medicine in cardiovascular disease. This figure was created with Biorender.com. All rights and ownership of BioRender content are reserved by BioRender

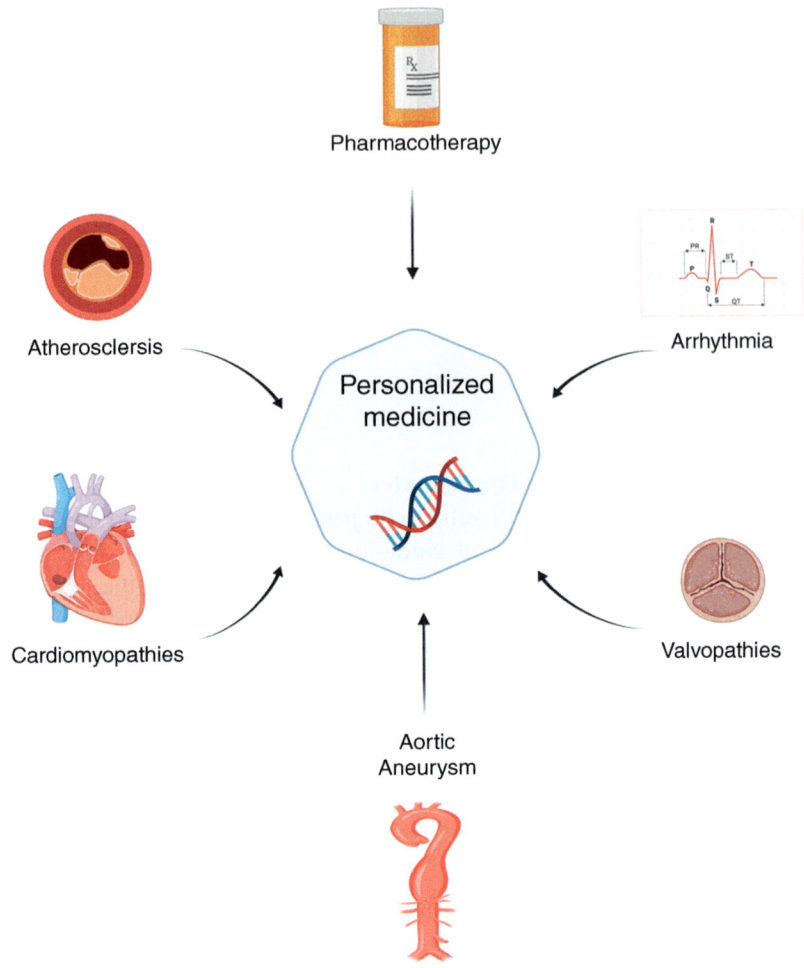

tion and treatment of several cardiologic conditions (Fig. 4.1).

## 4.2 Drugs

### 4.2.1 Warfarin

Warfarin is an anticoagulant that is often prescribed for the treatment and prevention of thromboembolic events in people with prosthetic heart valves, atrial fibrillation, venous thrombosis, and a history of stroke. Warfarin dose requirements, drug response, and risk of bleeding are influenced by environmental factors (such as vitamin K consumption, age, gender, and concurrent medications) and by genetic variations [5].

*VKORC1*, *CYP2C9*, and *CYP4F2* are considered the main genes that may influence warfarin metabolism and cause genetic variations.

### 4.2.2 *VKORC1*

The *VKORC1* gene encodes the target enzyme of the warfarin drug, the vitamin K epoxide reductase enzyme, which is responsible for reducing vitamin K epoxide to the active form [6]. A common non-coding variant of *VKORC1* that occurs in the promoter region of the gene, c.-1639G>A (rs9923231) polymorphism, affects protein expression and is associated with warfarin sensitivity and lower dose requirements. Patients who are carrying one or two "A" alleles at -1639 require lower warfarin doses than -1639G/G homozygotes [1, 7, 8]. The c.-1639G>A allele frequency shows a discrepancy among different ethnic groups and is more common among Asians, Caucasians, and African Americans [2, 9, 10]. Besides, other less common coding *VKORC1* polymorphisms (such as Asp36Tyr) are associated with warfarin resistance and higher dose requirements [4, 11].

### 4.2.3   *CYP2C9*

*CYP2C9* is a member of the cytochrome P450 superfamily (CYP450) that metabolizes the more potent warfarin S-enantiomer. *CYP2C9*1* is the wild-type allele in the "normal metabolizer" phenotype (those with normal enzyme activity and metabolism). Individuals carrying two well-characterized variant alleles, *CYP2C9*2* and *CYP2C9*3*, are known to be more sensitive to warfarin, require lower doses to achieve the therapeutic range, are at a higher risk of bleeding, and take longer to achieve a stable INR compared to normal metabolizers [12, 13]. The maintenance dose requirements of warfarin in patients with *1*1, *2*2, and *3*3 genotypes are reported as 5.28 mg/day, 3.04 mg/day, and 0.5 mg/day, respectively [14]. Other *CYP2C9* variants (*CYP2C9*5, *6, 8*, and *11*), which are more common among African Americans, are also associated with decreased enzyme activity and dose variability [15].

### 4.2.4   *CYP4F2*

*CYP4F2* is the vitamin K oxidase enzyme and acts as an important counterpart to *VKORC1* (vitamin K reductase enzyme), limiting vitamin K accumulation in the liver [16]. A known variant of *CYP4F2*3* (c.1297C>T, rs2108622) has been shown to affect enzyme activity and dose requirements of warfarin [17]. Caucasian individuals who carry two "T" alleles require a higher dosage of warfarin (1 mg/day) compared to those with two "C" alleles, which is explained by the reduced function of the enzyme in those with "T"

alleles [18]. Thus, including this *CYP4F2* variant in warfarin dosing models is helpful in dose prediction in Asians and Europeans, but not in African Americans [19–21].

### 4.2.5   P2Y12 Inhibitors

Clopidogrel is a prodrug, and genetic variants influence the catalytic activity of the CYP P450 isoforms (such as *CYP2C19*, *CYP1A2*, *CYP2B6*, *CYP2C9*, and *CYP3A*) and affect the efficiency of active metabolite generation [22]. The most common *CYP2C19* loss-of-function alleles are *CYP2C19*2* (G681A) and *CYP2C19*3* (G636A), and the most common allele that results in increased enzyme activity is CYP2C19*17 [23]. Therefore, based on the *CYP2C19* genotypes, patients are categorized as ultrarapid metabolizers (*1/*17, *17/*17), extensive metabolizers (*1/*1), intermediate metabolizers (*1/*2, *1/*3, *2/*17), and poor metabolizers (*2/*2, *2/*3, *3/*3) [24]. The ABCB1 gene polymorphisms are also known to be associated with clinical outcomes in clopidogrel-treated patients [25]; however, the association has been inconsistent across studies, with several studies finding no relationship between *ABCB1* variants and the antiplatelet effect of clopidogrel [26]. Prasugrel and ticagrelor are both stronger P2Y12 inhibitors than clopidogrel and lower platelet reactivity more effectively, irrespective of the *CYP2C19* genotype [27, 28]. Moreover, polymorphisms of the other isoforms of the CYP450 system appear to not influence prasugrel pharmacokinetics or pharmacodynamics [28].

### 4.2.6   Statin

Statins, HMG-CoA reductase inhibitors, act by inhibiting cholesterol biosynthesis and increasing low-density lipoprotein cholesterol (LDL-C) uptake by hepatocytes. *SLCO1B1* and *ABCB1* are proteins that play a role in the transportation of statins. The *SLCO1B1* 521C (rs4149056) variant is associated with a reduction of the lipid-lowering effect of simvastatin, atorvastatin, lovastatin, and pravastatin. Three ABCB1 gene polymorphisms

(1236T, 2677T, and 3435T) have been linked to statin pharmacokinetics and toxicity. HMG-CoA reductase is an important enzyme in cholesterol synthesis and is inhibited by statins within hepatocytes. The H7 haplotype of HMG-CoA reductase is associated with decreased lipid-lowering response to statins [29]. Polymorphisms in the *CYP3A* gene, such as *CYP3A4*22* (rs35599367) and *CYP3A5*3* (rs776746), have been shown to reduce CYP3A4 enzyme levels and activity, as well as to affect the pharmacokinetics of simvastatin, atorvastatin, and lovastatin [30, 31].

## 4.3 Cardiomyopathies

### 4.3.1 Hypertrophic Cardiomyopathy (HCM)

HCM is one of the common hereditary cardiac diseases, which is associated with two main pathogeneses; the first one is defects in myocardial filaments, associated with sarcomeric genes, and the second one is metabolic and infiltrative disorders [32]. The gene variants that are associated with HCM are the *MYH7* gene, which encodes the myosin heavy chain [33], *TNNsT2* which encodes cardiac troponin T [34], *MYBPC3* which encodes myosin-binding protein C [35], *TNNI3* which encodes Cardiac troponin I [36], and *FHOD3* which encodes "Formin homology 2 domains containing 3" [37]. Moreover, there are some syndromic genes without isolated left ventricular hypertrophy, including the autosomal recessive *GAA* gene as Pompe disease [38], and X-linked *GLA*, which presents as Anderson-Fabry disease [39]. Genotype-positive patients have been shown to present with illness approximately 10 years earlier, to have a greater maximum left-ventricular wall thickness, and to have a higher proportion of family history of HCM and sudden cardiac death than others [40].

### 4.3.2 Dilated Cardiomyopathy (DCM)

A strong familial component has been reassuringly confirmed in DCM [41]. About 111 genes are associated with DCM. The most associated gene is *TTN*, which encodes Titin, the largest structural protein of the heart [42]. Another gene variant that is associated with approximately 5% of the causes of DCM is *LMNA* missense and truncating mutations [43]. *LMNA* mutations are the main genetic cause of arrhythmogenic DCM.

### 4.3.3 Restrictive Cardiomyopathy (RCM)

RCM, one of the rarest and poor-prognosis cardiac disorders, is characterized by a normal-sized left ventricle with a hypertrophic atrium. Amyloidosis, as an infiltrative disorder, is the most common cause of RCM. *TTR* gene variants and *APOA1* are the main genetic perturbations in amyloidosis [44]. There is a lack of adequate data about non-infiltrative RCM genes; however, *TNNI3, TNNT2, TNNC1, TPM1, TTN, MYH7, MYL2, MYBPC3, MPN, DES, FLNC, LMNA*, and *BAG3* were labeled as associated genes in RCM [45, 46]. Most of these genes encode sarcomeric proteins. Moreover, *CRYAB*, which encodes heat-shock proteins (such as crystallin B and *BAG3*), is also reported in some studies [45, 47].

## 4.4 Thoracic Aortic Aneurysm/ Aortic Dissection (TAAD)

Several causal genes have been identified in syndromic and non-syndromic TAAD. Variants in the smooth muscle contractile (SMC) genes, including *ACTA2, MYH11, MYLK*, and *PRKG1*, have been associated with non-syndromic TAAD [48]. Syndromic TAAD is associated with several connective tissue disorders and their corresponding genes, including Marfan syndrome (*FBN1*), Loeys-Dietz syndrome (*TGFBR1, TGFBR2, SMAD3*, and *TGFB2*), Ehlers-Danlos syndrome (*COL1A1, COL1A2, COL3A1, COL5A1*, and *COL5A2*), arterial tortuosity syndrome (*SLC2A10*), and Shprintzen-Goldberg syndrome (*SKI*) [49]. Marfan syndrome patients with *FBN1* mutations have a low risk for acute aortic dissections at diameters less than 5.5 cm and for aneurysms of other arteries [50].

Common genetic variants at 15q21.1, in the *FBN1* gene, are associated with an increased risk of TAAD in the general population and are common pathogeneses of aortic disease in Marfan syndrome and sporadic TAAD [51]. Loeys-Dietz syndrome patients with *TGFBR1* and *TGFBR2* mutations are at higher risk of aortic dissections at aortic diameters less than 5.0 cm, and these patients have aneurysms and dissections of other arteries. Furthermore, studies have shown that *TGFBR1* mutation carriers may have a lower risk of aortic dissection with minimal enlargement than *TGFBR2* mutation carriers [52]. *MYLK* encodes the Ca2+/calmodulin-dependent myosin light-chain kinase, which phosphorylates the regulatory light chain in smooth muscle cells to initiate contraction [53]. *MYLK* missense variants were shown to be associated with earlier-onset aortic events compared to haploinsufficient variants [54].

## 4.5 Valvopathies

Aortic stenosis (AS) is the narrowing of the aortic valves that leads to obstruction of blood flow from the left ventricle (LV) to the aorta. The incidence of AS is increasing with the aging population. Today, AS is not considered a passive degenerative disease anymore. It is associated with a dynamic, complex, and highly regulated pathobiological process that leads to a multitude of events [55]. The characterization of the whole protein complements of the genome, which is termed "proteome," is the major goal of proteomics that could improve the patient's management. Analysis of the cell or tissue is a suitable platform as they are eventually the targets for novel medications and should provide important evidence for treatment discovery.

As a result, lipoproteins and oxidized phospholipids play a significant role in AS that generates inflammation, apoptosis, and calcification of the aortic valve [56]. *LPA* genetic variants linked to Lp(a) levels are significantly linked to aortic valve calcification and incident AS [57]. Accordingly, to manage the progression of AS, aiming lipoprotein(a) is a potential therapeutic target. The other potential mechanisms are:

1. Calcium deposition: which includes calcium, phosphate, vitamin D, fibroblast growth factor 23 (FGF-23), and PTH; the vitamin D/PTH axis biomarkers are the most verified factors [58]. The N-terminal propeptide of human procollagen type I (PINP), beta carboxy-terminal cross-linking telopeptide of type I collagen (β-CTx), osteocalcin, osteopontin, osteoprotegerin, and fetuin-A are the other suggested factors [59–62].

2. Inflammation: limited factors are associated with inflammation, which leads to AS. Remarkably, in contrast with CAD, C-reactive protein (CRP) is not associated with the progression of calcified aortic valve disease [63].

3. Cardiac remodeling, B-type natriuretic peptide (BNP), and cardiac troponin are potentially informative about the myocardial consequences of AS. Higher NT proBNP was associated with a higher grade of AS severity and NYHA class [64]. Cardiac troponin was identified as a separate variable associated with mid-wall fibrosis of the myocardium as part of a clinical risk score that predicts cardiovascular events in asymptomatic AS [65]. Biomarkers of extracellular matrix remodeling such as Fibulin-1 are significantly and inversely correlated with AVA index [66].

Personalized medicine contains a multimodal approach that might be especially useful for decision-making in patients with asymptomatic AS rather than patients with AS. Defining which patients could benefit from each therapeutic strategy would be possible with PM, for example, utilizing a transcatheter instead of the surgical aortic valve.

### 4.5.1 Mitral Valve Replacement

In patients with significant mitral regurgitation (MR) due to floppy mitral valve (FMV)/mitral valve prolapse (MVP), mitral valve replacement is crucial. Due to the significant variability in the size of the mitral annulus, one ring size can't fit all. The mitral "personalized ring" is a novel device constructed intraoperatively [67]. These "personalized rings" provide excellent support of the mitral annulus, which avoids annular dilatation and paravalvular leak.

## 4.6 Arrhythmia

### 4.6.1 Long QT Syndrome (LQTS)

LQTS is defined as QTc ≥480 ms in an asymptomatic patient or a QTc ≥460 ms in the presence of unexplained syncope [68]. Patients with LQTS are at high risk of arrhythmogenic syncope, polymorphous ventricular tachycardia (torsade de pointes), and sudden arrhythmic death [69]. LQT type 1 is caused by loss of function mutation in the *KCNQ1* gene which encodes the α-subunit of the slow rectifier current ($I_{KS}$) [70]. LQT type 2 arises from loss-of-function mutations in *KCNH2*, which encodes the α-subunit of the rapid rectifier current ($I_{Kr}$) [71]. In contrast, a gain of function in *SCN5A* will cause LQT type 3, which amplified late sodium current ($I_{Na}$) [72]. Based on LQTS genotyping studies, the best therapeutic option in LQT types 1, 2, and 3 has shown to be β-blockers [73, 74].

### 4.6.2 Brugada Syndrome (BrS)

Inward sodium current impairment compared with the transient outward potassium current ($I_{to}$) in the right ventricular outflow tract is the key pathogenesis of BrS [75]. The most common genetic mutation in BrS, which could be detected in 21% of the patients, is the loss of function in the *SCN5A* gene. Loss-of-function mutations in *SCN5A* reduce the overall available sodium current (INa) through either (1) impaired intracellular trafficking of the ion channel to the plasma membrane or (2) through altered gating properties of the channel [76]. *CACNA1C, GPD1L, HEY2, PKP2, RANGRF, SCN10A, SCN1B, SCN2B, SCN3B, SLMAP,* and *TRPM4* are some other rare genes associated with BrS [77]. *SCN10A*, which encodes α-subunit Nav1.8 sodium channel, is one of the most novel mutations and is responsible for 5 to 16 percent of BrS [78, 79].

### 4.6.3 Short QT Syndrome (SQTS)

SQTS is defined as QTc ≤330 ms, or QTc interval <360 ms, and at least one of the following conditions: history of cardiac arrest or syncope, family history of sudden cardiac death (SCD) at age 40 or younger, or family history of SQTS [80]. Potassium and calcium channelopathies are the main pathophysiology in SQTS [26, 27]. Gain of function mutations in *KCNH2, KCNQ1,* and *KCNJ2* genes (associated with potassium channels) are responsible for SQT type 1, 2, and 3, respectively [81, 82]. Loss of function in *CACNA1C, CACNB2,* and *CACNA2D1* (associated with calcium channels) leads to SQT 4, 5, and 6, respectively [83, 84].

### 4.6.4 Idiopathic Ventricular Fibrillation (IVF)

IVF is defined as resuscitated ventricular fibrillation (VF), which had no other causes for VF, that is, metabolic, toxicological, cardiac (including other channelopathies and structural heart disease), respiratory, and infectious causes [68]. IVF is responsible for 6.8% of sudden cardiac death causes [85]. IVF pathophysiology is mainly due to an abnormality affecting the microstructural myocardial or Purkinje system [86]. Several genes have been found in association with IVF, *DPP6* was reported in Dutch families [87], *CALM1* was reported in a Moroccan family [88], and *RYR2* causes a leaky channel at diastolic levels of calcium under non-stress conditions [89].

### 4.6.5 Catecholaminergic Polymorphic Ventricular Tachycardia (CPVT)

The main clinical manifestation of CPVT is episodic syncope occurring during exercise or acute emotion in individuals without structural cardiac abnormalities [90]. CPVT1 is caused by a mutation in the *RYR2* gene, which encodes the cardiac ryanodine receptor and accounts for 65% of the CVPT cases [91]. *RYR2* gene affects intracellular calcium hemostasis and excitation-contraction coupling of the heart [92]. Mutation in the *CASQ2* gene, which encodes cardiac calsequestrin (a calcium-buffering protein within the sarcoplasmic reticulum), accounts for 2–5% of the CPVT cases [93]. Some other genes associated with CPVT are *TECLR* [94], *TRDN* [95], *CALM* [96], and *CALM2* [97]. *ANK2* and *KCNJ2* may phenocopy CPVT; hence they are associated with LQT4 and LQT7, respectively [98, 99]. However, no specific gene could be found for almost one-third of CPVT cases [100].

### 4.6.6 Progressive Cardiac Conduction System Disease (PCCD)

PCCD is defined as impulse conduction progressive delay through the His-Purkinje system with right or left bundle branch block (RBBB or LBBB) [101]. The first reported gene associated with PCCD was *SCN5A*, which encodes the cardiac sodium channel Na$_v$ 1.5 [102]. *SCN5A* mutations could also be found in BrS type 1; thus, there is a significant overlap between BrS and PCCD. Individuals carrying this mutation may manifest isolated forms of each BrS and PCCD or coexisted forms [103]. Mutations in *TRPM4* gene, which encodes a Ca$^{2+}$-activated but Ca$^{2+}$-impermeable cation channel [104], are associated with PCCD as well as familial AV block and RBBB [105]. PCCD may be associated with HCM in the presence of mutations in *PRKAG2*, *LAMP2*, and *GLA*; also it may be accompanied by DCM in the occurrence of *LMNA*, *DES*, and *TNNI3K* alterations [106].

## 4.7 Coronary Artery Disease (CAD)

### 4.7.1 Genes and Mechanism

In addition to several traditional risk factors (such as smoking, hypertension, diabetes, dyslipidemia, and obesity), a strong genetic basis had been also identified for CAD. According to early GWA studies [107, 108], variants in two loci (*LTA* and *LGALS2*) are associated with pathogenesis and increased risk of myocardial infarction (MI). However, later studies failed to show such association between polymorphisms in *LTA* and *LGALS2* and myocardial infarction [109]. In 2007, GWA studies identified SNPs at the 9p21.3 locus, which is located near the *CDKN2A* and *CDKN2B* genes and is associated with a 30–40% increased risk of CAD [110, 111].

GWA studies for plasma lipoprotein traits have identified several common single nucleotide polymorphism (SNP) variants that are strongly associated with plasma LDL. Common variants in genes associated with LDL-C levels (*PCSK9*, *LDL-R*, *APOB*, *APOE*, *SORT1*, *ABCG5-ABCG8*, *ABO*, *LPA*, and *NPC1L1*), genes associated with triglyceride levels (*LPL*, *APOA5*, *ASGR1*, *ANGPTL4*, *APOC3,* and *TRIB1*), and the gene encoding cholesteryl ester transfer protein (CETP), which is associated with HDL-C levels, have been linked to CAD [112, 113]. SNPs on chromosome 1P13 have a strong association with LDL and have also been independently linked to CAD and MI [110, 114]. Not all mutations are associated with an increased risk of CAD, in some cases; inactivating mutations may decrease CAD risk in conclusion. *PCSK9*, *NPC1L1*, and *ASGR1* mutations result in CAD risk reduction by lowering LDL cholesterol levels [11, 12, 115]. Lipoprotein lipase (*LPL*) hydrolyses lipoprotein-bound triglycerides and reduces triglyceride levels consequently. *LPL* loss of function is associated with an increased risk of CAD [116]. Apolipoprotein A5 (*APOA5*) increases LPL activity [116]. In contrast, apolipoprotein C-III (*APOC3*) and angiopoietin-like 4 (*ANGPTL4*) reduce LPL activity, and they are associated with CAD [117, 118]. *APOA5* mutations increase plasma triglyceride levels [116]; nonetheless, *APOC3* loss-of-function has opposite effects, which causes a reduction in plasma triglycerides levels [117].

### 4.7.2 Premature CAD

GWA studies have identified a considerable number of genetic variants that are associated with premature CAD. Genetic variants in genes such as *PCSK9*, *LDL-R*, and *NPC1L1* contribute to premature CAD either directly or via traditional cardiovascular risk factors. Variants in locus 9p21.3, which is located on chromosome 9, are also associated with the risk of developing premature CAD [119]. It has been shown that a mutation in LDL receptor (LDLR) may lead to LDL metabolism dysfunction and thus increase the risk of premature CAD [120]. *LDLR* plays an important role in CAD pathogenesis by increasing LDL cholesterol and triglyceride-rich lipoproteins levels [121].

### 4.7.3 Vascular Inflammation and Remodeling

The encoding genes of cytokines (*CXCL12*) [112] and interleukin 6 (*IL6*) [122] are associated with CAD by vascular inflammation. *SH2B3* is one of the novel mutations associated with an increased risk of CAD [122]. *SH2B3* mutations trigger an elevation in numerous inflammatory mediators in left ventricle tissues including NRLP12, CCR2, and IFNγ [123]. There are two types of vascular remodeling, constrictive remodeling and expansive remodeling. Constrictive remodeling produces more stable plaque and narrow lumen, in contrast, expansive remodeling causes less stable plaque with no narrowing effect on the lumen [124]. *ADAMTS7* is one of the novel genes associated with CAD and plaque formation, but not plaque rupture [125]. *MIA3* is another gene associated with CAD, which regulates the levels of large proteins such as collagen VII [126].

### 4.8 Hypertension

Hypertension is one of the major risk factors for CAD. Based on CHARGE Consortium, *ATP2B1*, *CYP17A1*, *PLEKHA7*, and *SH2B3* were associated with systolic blood pressure (SBP); *ATP2B1*, *CACNB2*, *CSK-ULK3*, *SH2B3*, *TBX3-TBX5*, and *ULK4DBP* were associated

with diastolic blood pressure (DBP); and *ATP2B1* was labeled for hypertension [127]. According to another large-scale study, *CYP17A1*, *CYP1A2*, *FGF5*, *SH2B3*, *MTHFR*, *c10orf107*, *ZNF652*, and *PLCD3* genes caused hypertension [128].

### 4.9 Recognizing Ethical Issues and How to Deal with Them

Many patients are aware of the benefits of PM although their knowledge of its potential appears to be limited [129]. Patients in oncology request information about PM more frequently than patients with other diseases [130]. Even if patients are aware of the phrase "personalized medicine," some of them don't understand the concept of PM [131], which may affect the participation of patients in medical decision-making.

Professionals also describe a lack of knowledge about PM. According to the conducted studies, cardiologists have the lowest information about PM among family physicians, cardiologists, and oncologists [130].

One of the major ethical concerns in this field is data confidentiality not being guaranteed properly [132]. Besides sharing data with the legal system, patients are concerned about data sharing with families in cases where information about a genetic disposition needs to be shared with all at-risk family members.

Test results or the testing process itself can also cause harm for patients. This harm can be caused by professionals' misinterpretation of the results or making the wrong therapeutic decisions [133]. Besides mentioned issues, the psychological burden from knowing or expecting the assessment results is considerably high. Accordingly, harm to benefits must be evaluated in every patient.

In contrast with clinical practice, in which test results are only beneficial when they provide reliable and actionable evidence that can be used for clinical decisions, there is a lack of evidence in PM results and practice guidelines in this field. The cost-benefit ratio of PM is also questionable whether other treatment interventions would not have a superior benefit. PM costs are supposed as being massive and caused by a much minor proportion of the total patient population [134].

## 4.10  Digital Twin "Prospective of Precision Medicine in Cardiology"

The idea of digital twins was initially discussed in David Gelernter's book in the early 1990s [135]. A digital twin is a digital imitation or representation of a physical object, process, or service, but also beyond that. In other words, it is a virtual prototype (data plus algorithms) that will dynamically connect the physical and digital worlds and that will utilize modern technologies, such as smart sensors, data analytics, and artificial intelligence (AI), to monitor system performance, detect and prevent failures, and explore new advancements. A digital twin is intended to make a virtual representation of a physical object, test it, and optimize it in the virtual space, until the virtual representation meets the desired performance, at which point it can be built or enhanced (if already built) in the real world [136]. Collecting real-time data streams from linked clinical, health, and other sensors and combining these mass data with advanced data analytics, cloud computing, and artificial intelligence (including machine learning) will generate highly potent networked computational resources, which could be used in real-world decision-making. Precision cardiology could be established by the utilization of cardiac digital twins (CDT) [137]. This cardiovascular model will maximize the interaction between anatomical and physiologic understanding of the cardiovascular system. Treatment and prevention of cardiovascular disease will be based on precise predictions of both the underlying causes of disease and the pathways; hence, these predictions will be promising with CDT utilization.

## References

1. Vogenberg FR, Isaacson Barash C, Pursel M. Personalized medicine: part 1: evolution and development into theranostics. P T. 2010;35(10):560–76.
2. Leopold JA, Loscalzo J. Emerging role of precision medicine in cardiovascular disease. Circ Res. 2018;122(9):1302–15. https://doi.org/10.1161/circresaha.117.310782.
3. GBD 2015 Mortality and Causes of Death Collaborators. Global, regional, and national life expectancy, all-cause mortality, and cause-specific mortality for 249 causes of death, 1980-2015: a systematic analysis for the Global Burden of Disease Study 2015. Lancet. 2016;388(10053):1459–544. https://doi.org/10.1016/s0140-6736(16)31012-1.
4. Goetz LH, Schork NJ. Personalized medicine: motivation, challenges, and progress. Fertil Steril. 2018;109(6):952–63. https://doi.org/10.1016/j.fertnstert.2018.05.006.
5. Wadelius M, Pirmohamed M. Pharmacogenetics of warfarin: current status and future challenges. Pharmacogenomics J. 2007;7(2):99–111. https://doi.org/10.1038/sj.tpj.6500417.
6. Mozaffarian D, et al. Heart disease and stroke statistics--2015 update: a report from the American Heart Association. Circulation. 2015;131(4):e29–322. https://doi.org/10.1161/cir.0000000000000152.
7. Currie G, Delles C. Precision medicine and personalized medicine in cardiovascular disease. Adv Exp Med Biol. 2018;1065:589–605. https://doi.org/10.1007/978-3-319-77932-4_36.
8. Hamburg MA, Collins FS. The path to personalized medicine. N Engl J Med. 2010;363(4):301–4. https://doi.org/10.1056/NEJMp1006304.
9. Davey Smith G, Ebrahim S. 'Mendelian randomization': can genetic epidemiology contribute to understanding environmental determinants of disease?*. Int J Epidemiol. 2003;32(1):1–22. https://doi.org/10.1093/ije/dyg070.
10. Emdin CA, Khera AV, Kathiresan S. Mendelian randomization. JAMA. 2017;318(19):1925–6. https://doi.org/10.1001/jama.2017.17219.
11. Cohen JC, et al. Sequence variations in PCSK9, low LDL, and protection against coronary heart disease. N Engl J Med. 2006;354(12):1264–72. https://doi.org/10.1056/NEJMoa054013.
12. Stitziel NO, et al. Inactivating mutations in NPC1L1 and protection from coronary heart disease. N Engl J Med. 2014;371(22):2072–82. https://doi.org/10.1056/NEJMoa1405386.
13. Peloso GM, et al. Rare protein-truncating variants in APOB, lower low-density lipoprotein cholesterol, and protection against coronary heart disease. Circ Genom Precis Med. 2019;12(5):e002376. https://doi.org/10.1161/circgen.118.002376.
14. Ansell J, et al. Pharmacology and management of the vitamin K antagonists: American College of

Chest Physicians Evidence-Based Clinical Practice Guidelines (8th Edition). Chest. 2008;133(6 Suppl):160s–98s. https://doi.org/10.1378/chest.08-0670.

15. Kasner SE, et al. Warfarin dosing algorithms and the need for human intervention. Am J Med. 2016;129(4):431–7. https://doi.org/10.1016/j.amjmed.2015.11.012.

16. Ma Z, et al. Clinical model for predicting warfarin sensitivity. Sci Rep. 2019;9(1):12856. https://doi.org/10.1038/s41598-019-49329-0.

17. Danese E, et al. Impact of the CYP4F2 p.V433M polymorphism on coumarin dose requirement: systematic review and meta-analysis. Clin Pharmacol Ther. 2012;92(6):746–56. https://doi.org/10.1038/clpt.2012.184.

18. Caldwell MD, et al. CYP4F2 genetic variant alters required warfarin dose. Blood. 2008;111(8):4106–12. https://doi.org/10.1182/blood-2007-11-122010.

19. Aithal GP, et al. Association of polymorphisms in the cytochrome P450 CYP2C9 with warfarin dose requirement and risk of bleeding complications. Lancet. 1999;353(9154):717–9. https://doi.org/10.1016/s0140-6736(98)04474-2.

20. Mega JL, et al. Genetics and the clinical response to warfarin and edoxaban: findings from the randomised, double-blind ENGAGE AF-TIMI 48 trial. Lancet. 2015;385(9984):2280–7. https://doi.org/10.1016/s0140-6736(14)61994-2.

21. Lee CR, Goldstein JA, Pieper JA. Cytochrome P450 2C9 polymorphisms: a comprehensive review of the in-vitro and human data. Pharmacogenetics. 2002;12(3):251–63. https://doi.org/10.1097/00008571-200204000-00010.

22. Jiang X-L, et al. Clinical pharmacokinetics and pharmacodynamics of clopidogrel. Clin Pharmacokinet. 2015;54(2):147–66. https://doi.org/10.1007/s40262-014-0230-6.

23. Paré G, et al. Effects of CYP2C19 genotype on outcomes of clopidogrel treatment. N Engl J Med. 2010;363(18):1704–14. https://doi.org/10.1056/NEJMoa1008410.

24. Scott SA, et al. Clinical pharmacogenetics implementation consortium guidelines for CYP2C19 genotype and clopidogrel therapy: 2013 update. Clin Pharmacol Ther. 2013;94(3):317–23. https://doi.org/10.1038/clpt.2013.105.

25. Biswas M, et al. Effects of the ABCB1 C3435T single nucleotide polymorphism on major adverse cardiovascular events in acute coronary syndrome or coronary artery disease patients undergoing percutaneous coronary intervention and treated with clopidogrel: a systematic review and meta-analysis. Expert Opin Drug Saf. 2020;19(12):1605–16. https://doi.org/10.1080/14740338.2020.1836152.

26. Price MJ, Tantry US, Gurbel PA. The influence of CYP2C19 polymorphisms on the pharmacokinetics, pharmacodynamics, and clinical effectiveness of P2Y12 inhibitors. Rev Cardiovasc Med. 2011;12(1):1–12. https://doi.org/10.3909/ricm0590.

27. Tantry US, et al. First analysis of the relation between CYP2C19 genotype and pharmacodynamics in patients treated with ticagrelor versus clopidogrel. Circ Cardiovasc Genet. 2010;3(6):556–66. https://doi.org/10.1161/CIRCGENETICS.110.958561.

28. Varenhorst C, et al. Genetic variation of CYP2C19 affects both pharmacokinetic and pharmacodynamic responses to clopidogrel but not prasugrel in aspirin-treated patients with coronary artery disease. Eur Heart J. 2009;30(14):1744–52. https://doi.org/10.1093/eurheartj/ehp157.

29. Zhang JE, et al. Effects of CYP4F2 genetic polymorphisms and haplotypes on clinical outcomes in patients initiated on warfarin therapy. Pharmacogenet Genomics. 2009;19(10):781–9. https://doi.org/10.1097/FPC.0b013e3283311347.

30. Perera MA, et al. Genetic variants associated with warfarin dose in African-American individuals: a genome-wide association study. Lancet. 2013;382(9894):790–6. https://doi.org/10.1016/s0140-6736(13)60681-9.

31. Kitzmiller JP, et al. CYP3A4*22 and CYP3A5*3 are associated with increased levels of plasma simvastatin concentrations in the cholesterol and pharmacogenetics study cohort. Pharmacogenet Genomics. 2014;24(10):486–91. https://doi.org/10.1097/fpc.0000000000000079.

32. Mazzarotto F, et al. Contemporary insights into the Genetics of hypertrophic cardiomyopathy: toward a new era in clinical testing? J Am Heart Assoc. 2020;9(8):e015473. https://doi.org/10.1161/JAHA.119.015473.

33. Geisterfer-Lowrance AA, et al. A molecular basis for familial hypertrophic cardiomyopathy: a beta cardiac myosin heavy chain gene missense mutation. Cell. 1990;62(5):999–1006. https://doi.org/10.1016/0092-8674(90)90274-i.

34. Watkins H, et al. A disease locus for familial hypertrophic cardiomyopathy maps to chromosome 1q3. Nat Genet. 1993;3(4):333–7. https://doi.org/10.1038/ng0493-333.

35. Carrier L, et al. Mapping of a novel gene for familial hypertrophic cardiomyopathy to chromosome 11. Nat Genet. 1993;4(3):311–3. https://doi.org/10.1038/ng0793-311.

36. Kimura A, et al. Mutations in the cardiac troponin I gene associated with hypertrophic cardiomyopathy. Nat Genet. 1997;16(4):379–82. https://doi.org/10.1038/ng0897-379.

37. Ochoa JP, et al. Formin homology 2 domain containing 3 (FHOD3) is a genetic basis for hypertrophic cardiomyopathy. J Am Coll Cardiol. 2018;72(20):2457–67. https://doi.org/10.1016/j.jacc.2018.10.001.

38. Martiniuk F, et al. Identification of the base-pair substitution responsible for a human acid alpha glucosidase allele with lower "affinity" for glycogen (GAA 2) and transient gene expression in deficient cells. Am J Hum Genet. 1990;47(3):440–5.

39. Davies JP, Winchester BG, Malcolm S. Mutation analysis in patients with the typical form of Anderson-Fabry disease. Hum Mol Genet. 1993;2(7):1051–3. https://doi.org/10.1093/hmg/2.7.1051.

40. Ho CY, et al. Genotype and lifetime burden of disease in hypertrophic cardiomyopathy: insights from the Sarcomeric human cardiomyopathy registry (SHaRe). Circulation. 2018;138(14):1387–98. https://doi.org/10.1161/circulationaha.117.033200.

41. Pinto YM, et al. Proposal for a revised definition of dilated cardiomyopathy, hypokinetic non-dilated cardiomyopathy, and its implications for clinical practice: a position statement of the ESC working group on myocardial and pericardial diseases. Eur Heart J. 2016;37(23):1850–8. https://doi.org/10.1093/eurheartj/ehv727.

42. Herman DS, et al. Truncations of titin causing dilated cardiomyopathy. N Engl J Med. 2012;366(7):619–28. https://doi.org/10.1056/NEJMoa1110186.

43. Parks SB, et al. Lamin A/C mutation analysis in a cohort of 324 unrelated patients with idiopathic or familial dilated cardiomyopathy. Am Heart J. 2008;156(1):161–9. https://doi.org/10.1016/j.ahj.2008.01.026.

44. Muchtar E, Blauwet LA, Gertz MA. Restrictive cardiomyopathy: Genetics, pathogenesis, clinical manifestations, diagnosis, and therapy. Circ Res. 2017;121(7):819–37. https://doi.org/10.1161/circresaha.117.310982.

45. Cimiotti D, et al. Genetic restrictive cardiomyopathy: causes and consequences—an integrative approach. Int J Mol Sci. 2021;22(2):558.

46. Kostareva A, et al. Genetic Spectrum of idiopathic restrictive cardiomyopathy uncovered by next-generation sequencing. PLoS One. 2016;11(9):e0163362. https://doi.org/10.1371/journal.pone.0163362.

47. Brodehl A, et al. The novel αB-crystallin (CRYAB) mutation p.D109G causes restrictive cardiomyopathy. Hum Mutat. 2017;38(8):947–52. https://doi.org/10.1002/humu.23248.

48. Renard M, et al. Clinical validity of genes for heritable thoracic aortic aneurysm and dissection. J Am Coll Cardiol. 2018;72(6):605–15. https://doi.org/10.1016/j.jacc.2018.04.089.

49. Lindsay ME, Dietz HC. Lessons on the pathogenesis of aneurysm from heritable conditions. Nature. 2011;473(7347):308–16. https://doi.org/10.1038/nature10145.

50. Milewicz DM, Dietz HC, Miller DC. Treatment of aortic disease in patients with Marfan syndrome. Circulation. 2005;111(11):e150–7. https://doi.org/10.1161/01.cir.0000155243.70456.f4.

51. LeMaire SA, et al. Genome-wide association study identifies a susceptibility locus for thoracic aortic aneurysms and aortic dissections spanning FBN1 at 15q21.1. Nat Genet. 2011;43(10):996–1000. https://doi.org/10.1038/ng.934.

52. Tran-Fadulu V, et al. Analysis of multigenerational families with thoracic aortic aneurysms and dissections due to TGFBR1 or TGFBR2 mutations. J Med Genet. 2009;46(9):607–13. https://doi.org/10.1136/jmg.2008.062844.

53. Wang L, et al. Mutations in myosin light chain kinase cause familial aortic dissections. Am J Hum Genet. 2010;87(5):701–7. https://doi.org/10.1016/j.ajhg.2010.10.006.

54. Wallace SE, et al. MYLK pathogenic variants aortic disease presentation, pregnancy risk, and characterization of pathogenic missense variants. Genet Med. 2019;21(1):144–51. https://doi.org/10.1038/s41436-018-0038-0.

55. Messika-Zeitoun D, et al. Aortic valve calcification: determinants and progression in the population. Arterioscler Thromb Vasc Biol. 2007;27(3):642–8. https://doi.org/10.1161/01.ATV.0000255952.47980.c2.

56. Yu B, et al. Pathological significance of lipoprotein(a) in aortic valve stenosis. Atherosclerosis. 2018;272:168–74. https://doi.org/10.1016/j.atherosclerosis.2018.03.025.

57. Thanassoulis G, et al. Genetic associations with Valvular calcification and aortic stenosis. N Engl J Med. 2013;368(6):503–12. https://doi.org/10.1056/NEJMoa1109034.

58. Yang ZK, et al. Mineral metabolism disturbances are associated with the presence and severity of calcific aortic valve disease. J Zhejiang Univ Sci B. 2015;16(5):362–9. https://doi.org/10.1631/jzus.B1400292.

59. Cho HJ, Cho HJ, Kim HS. Osteopontin: a multifunctional protein at the crossroads of inflammation, atherosclerosis, and vascular calcification. Curr Atheroscler Rep. 2009;11(3):206–13. https://doi.org/10.1007/s11883-009-0032-8.

60. Kiechl S, et al. Osteoprotegerin is a risk factor for progressive atherosclerosis and cardiovascular disease. Circulation. 2004;109(18):2175–80. https://doi.org/10.1161/01.cir.0000127957.43874.bb.

61. Civitelli R, Armamento-Villareal R, Napoli N. Bone turnover markers: understanding their value in clinical trials and clinical practice. Osteoporos Int. 2009;20(6):843–51. https://doi.org/10.1007/s00198-009-0838-9.

62. Schafer C, et al. The serum protein alpha 2-Heremans-Schmid glycoprotein/fetuin-A is a systemically acting inhibitor of ectopic calcification. J Clin Invest. 2003;112(3):357–66. https://doi.org/10.1172/jci17202.

63. Novaro GM, et al. Clinical factors, but not C-reactive protein, predict progression of calcific aortic-valve disease: the cardiovascular health study. J Am Coll Cardiol. 2007;50(20):1992–8. https://doi.org/10.1016/j.jacc.2007.07.064.

64. Cimadevilla C, et al. Prognostic value of B-type natriuretic peptide in elderly patients with aortic valve stenosis: the COFRASA-GENERAC study. Heart. 2013;99(7):461–7. https://doi.org/10.1136/heartjnl-2012-303284.

65. Chin CW, et al. A clinical risk score of myocardial fibrosis predicts adverse outcomes in aortic stenosis. Eur Heart J. 2016;37(8):713–23. https://doi.org/10.1093/eurheartj/ehv525.

66. Kruger R, et al. Extracellular matrix biomarker, fibulin-1, is closely related to NT-proBNP and soluble urokinase plasminogen activator receptor in patients with aortic valve stenosis (the SEAS study). PLoS One. 2014;9(7):e101522. https://doi.org/10.1371/journal.pone.0101522.

67. Pitsis A, et al. Mitral valve repair: moving towards a personalized ring. J Cardiothorac Surg. 2019;14(1):108. https://doi.org/10.1186/s13019-019-0926-7.

68. Priori SG, et al. 2015 ESC guidelines for the management of patients with ventricular arrhythmias and the prevention of sudden cardiac death: the task force for the Management of Patients with ventricular arrhythmias and the prevention of sudden cardiac death of the European Society of Cardiology (ESC). Endorsed by: Association for European Paediatric and Congenital Cardiology (AEPC). Eur Heart J. 2015;36(41):2793–867. https://doi.org/10.1093/eurheartj/ehv316.

69. Goldenberg I, Zareba W, Moss AJ. Long QT syndrome. Curr Probl Cardiol. 2008;33(11):629–94. https://doi.org/10.1016/j.cpcardiol.2008.07.002.

70. Heijman J, et al. Dominant-negative control of cAMP-dependent IKs upregulation in human long-QT syndrome type 1. Circ Res. 2012;110(2):211–9. https://doi.org/10.1161/circresaha.111.249482.

71. Sanguinetti MC, et al. A mechanistic link between an inherited and an acquired cardiac arrhythmia: HERG encodes the IKr potassium channel. Cell. 1995;81(2):299–307. https://doi.org/10.1016/0092-8674(95)90340-2.

72. Kambouris NG, et al. Phenotypic characterization of a novel long-QT syndrome mutation (R1623Q) in the cardiac sodium channel. Circulation. 1998;97(7):640–4. https://doi.org/10.1161/01.cir.97.7.640.

73. Abu-Zeitone A, et al. Efficacy of different beta-blockers in the treatment of long QT syndrome. J Am Coll Cardiol. 2014;64(13):1352–8. https://doi.org/10.1016/j.jacc.2014.05.068.

74. Wilde AA, et al. Clinical aspects of type 3 long-QT syndrome: an international multicenter study. Circulation. 2016;134(12):872–82. https://doi.org/10.1161/circulationaha.116.021823.

75. Antzelevitch C, et al. J-wave syndromes expert consensus conference report: emerging concepts and gaps in knowledge. J Arrhythmia. 2016;32(5):315–39. https://doi.org/10.1016/j.joa.2016.07.002.

76. Kapplinger JD, et al. An international compendium of mutations in the SCN5A−encoded cardiac sodium channel in patients referred for Brugada syndrome genetic testing. Heart Rhythm. 2010;7(1):33–46. https://doi.org/10.1016/j.hrthm.2009.09.069.

77. Brugada J, et al. Present status of Brugada syndrome: JACC state-of-the-art review. J Am Coll Cardiol. 2018;72(9):1046–59. https://doi.org/10.1016/j.jacc.2018.06.037.

78. Hu D, et al. Mutations in SCN10A are responsible for a large fraction of cases of Brugada syndrome. J Am Coll Cardiol. 2014;64(1):66–79. https://doi.org/10.1016/j.jacc.2014.04.032.

79. Behr ER, et al. Role of common and rare variants in SCN10A: results from the Brugada syndrome QRS locus gene discovery collaborative study. Cardiovasc Res. 2015;106(3):520–9. https://doi.org/10.1093/cvr/cvv042.

80. Priori SG, et al. Executive summary: HRS/EHRA/APHRS expert consensus statement on the diagnosis and management of patients with inherited primary arrhythmia syndromes. Heart Rhythm. 2013;10(12):e85–108. https://doi.org/10.1016/j.hrthm.2013.07.021.

81. Bellocq C, et al. Mutation in the KCNQ1 gene leading to the short QT-interval syndrome. Circulation. 2004;109(20):2394–7. https://doi.org/10.1161/01.cir.0000130409.72142.fe.

82. Wilde AA, Behr ER. Genetic testing for inherited cardiac disease. Nat Rev Cardiol. 2013;10(10):571–83. https://doi.org/10.1038/nrcardio.2013.108.

83. Templin C, et al. Identification of a novel loss-of-function calcium channel gene mutation in short QT syndrome (SQTS6). Eur Heart J. 2011;32(9):1077–88. https://doi.org/10.1093/eurheartj/ehr076.

84. Antzelevitch C, et al. Loss-of-function mutations in the cardiac calcium channel underlie a new clinical entity characterized by ST-segment elevation, short QT intervals, and sudden cardiac death. Circulation. 2007;115(4):442–9. https://doi.org/10.1161/circulationaha.106.668392.

85. Waldmann V, et al. Characteristics and clinical assessment of unexplained sudden cardiac arrest in the real-world setting: focus on idiopathic ventricular fibrillation. Eur Heart J. 2018;39(21):1981–7. https://doi.org/10.1093/eurheartj/ehy098.

86. Haïssaguerre M, et al. Idiopathic ventricular fibrillation: role of Purkinje system and microstructural myocardial abnormalities. JACC Clin Electrophysiol. 2020;6(6):591–608. https://doi.org/10.1016/j.jacep.2020.03.010.

87. Alders M, et al. Haplotype-sharing analysis implicates chromosome 7q36 harboring DPP6 in familial idiopathic ventricular fibrillation. Am J Hum Genet. 2009;84(4):468–76. https://doi.org/10.1016/j.ajhg.2009.02.009.

88. Marsman RF, et al. A mutation in CALM1 encoding calmodulin in familial idiopathic ventricular fibrillation in childhood and adolescence. J Am Coll Cardiol. 2014;63(3):259–66. https://doi.org/10.1016/j.jacc.2013.07.091.

89. Beach LY, et al. Idiopathic ventricular fibrillation in a 29-year-old man. Circulation. 2017;136(1):112–4. https://doi.org/10.1161/CIRCULATIONAHA.117.029120.

90. Pflaumer A, Davis AM. Guidelines for the diagnosis and management of Catecholaminergic polymorphic ventricular tachycardia. Heart Lung Circ. 2012;21(2):96–100. https://doi.org/10.1016/j.hlc.2011.10.008.

91. Ackerman MJ, et al. HRS/EHRA expert consensus statement on the state of genetic testing for the channelopathies and cardiomyopathies this document was developed as a partnership between the Heart Rhythm Society (HRS) and the European heart rhythm association (EHRA). Heart Rhythm. 2011;8(8):1308–39. https://doi.org/10.1016/j.hrthm.2011.05.020.

92. Laitinen PJ, et al. Mutations of the cardiac ryanodine receptor (RyR2) gene in familial polymorphic ventricular tachycardia. Circulation. 2001;103(4):485–90. https://doi.org/10.1161/01.cir.103.4.485.

93. Lahat H, et al. A missense mutation in a highly conserved region of CASQ2 is associated with autosomal recessive catecholamine-induced polymorphic ventricular tachycardia in Bedouin families from Israel. Am J Hum Genet. 2001;69(6):1378–84. https://doi.org/10.1086/324565.

94. Bhuiyan ZA, et al. A novel early onset lethal form of catecholaminergic polymorphic ventricular tachycardia maps to chromosome 7p14-p22. J Cardiovasc Electrophysiol. 2007;18(10):1060–6. https://doi.org/10.1111/j.1540-8167.2007.00913.x.

95. Roux-Buisson N, et al. Absence of triadin, a protein of the calcium release complex, is responsible for cardiac arrhythmia with sudden death in human. Hum Mol Genet. 2012;21(12):2759–67. https://doi.org/10.1093/hmg/dds104.

96. Nyegaard M, et al. Mutations in calmodulin cause ventricular tachycardia and sudden cardiac death. Am J Hum Genet. 2012;91(4):703–12. https://doi.org/10.1016/j.ajhg.2012.08.015.

97. Makita N, et al. Novel calmodulin mutations associated with congenital arrhythmia susceptibility. Circ Cardiovasc Genet. 2014;7(4):466–74. https://doi.org/10.1161/circgenetics.113.000459.

98. Mohler PJ, et al. Ankyrin-B mutation causes type 4 long-QT cardiac arrhythmia and sudden cardiac death. Nature. 2003;421(6923):634–9. https://doi.org/10.1038/nature01335.

99. Plaster NM, et al. Mutations in Kir2.1 cause the developmental and episodic electrical phenotypes of Andersen's syndrome. Cell. 2001;105(4):511–9. https://doi.org/10.1016/s0092-8674(01)00342-7.

100. Priori SG, et al. HRS/EHRA/APHRS expert consensus statement on the diagnosis and management of patients with inherited primary arrhythmia syndromes: document endorsed by HRS, EHRA, and APHRS in May 2013 and by ACCF, AHA, PACES, and AEPC in June 2013. Heart Rhythm. 2013;10(12):1932–63. https://doi.org/10.1016/j.hrthm.2013.05.014.

101. Baruteau AE, Probst V, Abriel H. Inherited progressive cardiac conduction disorders. Curr Opin Cardiol. 2015;30(1):33–9. https://doi.org/10.1097/hco.0000000000000134.

102. Schott JJ, et al. Cardiac conduction defects associate with mutations in SCN5A. Nat Genet. 1999;23(1):20–1. https://doi.org/10.1038/12618.

103. Asatryan B, Medeiros-Domingo A. Emerging implications of genetic testing in inherited primary arrhythmia syndromes. Cardiol Rev. 2019;27(1):23–33. https://doi.org/10.1097/crd.0000000000000203.

104. Nilius B, et al. Voltage dependence of the Ca$^{2+}$-activated Cation Channel TRPM4 *. J Biol Chem. 2003;278(33):30813–20. https://doi.org/10.1074/jbc.M305127200.

105. Stallmeyer B, et al. Mutational spectrum in the Ca(2+)-activated cation channel gene TRPM4 in patients with cardiac conductance disturbances. Hum Mutat. 2012;33(1):109–17. https://doi.org/10.1002/humu.21599.

106. Asatryan B, Medeiros-Domingo A. Molecular and genetic insights into progressive cardiac conduction disease. EP Europace. 2019;21(8):1145–58. https://doi.org/10.1093/europace/euz109.

107. Ozaki K, et al. Functional variation in LGALS2 confers risk of myocardial infarction and regulates lymphotoxin-alpha secretion in vitro. Nature. 2004;429(6987):72–5. https://doi.org/10.1038/nature02502.

108. Ozaki K, et al. Functional SNPs in the lymphotoxin-alpha gene that are associated with susceptibility to myocardial infarction. Nat Genet. 2002;32(4):650–4. https://doi.org/10.1038/ng1047.

109. Kimura A, et al. Lack of association between LTA and LGALS2 polymorphisms and myocardial infarction in Japanese and Korean populations. Tissue Antigens. 2007;69:265–9. https://doi.org/10.1111/j.1399-0039.2006.00798.x.

110. Samani NJ, et al. Genomewide association analysis of coronary artery disease. N Engl J Med. 2007;357(5):443–53. https://doi.org/10.1056/NEJMoa072366.

111. McPherson R, et al. A common allele on chromosome 9 associated with coronary heart disease. Science. 2007;316(5830):1488–91. https://doi.org/10.1126/science.1142447.

112. Schunkert H, et al. Large-scale association analysis identifies 13 new susceptibility loci for coronary artery disease. Nat Genet. 2011;43(4):333–8. https://doi.org/10.1038/ng.784.

113. Cheng CY, et al. New loci and coding variants confer risk for age-related macular degeneration in East Asians. Nat Commun. 2015;6:6063. https://doi.org/10.1038/ncomms7063.

114. Myocardial Infarction Genetics, C, et al. Genome-wide association of early-onset myocardial infarction with single nucleotide polymorphisms and copy number variants. Nat Genet. 2009;41(3):334–41. https://doi.org/10.1038/ng.327.

115. Nioi P, et al. Variant ASGR1 associated with a reduced risk of coronary artery disease. N

Engl J Med. 2016;374(22):2131–41. https://doi.org/10.1056/NEJMoa1508419.

116. Khera AV, et al. Association of Rare and Common Variation in the lipoprotein lipase gene with coronary artery disease. JAMA. 2017;317(9):937–46. https://doi.org/10.1001/jama.2017.0972.

117. Crosby J, et al. Loss-of-function mutations in APOC3, triglycerides, and coronary disease. N Engl J Med. 2014;371(1):22–31. https://doi.org/10.1056/NEJMoa1307095.

118. Lim GB. Polymorphisms in ANGPTL4 link triglycerides with CAD. Nat Rev Cardiol. 2016;13(5):245. https://doi.org/10.1038/nrcardio.2016.46.

119. Guo J, et al. Association between 9p21.3 genomic markers and coronary artery disease in East Asians: a meta-analysis involving 9,813 cases and 10,710 controls. Mol Biol Rep. 2013;40(1):337–43. https://doi.org/10.1007/s11033-012-2066-1.

120. Mabuchi H. Half a century Tales of familial hypercholesterolemia (FH) in Japan. J Atheroscler Thromb. 2017;24(3):189–207. https://doi.org/10.5551/jat.RV16008.

121. Do R, et al. Exome sequencing identifies rare LDLR and APOA5 alleles conferring risk for myocardial infarction. Nature. 2015;518(7537):102–6. https://doi.org/10.1038/nature13917.

122. Deloukas P, et al. Large-scale association analysis identifies new risk loci for coronary artery disease. Nat Genet. 2013;45(1):25–33. https://doi.org/10.1038/ng.2480.

123. Flister MJ, et al. SH2B3 is a genetic determinant of cardiac inflammation and fibrosis. Circ Cardiovasc Genet. 2015;8(2):294–304. https://doi.org/10.1161/circgenetics.114.000527.

124. Smits PC, et al. Coronary artery disease: arterial remodelling and clinical presentation. Heart. 1999;82(4):461–4. https://doi.org/10.1136/hrt.82.4.461.

125. Reilly MP, et al. Identification of ADAMTS7 as a novel locus for coronary atherosclerosis and association of ABO with myocardial infarction in the presence of coronary atherosclerosis: two genome-wide association studies. Lancet. 2011;377(9763):383–92. https://doi.org/10.1016/s0140-6736(10)61996-4.

126. Roberts R. A genetic basis for coronary artery disease. Trends Cardiovasc Med. 2015;25(3):171–8. https://doi.org/10.1016/j.tcm.2014.10.008.

127. Levy D, et al. Genome-wide association study of blood pressure and hypertension. Nat Genet. 2009;41(6):677–87. https://doi.org/10.1038/ng.384.

128. Newton-Cheh C, et al. Genome-wide association study identifies eight loci associated with blood pressure. Nat Genet. 2009;41(6):666–76. https://doi.org/10.1038/ng.361.

129. Bombard Y, et al. The value of personalizing medicine: medical oncologists' views on gene expression profiling in breast cancer treatment. Oncologist. 2015;20(4):351–6. https://doi.org/10.1634/theoncologist.2014-0268.

130. Bonter K, et al. Personalised medicine in Canada: a survey of adoption and practice in oncology, cardiology and family medicine. BMJ Open. 2011;1(1):e000110. https://doi.org/10.1136/bmjopen-2011-000110.

131. Gray SW, et al. Attitudes of patients with cancer about personalized medicine and somatic genetic testing. J Oncol Pract. 2012;8(6):329–35. https://doi.org/10.1200/jop.2012.000626. 2 p following 335

132. Hammack CM, Brelsford KM, Beskow LM. Thought leader perspectives on participant protections in precision medicine research. J Law Med Ethics. 2019;47(1):134–48. https://doi.org/10.1177/1073110519840493.

133. Pearce C, et al. Delivering genomic medicine in the United Kingdom National Health Service: a systematic review and narrative synthesis. Genet Med. 2019;21(12):2667–75. https://doi.org/10.1038/s41436-019-0579-x.

134. Di Paolo A, et al. Personalized medicine in Europe: not yet personal enough? BMC Health Serv Res. 2017;17(1):289. https://doi.org/10.1186/s12913-017-2205-4.

135. Gelernter D. Mirror Worlds: or the Day Software Puts the Universe in a Shoebox...How It Will Happen and What It Will Mean.

136. Bruynseels K, Santoni de Sio F, van den Hoven J. Digital twins in health care: ethical implications of an emerging engineering paradigm. Front Genet. 2018;9:31. https://doi.org/10.3389/fgene.2018.00031.

137. Corral-Acero J, et al. The 'Digital Twin' to enable the vision of precision cardiology. Eur Heart J. 2020;41(48):4556–64. https://doi.org/10.1093/eurheartj/ehaa159.

# Precision Medicine in Endocrinology Practice

# 5

Hamid Reza Aghaei Meybodi (iD),
Mandana Hasanzad (iD), Negar Sarhangi (iD),
and Bagher Larijani (iD)

## What Will You Learn in This Chapter?

Precision (personalized) medicine can potentially use some omics data to help disease management and improve clinical outcomes. As we are in a time in which technological advances are combined with increased knowledge of bioinformatics, the comprehensive investigation of the molecular basis of many common diseases such as diabetes will be possible.

Disease management in each patient is investigated according to the genomic profile and system biology to find a better solution in prediction, prevention, and personalized treatment.

H. R. Aghaei Meybodi (✉) · N. Sarhangi
Personalized Medicine Research Center,
Endocrinology and Metabolism Clinical Sciences
Institute, Tehran University of Medical Sciences,
Tehran, Iran
e-mail: hraghai@tums.ac.ir

M. Hasanzad
Personalized Medicine Research Center,
Endocrinology and Metabolism Clinical Sciences
Institute, Tehran University of Medical Sciences,
Tehran, Iran

Medical Genomics Research Center, Tehran Medical
Sciences, Islamic Azad University, Tehran, Iran

B. Larijani
Endocrinology and Metabolism Research Center,
Endocrinology and Metabolism Clinical Sciences
Institute, Tehran University of Medical Sciences,
Tehran, Iran

Accordingly, this chapter is an introduction to precision medicine in endocrinology. At the beginning of this chapter, precision medicine in diabetes has been introduced. Patients with diabetes might benefit from classification based on the cause of diabetes, pathophysiology, and historical background, allowing for the selection of the best treatments. The role of precision medicine in other endocrine conditions like osteoporosis and thyroid cancer will be addressed in the present chapter.

## Rationale and Importance

The precision medicine approach can be utilized for individuals with different types of endocrine diseases which provides improved clinical outcomes and cost-effectiveness in the healthcare system.

Early prevention, early diagnosis, and treatment programs in the precision medicine approach are integrating genomic-based data and clinical information of the patient.

Even with clinical guidelines, it is impossible to predict which individuals will benefit the most from a certain diabetes treatment or who will be more prone to toxicity. Currently, clinicians encounter the decision of prescribing a drug for type 2 diabetes as they have a variety of choices, all of which appear to be equally likely to provide benefit, and there is little specific data available on whether this particular patient is more or less likely than the average to benefit from any of the drugs or whether the therapeutic effect will be maintained. As a result, the promise of pharmacogenomics is to help in the selection of the best antidiabetic pharmacological therapy for the patients.

## 5.1 Precision Diabetes Medicine

Patient-centered medicine is a treatment strategy that takes into account the patient's genetic profile as well as their desires, attitudes, behaviors, awareness, and social context, whereas precision medicine is a healthcare model that goes beyond omics and is based on data, analytics, and information. Patients and providers are linked by personalized medicine, whereas patients, providers, clinical laboratories, and researchers are linked by the precision medicine ecosystem [1].

Precision medicine is an emerging approach to illness, treatment, and prevention that focuses on the individual heterogeneity in genes, environment, and lifestyle, according to the National Institutes of Health (NIH) [2].

The application of "omics" data, including genomics, transcriptomics, proteomics, metabolomics, and other big data, may eventually make the promise of precision medicine a reality [3].

Although the concept of providing the right treatment for the right patient at the right time is not new, leveraging new technology by assessment of genetic data, using omics data for disease categorization, and applying predictive and preventive approaches to pathological conditions open the new and quick way to precision medicine.

Precision medicine promises to provide tailored treatment at the level of the individual patient.

As far as we have several concerns for polygenic forms of diabetes type 1 and type 2, precision diabetes medicine is still in its infancy.

A new concept of disease categorization has emerged in recent years based on a personalized medicine approach. Individuals with the same diagnostic criteria might present a variety of symptoms, have varied clinical outcomes, and respond to treatment interventions in different ways. As a result, the concept of precision medicine has developed further by scientific data, becoming precision medicine to create a new taxonomy for diabetes [4].

Diabetes mellitus is a complex disease, and in recent decades, the definition of diabetes in terms of underlying causes has expanded. Hyperglycemia is the endpoint of several pathophysiological processes that often develop over time that is caused by pancreatic beta-cell dysfunction and lead to secrete insufficient insulin for the target tissues.

Diabetes is a considerably more complex condition than the T1D and T2D classifications currently being used [5, 6]. T1D, which represents 10% of all diabetes cases, occurs as a result of autoimmune destruction of the insulin-producing beta-cells, with early onset in childhood, whereas T2D is caused by a combination of insulin resistance and an insulin-secretory deficiency. Therefore, T1DM diagnosis is crucial for survival.

Clinical research into various processes that contribute to diabetes pathophysiology proposes that individuals with diabetes may have a combination of biological changes and hybrid forms.

The development of a range of hybrid types of diabetes, such as latent autoimmune disease of adults (LADA) and ketosis-prone diabetes (KPD), suggests molecular changes role [7].

T1DM requires insulin therapy multiple times per day for whole life because endogenous insulin secretion is nearly non-existent.

T2D patients range from those with a mainly insulin-resistance phenotype but enough β-cell reserve to continue insulin-independent to those who may need insulin treatment early in the disease's course.

There are also various types of monogenic diabetes caused by highly penetrant single gene deficiencies, the most prevalent of which are maturity-onset diabetes of the young (MODY), neonatal diabetes mellitus (NDM), and congenital hyperinsulinemic hypoglycemia (CHH). Precision diabetes has already been employed in clinical settings with several types of diabetes [8].

Monogenic diabetes is inherited or can result from de novo loss of activity in a single gene, which accounts for 2–3% of all diabetes diagnosed in children and young adults. Hyperglycemia can also result from an inherited or de novo loss of activity in a single gene.

Precision medicine in diabetes promises several applications which can be categorized as polygenic and monogenic diabetes.

Precision medicine of T1DM is based on making a precise diagnosis of T1DM with clear evidence of the islet cell autoantibodies and β-cell dysfunction as measured by levels of C peptide and prescribing the proper doses of insulin.

The prediction of T2DM development, personalized treatment, prediction of response to therapy, and prediction of micro- and macrovascular complications will be achievable by precision medicine in T2DM.

## 5.2 Some Lessons from the American Diabetes Association (ADA) and the European Association for the Study of Diabetes (EASD) Consensus Report for Precision Medicine in Diabetes

According to the Consensus Report From the American Diabetes Association (ADA) and the European Association for the Study of Diabetes (EASD), precision diabetes medicine is a method of integrating multidimensional data and considering the individual characteristics to improve diabetes diagnosis, prediction, prevention, and precision treatment (Fig. 5.1) [9].

Our inability to determine the pathophysiological mechanisms that cause diabetes in patients, as well as our failure to thoroughly know the numerous molecular and environmental

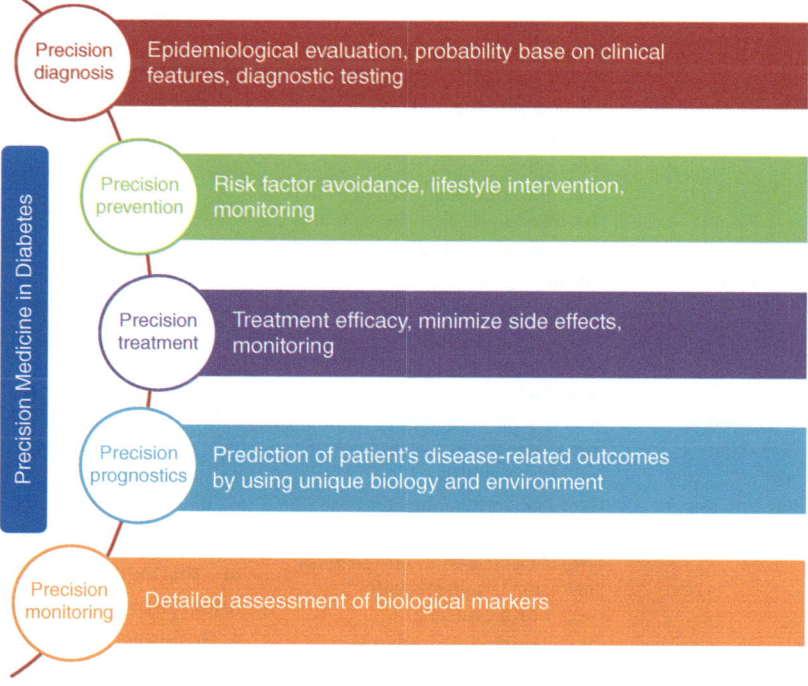

**Fig. 5.1** Precision medicine diabetes approach [9]

Precision Medicine in Diabetes

**Precision diagnosis** Epidemiological evaluation, probability base on clinical features, diagnostic testing

**Precision prevention** Risk factor avoidance, lifestyle intervention, monitoring

**Precision treatment** Treatment efficacy, minimize side effects, monitoring

**Precision prognostics** Prediction of patient's disease-related outcomes by using unique biology and environment

**Precision monitoring** Detailed assessment of biological markers

processes that underlay diabetes, limits our ability to prevent and manage the disease, although there are a lot of barriers for precision medicine diabetes implementation (Table 5.1).

Precision diabetes medicine includes precision diagnosis, precision therapeutics, precision prevention, precision treatment, precision prognostics, and precision monitoring.

*Precision diagnosis* involves refining the diagnostic features of diabetes for treatment optimization or improving prognostic clarity using information about a person's unique biology and environment. Successful precision treatment often requires careful diagnosis, and precision diagnostics can also include classification into diagnostic subtypes.

*Precision therapeutics* is a medical approach using omics and environmental information for disease prevention or treatment.

*Precision prevention* uses omics and environment information to determine their potential response to health interventions and risk factors and monitor disease susceptibility.

**Table 5.1** Precision medicine diabetes challenges [9]

| Precision medicine diabetes approach | Implementation barriers |
|---|---|
| Precision diagnosis | Fasting C-peptide measurement is not regularly conducted in clinical practice |
| | Clustering biomarkers change over time |
| | The clustering approach does not have good predictive accuracy |
| Precision prevention | Availability of appropriate technologies |
| | Accessibility to precision prevention programs |
| | Willingness to embrace the approach |
| Precision therapeutics | Complications and drug outcomes will differ because of ethnic variability |
| | Need powerful computational servers for analysis |
| | Novel clinical trials designed for precision treatments with real-world evidence |
| | Require to train healthcare providers to translate big data into a meaningful data |

*Precision treatment* uses genomic data to guide the choice of efficient therapy to achieve the preferred therapeutic outcome and to reduce adverse drug reactions.

*Precision prognostics* improve the accuracy of predicting a patient's disease-related outcomes based on genetic information.

*Precision monitoring* is a comprehensive analysis of biomarkers such as continuous glucose monitoring, psychophysiological stress, etc., which can be achieved by digital apps or specialized sensors.

## 5.3 Precision Diagnostics in Diabetes

Precision diabetes medicine is the monitoring of disease progression using advanced technologies or considering how patient features affect the reliability of assays.

Precision diagnostics use strategies to subclassify patients to successfully apply the precision medicine approach. The precision diagnostic approach should have specific criteria, including robustness, potential to define the new taxonomy and subclassification, availability, cost-effectiveness, accredited related agency, and easily performed.

The best diagnosis and outcomes are obtained by integrating all diagnostic modalities rather than depending solely on prior prevalence, clinical symptoms, or medical reports.

## 5.4 Precision Diagnostics in T1D and T2DM

There is currently no immune-based test that is reproducible and reliable enough to be utilized as a diagnostic tool.

The first islet autoantibody appearance, as well as the type of autoantibody, may be crucial in distinguishing T1D etiological subtypes. Integration of immunological biomarker data with ß-cell secretion assessments provides a tailored alternative to the present disease classification system [10].

Recent biomarker studies show new prospects for patient stratification based on personalized characteristics such as age and specific immunological phenotypes, which combine active disease timing with targeted therapy options [11].

The great majority of T1D genetic risk is now recognized, and the sensitivity and specificity of a T1D genetic risk score (T1D-GRS) are both above 80%, although a high T1D-GRS will have a low positive predictive value in patient populations with a low overall prevalence of T1D, such as those over the age of 50 when diagnosed with diabetes. It will most likely be most beneficial when coupled with clinical characteristics and islet autoantibodies.

The precision approach for the diagnosis of diabetes is not more based on blood glucose levels, and glycated hemoglobin (HbA1c) is widely used.

The fact that the vast majority of T2D genome-wide association studies (GWAS) variations do not impact protein-coding sequences indicates that gene regulation plays a critical role in disease progression [12].

The current T2DM genetic data lack the prediction accuracy required to replace existing delineative techniques. Although the subcategorization of T2D using genetic data is interesting about the etiological processes that influence the disease, the methods outlined so far are not designed to be used to subclassify a T2D diagnosis, nor are the existing genetic data sufficient for this purpose for the majority of individuals with T2D [13].

The current genetic risk variants at these loci have only minor effects on disease predisposition (10–15% of overall disease risk). The knowledge gained has prepared the path for revealing the disease's molecular taxonomy and the potential recognition for new therapeutic approaches [14].

Due to genotyping and sequencing research in a variety of populations, ethnic-specific alleles are emerging to highlight illness heterogeneity.

National guidelines keep track of diabetes classification, but this classification is still not updated in two decades, and very few efforts have been made to investigate the heterogeneity of type 2 diabetes. Current treatment guidelines are limited in that they respond to poor metabolic control once it has developed, but cannot predict which patients will require more intensive treatment.

Since target tissues appear to remember poor metabolic control decades later that is called metabolic memory, evidence suggests that early treatment is critical for preventing life-shortening complications [13, 15].

A precise classification could be a useful tool for determining those at the highest risk of complications at diagnosis and enabling personalized treatment regimens, similar to how genetic diagnosis of monogenic diabetes guides practitioners to optimal treatment [16].

Currently, the important impact of precision diagnosis in clinical diabetes medicine is the classification of T1D versus T2D, the two most prevalent ones with different etiologies and different treatment needs [17].

T2D-associated genetic variations can also be used as biomarkers in unsupervised classification approaches or aggregated into biologically relevant polygenic risk scores to further characterization certain T2D subtypes (GRS). The results of unsupervised clustering analysis of 37 identified T2D susceptibility loci revealed that T2D risk loci can be classified into many groups.

The results of unsupervised clustering analysis of 37 identified T2D susceptibility loci revealed that T2D risk loci can be classified into different groups: (1) insulin processing, (2) insulin secretion, (3) insulin sensitivity, and (4) insulin processing and secretion without a detectable change in fasting glucose levels [18].

As the number of T2DM-associated variants increases, GRS construction provides a quantitative measure of T2DM genetic susceptibility that can help in the recognition of the subgroups. Another approach to categorizing diabetes and providing biological insights into early metabolic changes is through specific metabolomics profiling [19, 20].

In a data-driven cluster analysis in 9000 Swedish individuals with newly diagnosed diabetes, clusters were based on six variables including age at diagnosis, body mass index (BMI), glutamate decarboxylase antibodies, HbA1c, and homeostatic model assessment of β-cell function and insulin resistance. Five subgroups or clusters of diabetics were developed, each with a different disease progression and risk of diabetic complications. These subgroups are severe insulin-deficient diabetes (SIDD), severe insulin-resistant diabetes (SIRD), mild obesity-related diabetes (MOD), and mild age-related diabetes (MARD). These five clusters have distinct clinical characteristics and risk factors for developing diabetic complications [21]. This new classification could eventually aid in tailoring and targeting early treatment to patients who would benefit the most, consequently indicating the first step toward precision medicine in diabetes [22]. Although the proposed clusters differ in terms of diabetes progression and treatment response, models based on simple continuous clinical features are more useful for stratifying patients. Precision medicine in type 2 diabetes is more likely to be clinically useful if it is based on the use of specific phenotypic measures to predict specific outcomes rather than classifying patients into subgroups [9, 23].

More study in T1D and T2D is needed to characterize subtypes and determine the best interventional and therapeutic strategies due to the limited ideal testing and uncertainty in etiology.

## 5.5 Precision Therapeutics of Diabetes

Precision therapeutics can be categorized as precision prevention and precision treatment. Precise prevention or treatment is crucial for precision therapy.

Precision prevention ways in T2D are wider than T1D as tailoring approaches of lifestyle (e.g., diet) in T2D is more possible.

## 5.6 Precision Prevention

Precision prevention in T1D mostly involves optimizing evaluation methods, allowing for earlier diagnosis and treatment. T1D is characterized by insulin-producing pancreatic b-cell damage, dysfunction, and eventual destruction, which is assumed to be the outcome of an autoimmune process. The course of T1D has been classified into three stages [9, 24].

The current staging is based on the presence of islet autoantibodies and the level of blood glucose [25, 26]. Approximately half of the risk of T1D is attributed to hereditary factors, with over 30% of the genetic risk related to genes of the human leukocyte antigen (HLA) complex, as well as more than 50 non-HLA loci [13].

In genetically susceptible individuals who have not yet produced the autoantibodies, primary prevention is possible, and secondary prevention studies in T1D patients with stage 1 and stage 2 are under investigation [27]. In T1D patients, dietary supplementation may not be effective if the particular genetic profiles of children are not considered.

As a result, precision prevention in T1D is believed to involve the stratification of at-risk individuals as well as novel monitoring technology.

In recent years, T2D as a global public health concern has prompted many studies to assess interventional preventive approaches which are applied in prediabetes individuals. The levels of fasting blood glucose, 2-hour blood glucose, or HbA1c in prediabetes are below the diagnostic thresholds for diabetes. But this approach is not cost-effective [28, 29].

Preventive interventions focus on younger prediabetes individuals with risk factors including obesity, maternal history, positive family history, certain biomarker profiles such as genetics, etc.), which is a more cost-effective approach.

Large population-based studies and intervention trials significantly indicate that standard lifestyle modification strategies are similarly effective in preventing diabetes regardless of genetic risk. Increasing the knowledge of genetic risk factors of T2D may not induce using lifestyle modification methods [30–32].

## 5.7 Precision Treatment/ Pharmacogenomics

The clinical management of diabetes is now based on lowering plasma glucose levels to reduce the risk of developing long-term complications. There are currently 12 drug classes available for the treatment of type 2 diabetes: biguanides (metformin), thiazolidinediones (glitazones), sulfonylureas (SU), α-glucosidase inhibitors, meglitinides (glinides), DPP4 inhibitors (gliptins), incretin mimetics (GLP-1 receptor agonists), SGLT2 inhibitors (gliflozins), bile acid sequestrants, amylin mimetics, dopamine agonists, and insulin/insulin analogs [33].

The ADA Standards of Medical Care in Diabetes recommend individualized treatment to achieve the control of HbA1c levels quickly after diagnosis of type 2 diabetes mellitus [33]. Nevertheless, there is significant variation in response to various therapies, implying that treatment heterogeneity may indicate underlying biological variations. Twelve approved classes of antidiabetic drugs have been evaluated for preventive efficacy. Several studies have revealed widespread diversity in glycemic response toler-

ability, as well as several variable outcomes in patients treated with the same antidiabetic drugs [34].

Antihyperglycemic drugs have been rapidly developed over the last decade. One of the primary strategies for improving therapeutic quality may be an individualized strategy which is facilitated by pharmacogenetic investigations.

Thus, precision treatment can be defined as the use of patient information to direct the selection of effective therapy to achieve the desired therapeutic outcome which is also known as the pharmacogenomics approach.

The well-known example of the influence of genetics on precision treatment was seen in monogenic diabetes as single-gene mutations in a gene such as HNF1A can be causal of diabetes onset and targeted therapies [16].

It seems that genetic variants influencing OAD treatment outcome can be generally split into two groups:

classical pharmacogenomics genes that impact pharmacokinetics/pharmacodynamics (e.g., SLC47A1) and T2D risk genes (Fig. 5.2) [35].

Several numbers of genetic variations in T2D patients have been identified in genes closely

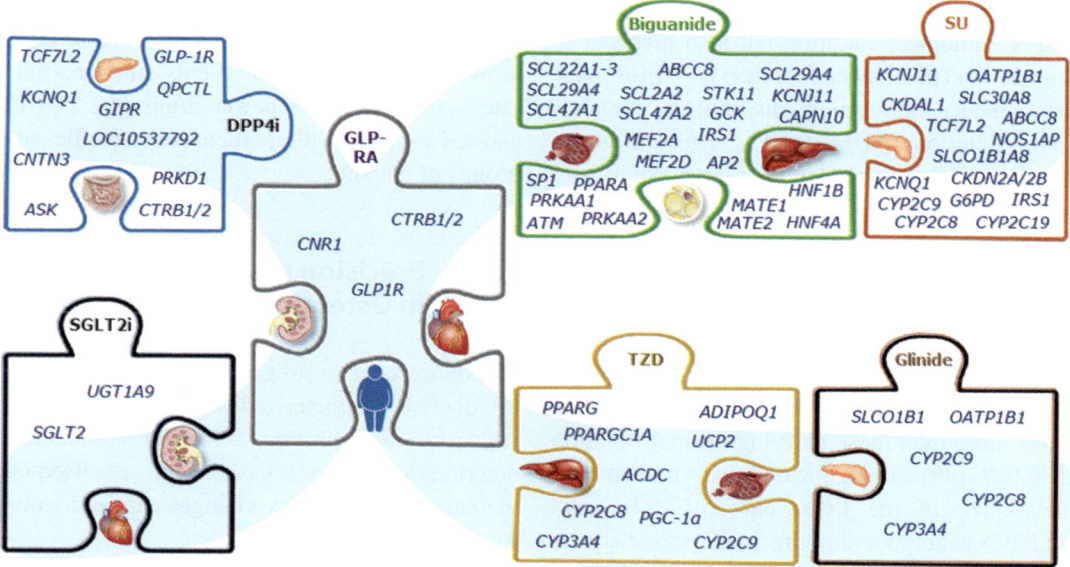

**Fig. 5.2** Pharmacogenetics of oral antidiabetic drugs [42, 53, 54]

related to the activity of metabolizing enzymes that metabolize oral antidiabetic drugs (OAD). Generally, drugs that improve insulin action are more effective in preventing diabetes than those that boost insulin secretion [36–46].

The emergence of GWAS in the last decade has gradually moved the genetics of T2D forward, transforming pharmacogenetics into pharmacogenomics.

In the last decade, the advent of GWAS has gradually shifted the genetics of T2D to a step forward, definitely turning pharmacogenetics into pharmacogenomics [47].

Generally, drug efficacy pharmacogenomics has been categorized into two types of variants: those that affect drug pharmacokinetics (PK) and those that affect drug pharmacodynamics (PD).

In T2D there are substantial numbers of etiological variations which can be divided into different categories as PD and PK. Actually, in the PD approach, we expect that addressing the genetic etiology of diabetes will lead to pharmacogenomics of that drug and the etiological pathway. Understanding the genetic architecture of treatment response in diabetes may yield insights into diabetes etiology [48]. But in the case of PK variations, genetics of drug transporters and metabolism play crucial role.

Therefore, genetic variation not only encompasses etiological variation but also involves in drug action (pharmacodynamics) and drug pharmacokinetics (absorption, distribution, metabolism, excretion [ADME]). The effect of these variations is from a moderate to large extent.

There are several genes related to different OADs that can indicate variations in drug outcomes such as HbA1c. For example, in sulfonylureas, about 8% of the white population with two loss-of-function variants in cytochrome P450 family 2 subfamily C member 9 (*CYP2C9*), the probability of achieving the HbA1c target is 3.4 times more than those carrying normal function *CYP2C9* [49]. Rosiglitazone can uptake and metabolize in the liver, but SLCO1B1 and CYP2C8 genotypes alter this process, and consequently HbA1c targets as much as 0.7% [50].

The treatment efficacy of most OADs depends on variations in classic pharmacogenetics genes and/or T2D susceptibility genes. The genetic variation profile in type 2 diabetes is significantly different across ethnicities; therefore, it is expected that drug responses will differ between populations. It is undeniable that we now have the potential to use clinical and genetic characteristics to predict who is more or less likely to benefit from treatment [37].

Genome wide association studies in precision diabetes medicine have been performed on well-known large clinical trial studies such as the Action to Control Cardiovascular Risk Factors in Diabetes (ACCORD) [51].

Diabetes incidence is increasing which results in higher healthcare costs, morbidity, mortality, and diabetes-related comorbidities. Several genetic loci associated with T2DM have been discovered using a variety of genomic technologies. However, more genetic investigations on people of diverse ethnicities are needed to complete the picture of T2DM susceptibility gene variations. Furthermore, replication studies on the reported gene variations using sophisticated sequencing technologies on various populations and subethnic groups are required to provide more compelling data for clinical translation [52–54].

In terms of pharmacological outcomes, there is an urgent need to explore beyond early glycemic response and study variation in response in terms of cardiovascular events and mortality rates, particularly for newer drugs like SGLT2i and GLP-1RA, with a focus on specific subgroups of patients.

## 5.8 Precision Medicine in Osteoporosis

Osteoporosis is "a progressive systematic skeletal disease characterized by weakened bones strength, low bone mass and microarchitectural deterioration of bone tissue, with a subsequent increase in degenerative changes and susceptibility to fracture" [55].

Osteoporosis is often characterized by low bone mineral density (BMD) and an increased risk of fracture. According to the International

Osteoporosis Foundation (IOF), osteoporosis causes more than 8.9 million fractures worldwide each year, with an osteoporosis fracture occurring every 3 seconds [56]. The most common fractures are in the hip, forearm, and vertebral [55].

Currently, osteoporosis diagnostic approaches such as bone density testing, bone turnover markers, or fracture algorithms such as the Fracture Risk Assessment Tool (FRAX) have not been able to meet all of the parameters for predicting individual fracture risk.

Many factors contribute to the risk of osteoporosis, including age, gender, nutrition, physical activity, medication usage, and menopausal status, but positive family history is one of the most important clinical risk factors, highlighting the importance of genetic factors in predicting disease susceptibility [57].

The genetics of osteoporosis is classified into two categories: disease susceptibility genetics and treatment response pharmacogenetics.

Numerous GWASs have introduced polymorphisms of more than 40 candidate genes, including LRP5, ESR1, OPG, SOST, RANK, and RANKL that are involved in bone quantitative and qualitative features. Easy-to-measure osteoporosis biomarkers are required as a significant tool for assessing both primary and secondary osteoporosis risks.

"The Food and Drug Administration defines biomarkers as a feature that is assessed as an indicator of normal biological and pathological processes, or pharmacologic responses to a therapeutic intervention" [58].

miRNAs have been described as playing an essential role in biological processes (BPs) including cell proliferation, differentiation, development, and regulation of bone homeostasis [59]. MiRNAs have emerged as critical regulators of genes involved in bone production and resorption, bone remodeling, and bone degeneration [60].

After standardizing the quantitative procedures of miRNA detection and validity, potential biomarkers for osteoporosis prediction, prevention, prognosis, and tailored treatment will be achievable in clinical practice.

The field of osteoporosis pharmacogenetics is still in its infancy, with only a few association studies published to date. Furthermore, no functional study is currently available [61, 62].

Three biological pathways are represented by these genes: the Wnt/bcatenin signaling pathway, the RANK/RANKL/osteoprotegerin (OPG) pathway, and the estrogen endocrine pathway [63].

Possible principles of personalized therapy are recommended based on the biological pathways in osteoporosis. Some experimental and preliminary studies in humans indicated the feasibility of personalized treatment of osteoporosis by targeting the genes involved in the biological pathways like LRP5 and SOST. Loss-of-function mutation in the ESR1 gene not only predicts an individual with osteoporosis but also gives resistance to estrogen replacement therapy (ERT). Inhibition of RANK/RANKL/OPG signaling in osteoporosis patients with susceptibility variants in *ANK*, *RANKL*, and *OPG* genes might be an ideal tailored therapeutic approach [64].

Antiresorptive drugs like estrogen replacement therapy, raloxifene, and bisphosphonates responses can be influenced by genetic variations in different genes like as estrogen receptors alpha (ERa), vitamin D receptor (VDR), and beta (ERb), and collagen I alpha 1 (COL1A1) [65–68].

Additionally, because many genes encoding drug-metabolizing enzymes, drug receptors, drug transporters, and drug targets are epigenetically controlled, epigenetic elements should be considered in the design of osteoporosis pharmacogenetic investigations.

## 5.9 Precision Medicine in Polycystic Ovary Syndrome

One of the most common endocrine disorders in women of reproductive age is polycystic ovary syndrome (PCOS). Affected women usually show metabolic abnormalities such as insulin resistance and glucose homeostasis dysregulation [69]. The symptoms of PCOS are heterogeneous, implying that the etiology may vary considerably between subsets of women with PCOS. According to the Rotterdam criteria, PCOS is characterized by polycystic ovarian morphology (PCOM), and it requires at least two of three key reproductive traits [70], while the NIH PCOS inclusion criteria are hyperandrogenism (HA) and ovulatory dysfunction (OD) [71]. It appears that the current diagnostic criteria do not recognize genetically distinct disease subtypes.

A combination of genetic background and environment is considered as a risk factor for PCOS [72]. Genome wide association studies (GWASs) have been found nearly 16 genetic loci in women of European and Han Chinese and Korean ancestry [73–77]. Approximately 200 genes have been identified to be related to PCOS and its clinical manifestations, including *FSHB*, *INSR*, *FBN3*, *TOX3*, *LHCGR*, and *GATA4/NEIL2* [78–80]. These loci have linked PCOS etiology to gonadotropin secretion and activity, androgen biosynthesis, ovarian aging, and metabolic control. The susceptibility loci identified in PCOS appeared to be shared by NIH criteria and self-reported diagnosis [74]. Genetic studies by Mendelian randomization analysis indicated that insulin resistance, sex hormone-binding globulin, age at menopause, and body mass index contribute to PCOS pathogenesis [74].

In a recent large meta-analysis of PCOS, the case and control groups were categorized based on NIH or Rotterdam criteria, or by self-report. The genetic architecture of PCOS and separate susceptibility loci related to the main symptoms of PCOS, HA, OD, and PCOM, across different criteria were investigated. Eleven previously reported loci are replicated in the present meta-analysis, and three novel loci near *ZBTB16*, *PLGRKT*, and *MAPRE1* are reported. These data indicate that the genetic architecture of the phenotypes characterized by the various PCOS diagnostic criteria was mostly the same. One locus indicated significant differences in association with diagnostic criteria. Across common variations at 13 loci, the genetic architecture of PCOS diagnosed by self-report or NIH or non-NIH or Rotterdam criteria was identical [81].

## 5.10 Thyroid Cancer

Thyroid cancer accounts for 586,000 cases worldwide, ranking it ninth in terms of incidence in 2020. The global incidence rate in women is 10.1 per 100,000, which is three times greater than in males, and the disease accounts for 1 out of every 20 malignancies diagnosed in women [82].

Thyroid cancer's pathogenesis is not well recognized. Ionizing radiation is the main well-established risk factor for thyroid cancer, particularly when exposure occurs in childhood, while there is evidence that other factors (excess body weight, larger height, hormonal exposures, and certain environmental toxins) may also play a role [83]. Over the last two decades, substantial advances in understanding the genetic and biologic aspects of the condition have resulted in the development of molecular targeted therapies, entering a new era of precision medicine for thyroid cancer care.

Differentiated thyroid cancer (DTC) is the most frequent malignancy in thyroid cancer (TC) (accounting for more than 90% of all thyroid cancers) [84], and it arises from the follicular cells (thyrocytes). Differentiated TC includes follicular TC (FTC), papillary TC (PTCs, up to 85%), and Hürthle cell TC. Medullary TC (MTC) representing 5% of TC develops from thyroid parafollicular C cells and can be sporadic (in 75% of cases) or inherited (in 25% of cases) [85, 86].

The World Health Organization suggested in 2017 to reclassify FTC into three subtypes based on clinical and biological characteristics: minimally invasive (miFTC), encapsulated angioinvasive (eaFTC), and widely invasive (wiFTC) [87]. Poorly differentiated thyroid carcinoma (PDTC, 5–10%) and anaplastic thyroid carcinoma (ATC, 2–3%) are rare tumors that exhibit aggressive behavior and have a short median survival time (5 years and 6 months, respectively) [88]. Anaplastic thyroid cancer (ATC), one of the most aggressive human malignancies, accounts for 1% of all TC, accounting for 15–40% of all TC deaths [86, 89]. PTC and FTC patients are treated with total thyroidectomy, and aggressive PTCs and FTCs are treated with successive radioactive iodine (RAI) remnant ablation 131I too [90, 91].

## 5.11 Molecular Genetics of Thyroid Cancer

Recent advances in sequencing techniques could pave the road for identifying somatic mutations and altered signal pathways as PTC progresses. Different TC subtypes are distinguished by distinct genetic changes, leading to a close genotype-phenotype association [92–95]. The majority of mutated genes in each of the four TC subtypes are involved in cell cycle regulation.

TC is a genetic disease with a low somatic mutation burden in each tumor. More than 90% of TCs have driver mutations, which are mutations that provide a selective growth advantage, promoting cancer development [96]. The majority of TC is caused by dysregulation of the mitogen-activated protein kinase (MAPK) and phosphatidylinositol-3 kinase (PI3K)/AKT signaling pathways. MAPK activation is thought to be critical for PTC development via point mutations in the *BRAF* and *RAS* genes or gene fusions of *RET/PTC* and *TRK* [88].

PTCs have a high frequency (up to 60%) of activating *BRAF* gene mutations (often BRAFV600E), *RAS* mutations, or *RET-PTC* fusion, while most FTCs have *PAX8-PPAR* fusion or *RAS* mutation but rarely have changed *BRAF*. However, *BRAF* and *RAS* mutations are common in PDTCs and ATCs, and the tumor suppressor *TP53* is frequently inactivated, particularly in ATCs [95, 97]. The *RET* proto-oncogene encodes a tyrosine kinase receptor, and its activation initiates intracellular signaling cascades that regulate gene expression and biological responses [88].

The RET/PTC fusion protein leads to maintains the tyrosine kinase domain to be intact and allows unlimited activation of the MAPK signaling pathway. Different types of RET/PTC rearrangements have been reported, including RET/PTC1 and RET/PTC3. The prevalence of *RET* rearrangements in PTC is widely varied [98–101]. RET/PTC1 has been associated with a better prognosis in some studies, whereas RET/PTC3 has been associated with a more aggressive and malignant phenotype [102, 103]. Patients with these rearrangements usually have a good prognosis due to their better response to radioactive iodine (RAI) therapy [104].

*BRAF*, a member of serine/threonine protein kinases, is mutated in 7% of all malignancies. The prevalence of *BRAF* mutations in PTC varies among several studies (29% to 83%) [88]. *BRAF* mutations activate downstream transcription factors, resulting in cell differentiation, proliferation, growth, and apoptosis [88].

*RAS* is a family member of GTP-binding proteins upstream of *BRAF*, and its function is through the MAPK and PI3K-AKT signaling pathways. *RAS* encodes HRAS, KRAS, and NRAS that play critical roles in cell development, differentiation, and survival. RAS is also found in other types of TCs like FTC and follicular-variant PTC (FVPTC) [88].

*TERT* encodes the telomerase reverse transcriptase, and two hotspot genetic mutations (C228T and C250T) have been identified. These mutations, which are found in approximately 10% of PTC, increase telomerase activity and telomere length maintenance in cancer cells [105].

The identification of promoter mutations in the telomerase reverse transcriptase (*TERT*) gene is an important breakthrough in the field of thyroid cancer research. *TERT* promoter mutations are associated with aggressive clinical characteristics and the radioiodine-refractory nature of thyroid cancer. The use of *TERT* promoter mutation as an excellent biomarker for risk stratification, diagnosis, prognosis, treatment decision-making, and follow-up design in TC is well established [106].

The most common mutations in FTC are in the *RAS* gene family (*HRAS*, *KRAS*, and *NRAS*). The *PAX8-PPARγ* fusion gene was identified in one-third of FTCs (12–53%) [107, 108]. *TERT* promoter mutations have been identified in around 15% of FTCs and have been related to the worst clinical and prognosis characteristics [109].

*BRAF* and H-, K-, and N-RAS mutations are less common in anaplastic thyroid cancer than in DTC, with a frequency of 19–45% and 9.5–27%, respectively [97, 110]. *TERT* promoter mutations, which occur in 43–73% of patients, and *TP53* mutations, which occur in 48–73% of cases, are the two most common mutations found in ATC [97, 110].

Because TP53 is only found in ATC, it may be regarded pathognomonic for this tumor and its extreme aggressiveness [97]. 29% of ATC had mutations in cell cycle control genes (*CDKN2A*, *CDKN2B*, and *CCNE1*) [97]. A substantial number of genetic mutations in ATC are those producing dysfunctions in the ERK1/2-MEK1/2 and PI3K-AKT signaling pathways [111]. The tumor suppressor gene *p53* mutation is uncommon in FTC and PTC but common in ATC which is ranging from 70% to 88% [112, 113]. Notably, four distinct subtypes of ATC molecular pattern have been suggested in ATC heterogeneous molecular status [114]:

1. Type 1 ATC, BRAF-positive ATC, may originate from PTC and with a genetic landscape similar to PTC. Mutations in *TP53* or *PIK3CA* are also common.
2. Type 2 ATC, NRAS-positive ATC, which may originate from FTC.
3. Type 3 ATC, with *RAS* mutations or more atypical ones (e.g., *PTEN*, *NF1*, and *RB1*). It is most likely caused by FTC or Hürthle cell cancer.
4. Mixed ATC with loss-of-function mutations in the cell-cycle regulation genes (*CDKN2A* and *CDKN2B*).

Proto-oncogene *RET* point mutations exert an important role in MTC oncogenesis and are very common in familial cases [115]. But in the sporadic form of MTC, RET and RAS (*HRAS* and *KRAS*) mutations are found in 44% and 13% of cases, respectively [116].

## 5.12 Precision Medicine for Thyroid Cancer

Following the release of the first draft of the human genome project, genotyping and genomics are becoming standard treatments for some types of cancer [117]. Comprehensive available data about thyroid cancer genetics can help increase the speed of precision medicine development for thyroid cancer.

"Precision medicine is defined by the US National Cancer Institute as a type of medicine that uses information about a person's genes, proteins, and environment to prevent, diagnose, and treat disease" [118].

Over the last decade, there has been a significant shift in the treatment of metastatic or advanced thyroid cancer. It transitioned from a one-size-fits-all approach cytotoxic chemotherapy for these types of patients to an era of personalized targeted therapy based on tumor type and genomic profile [119, 120]. The systemic therapy will depend on the genotype of the tumors.

Anaplastic thyroid carcinoma is one of the most aggressive cancers we are aware of. From the day of diagnosis to the day of death, the average survival time is 6 months. As a result, one's approach to managing this patient differs significantly from that of other well-differentiated tumors, which anaplastic is not [86].

The American Joint Committee on Cancer classified ATC as stage IV regardless of tumor size or the existence of lymph nodes and distant metastases [121]. ATC is caused by many genetic mutations that were already found in several molecular pathways and are linked to tumor growth.

According to ATA guidelines, paclitaxel or docetaxel, doxorubicin, and cisplatin or carboplatin are all effective in ATC; however, none of these medications can prolong survival in advanced ATC [122].

Because chemotherapy is not targetable, these drugs are ineffective against ATC, and mortality, and despite progress in diagnosis and treatment of cancers, ATC remains one of the most significant issues in clinical practice [123].

A large number of genetic alterations are associated with different molecular pathophysiological pathways of ATC. Recently, new treatments have been developed based on the known genetic mutations and pathways [122].

ATC is related to a significant number of genetic mutations especially dysfunctions in the ERK1/2-MEK1/2 and PI3K-AKT signaling pathways, and new drugs using these molecular pathways as targets have recently been developed [86]. New drugs which target different molecular pathways interestingly have been introduced as follows [86, 120, 124]:

Angiogenesis: vandetanib, lenvatinib, sorafenib, sunitinib, combretastatin, etc.

BRAF: vemurafenib, dabrafenib/trametinib

PPARγ agonists: efatutazone, pioglitazone, rosiglitazone

EGFR: gefitinib

Identification of novel therapeutic techniques is required to improve ATC patients' survival and quality of life.

The capability of targeted therapy by new drugs to combine with radiation and chemotherapy is now being investigated to solve the drug resistance problem in ATC patients [86].

Cancer treatment is only one aspect of cancer care. Precision oncology will continue to make scientific achievements, but to make real changes for patients, worldwide collaboration and a comprehensive approach to the patient that extends beyond the laboratory are essential.

## 5.13 Precision Medicine for Obesity

Obesity is a complex, multifactorial, and mostly preventable condition, and its growing prevalence has reached epidemic proportions with many associated comorbidities [125]. It is primarily caused by genetic, behavioral, socioeconomic, and environmental risk factors [126].

Obesity is difficult to prevent and treat because the origin, clinical presentation, and comorbidities of the disease are different from patient to patient.

Current treatment of obesity begins with lifestyle changes and proceeds to pharmacologic therapy, endoscopic devices, and/or bariatric surgery depending on the patient's response. Each of these treatment strategies has been demonstrated to have a wide range of weight loss outcomes, implying that the current strategy does not provide patients with obesity with the comprehensive and tailored care they require. Also, the

increased incidence of obesity combined with the absence of well-validated treatment alternatives makes necessary the investigation of new therapeutic strategies [127].

Contrary, precision medicine for obesity proposes a new paradigm in which the disease is stratified based on specific biological markers gathered predominantly from high-throughput or "omics assays" (e.g., genomics, epigenomics, transcriptomics, and microbiomics, among others), as well as from other clinical, physiological, and behavioral characteristics can increase therapeutic efficacy and tolerance [128]. Consequently, it will allow bridging current evidence-based medical practice and precision medicine.

Even though precision medicine has the potential to improve the prevention and treatment of common multifactorial diseases, it has yet been limited to being applied for obesity. Furthermore, these biological markers cannot only predict the risk of progression into developing other comorbidities but can potentially be used to predict the response to specific therapies as well [129].

Obesity is now classified based on the rate at which weight is gained, the distribution of fat, and the effects. The discovery and development of gene-targeted medicines, more novel therapeutic techniques, and drug discovery and development could all be aided to identify specific subgroups.

## 5.14 Obesity Classification

Obesity is typically classified according to BMI, which categorized individuals from overweight to morbid obesity.

- Severely underweight: BMI less than 16.5 kg/m$^2$
- Underweight: BMI less 18.5 kg/m$^2$
- Normal weight: BMI more than or equal to 18.5–24.9 kg/m$^2$
- Overweight: BMI more than or equal to 25 to 29.9 kg/m$^2$
- Obesity: BMI more than or equal to 30 kg/m$^2$

- Obesity class I: BMI 30 to 34.9 kg/m$^2$
- Obesity class II: BMI 35 to 39.9 kg/m$^2$
- Obesity class III: BMI greater than or equal to 40 kg/m$^2$ (also referred to as severe, extreme, or massive obesity)

A high BMI can indicate a high percentage of body fat [130].

The risk of metabolic abnormalities and consequences, which are usually associated with obesity, is not necessarily differentiated by BMI. Although obese patients have a higher risk of comorbidities than people of normal weight, some patients with obesity may have no metabolic problems [131].

Classification of obesity based on BMI causes limitations in guiding clinical decisions. However, recent research has revealed new insights about the identification of different subgroups among obese patients that have provided crucial information regarding the underlying pathophysiologic mechanisms derived from obesity and its complications and have paved the way for precision obesity management.

From a genetic point of view, obesity is classified into rare, severe, early-onset monogenic, and polygenic (or common) disorders, which are frequently separated as distinguished diseases [132]. However, gene discovery research reveals that both types of obesity have genetic and biological basis, implying that the central nervous system (CNS) and neuronal pathways play a crucial role in body weight by controlling the food intake [133].

Monogenic obesity is inherited in a Mendelian pattern that is the result of either small or large chromosomal deletions or single-gene defects. Contrary, polygenic obesity has a heritability pattern that is comparable to that of other complex traits and disorders and is the result of hundreds of polymorphisms, each of which has a minor effect [134].

## 5.15    Omics and Obesity

Obesity and metabolism research is increasingly relying on new technology to find processes in the development of obesity using multiple "omics" platforms. Genetic and epigenetic variants that translate into transcriptome, proteome, and metabolome modifications have been detected using a range of high-throughput profiling tools across biological tissues and fluids. These changes may play a role in disease pathophysiology, leading to the identification of biomarkers that could be used therapeutically to reveal novel pathways responsible for the development and advancement of the disease and could finally serve as targets for obesity prevention [135, 136]. "Omics" data have increased our understanding of the etiology of obesity and its pathophysiological links with chronic diseases, which may one day help us tackle obesity more effectively and individually. They have a lot of potential in terms of developing successful public health approaches that pave the way for patient stratification and precision prevention [137]. Although different genes, epigenetic factors, transcripts, proteins, and metabolites are identified as a first step toward understanding pathways and mechanisms, the ultimate goal is to use multi-omics data for health assessment and prediction.

### 5.15.1 Genomics

The heredity of human obesity is estimated to be from about 30% (in normal-weight subjects), 40–50% (variability in body weight status), to 60–80% (among the subpopulation of individuals with obesity and severe obesity) [138]. It is reported more than 300 single nucleotide variants for obesity according to the GWAS could pave the way for the development of new obesity prevention and treatment strategies. Since the progress of these approaches requires a thorough understanding of how a single nucleotide polymorphism (SNP) changes the expression of target genes and related phenotype, therefore, it needs to take more time.

Obesity is frequently divided into two types including polygenic (or common) obesity and rare, severe, early-onset monogenic obesity. Although often considered to be two distinct forms, gene discovery investigations reveal that both types of obesity share genetic and biological bases, implying that the brain plays a crucial role in body weight control [134]. Small or large chromosomal deletions or single-gene abnormalities are involved in monogenic obesity, which is inherited in a Mendelian model, whereas polygenic obesity is the result of hundreds of genetic variations that each have a small effect and follows a pattern of heritability that is similar to other complex traits [134].

As more obesity-related loci are being identified by GWAS, a growing number of these loci now contain genes that were first found in humans or animal models for early-onset obesity such as *MC4R* [139, 140], *BDNF* [141], *POMC* [142], *LEP* [143, 144], *LEPR* [145, 146], *NPY* [147], *SH2B1* [141, 148], *SIM1* [147], *NTRK2* [149], *PCSK1* [145], and *KSR2* [150]. The majority of these genes are involved in the leptin-melanocortin and BDNF-TrkB signaling pathways. As a result, whereas genetic deficiency in these pathways causes severe obesity, genetic variants in or near these genes that have a more exact impact on their expression can influence where an individual is in the normal BMI range.

The first known genes related with obesity were discovered as a result of the human genome sequencing and the realizing of the regulation of energy balance mediated by the leptin-melanocortin pathway. Some evidence suggests that using recombinant leptin and setmelanotide in patients with leptin and POMC deficiency, respectively, has resulted in considerable weight loss. This wide range of outcomes suggests that several SNPs could share a common mechanism of action [151, 152]. In this way, if these SNPs are discovered, they may be used to identify a subset of people with common obesity who could benefit from similar treatments [153].

Reports indicate that individuals with genetic susceptibilities to obesity are more predisposed to gain weight when they are exposed to environmental factors. Researchers have combined numerous SNPs into polygenic risk scores due to the small effect of genetic variations on obesity trait [132]. Low socioeconomic level, chronic psychosocial stress, decreased sleep duration, gender, higher consumption of sugar-containing beverages, increased fried food intake, and decreased physical activity, according to data, intensify the association of genetic risk scores with BMI [154, 155].

## 5.15.2 Transcriptomics

Transcriptomics is defined as the study of the transcriptome (all RNA transcripts, both coding and non-coding) in an individual or a group of cells [156, 157].

The research into the role of transcriptomics in obesity is progressing slowly. Small nucleolar RNAs have been linked to food intake and body weight in patients with Prader-Willi syndrome, according to several correlations and validation studies [158]. On the other hand, microRNAs have been correlated to adipogenesis and adipocyte differentiation metabolic pathways [159].

RNA sequencing allows researchers to quantify gene expression and gain a better knowledge of how cells work. Researchers have been able to assess gene expression and classify adipose tissue more extensively because of developments in sequencing technology [160].

The significance of adipose tissue in inflammation and the subsequent influence on obesity development have been confirmed by transcriptomics profiling of subcutaneous adipose tissue in patients receiving surgical therapy [161]. However, alterations in the transcriptome of adipose tissue following weight reduction therapies may not be sufficient to identify a therapeutic target or predict treatment success.

## 5.15.3 Metabolomics

Metabolites are fundamental components of biological activity, and alterations in their regulation can have therapeutic implications. These molecules can be either substrates or products of the complicated biochemical networks connecting cellular and systemic biological mechanisms [162]. A metabolome is a collection of whole metabolites in a cell that carries information about disease mechanisms.

Metabolomics is a new bioanalytical approach that characterizes and quantifies all small molecules in biological samples to monitor metabolites and their changes in response to different stimuli. As a result, metabolomics data can reveal what is really going on in a biological system, acting as a vital link between phenotype and genetics.

The metabolic differences are associated with an interaction between genetic and environmental factors [163].

A metabolic phenotype is an end result of systemic biological alternations and environmental factors that are influenced in complex ways by the integrity of genes, enzymes, and related proteins.

Metabolic profiling has a lot of potential in terms of detecting the condition since it provides more mechanistic data about how the disease progresses. Indeed, the genome implicitly predicts what might occur, but the metabolome specifies the process's endpoint based upon what happened previously [164].

Metabolomics has made great progress in the last decade in terms of offering a beneficial organized view into the mechanisms behind a variety of metabolic illnesses, such as obesity that have mostly resulted in the development of biomarkers and risk factors for these diseases [164].

Obesity is a whole-body illness that certainly involves metabolic changes, and related research has been able to discover different metabolomics patterns in healthy obese individuals and obese people with metabolic comorbidities such as cardiovascular disease, dyslipidemia, metabolic syndrome, and diabetes. Therefore, metabolomics analysis can improve our understanding of obesity and obesity-related disorders because it can identify small alterations in the metabolic network and also provides a basis for the prevention or treatment of obesity [165].

An aberrant metabolome is related to increased cardiovascular events in healthy subjects (obese and even lean persons) in comparison with participants matched for BMI and with opposite metabolomes [166].

Evidence shows that the metabolic profile of people who follow a low-calorie diet can predict how much weight they lose [167].

Although the metabolomics field is actively evolving as a branch of the precision medicine approach, it can identify clinically meaningful heterogeneity in obesity that could potentially lead to the identification of a metabolic fingerprint that cannot only help phenotype a patient but can also help select certain patients for certain specific therapies.

### 5.15.4 Microbiome

Among the risk factors contributing to obesity, alterations of the human gut microbiome, as one of the most remarkable discoveries of the last decades, have shown a vital role in obesity risk [168]. But the exact underlying mechanisms remain unknown. Diet, pre- and probiotics, antibiotics, and surgery can all regulate the gut microbiota, having the power to change the weight and metabolic profile in either direction [169]. Gut microbiome genes are complementary to the genetic of humans, which have distinct activities [170]. Nutritional, environmental, and host factors influence the gut microbiota that can create microbiome changes and metabolic abnormalities and consequently result in disease [171]. More precise knowledge of the complicated links relating to gut microbiome and obesity has been revealed as the result of sequencing technology progresses. The gut microbiota composition plays an important role in energy metabolism and regulating the host's ability to extract energy from food, which is becoming more connected to body weight and BMI alternations and also obesity cause [172]. The gut microbiota secreted metabolites that could be influenced hunger modulation with either straightly influencing the central nervous system or indirectly by changing hormone production [171]. More evidence showed that opportunistic infections in a transmissible obesity microbiome may play a role in the obesity progress by modifying gene expression in the host and insulin resistance inducing through metabolic endotoxemia or affecting the brain-gut axis [173]. The change in the composition of obese individuals' gut microbiota drew attention to the importance of microbiota in the treatment of obesity, and researchers eventually discovered a novel strategy to control obesity using various therapeutic precision medicine approaches [174]. According to this, there are two key tailored therapeutic approaches to obesity control that includes microbiota transplantation and probiotics prioritizing gut microbiota composition.

### 5.15.5 Pharmacogenomics

The US Food and Drug Administration (FDA) has approved five pharmaceuticals for the management of obesity: orlistat, lorcaserin, liraglutide, phentermine/topiramate, and bupropion/naltrexone.The combination of phentermine and topiramate, followed by lorcaserin and bupropion/naltrexone, appears to be the most successful treatment. Other factors to consider while managing excessive weight include comorbidities associated with obesity, pharmacological interactions, and the likelihood of unfavorable collateral effects, as well as personalized treatments based on genetic makeup [175].

Linkage and candidate gene investigations have also been utilized to identify and describe the roles of certain genes whose functional regulation is changed by pharmacological treatments or dietary methods and has an effect on weight loss. These genes are involved in food intake regulation (*MC3R, MC4R, POMC, LEPR, FTO*), lipid metabolism and adipogenesis (*PLIN1, APOA4, APOA5, LIPC, FABP2*), thermogenesis (*ADBR3, UCP1,* UCP2, *UCP3*), adipocytokine and hormone secretion (*LEP, ADIPOQ, IL6, RETN, ACE*), and insulin action and signaling (*IRS1, INSIG2, PPARG, TFAP2B, TCF7L2, CLOCK, GNAS*).

Genetic information may contribute to tailored obesity management based not only on human phenotypic traits but also on genetic and epigenetic markers, by taking into account current pharmacogenetics knowledge about novel anti-obesity drugs [176–178].

Pharmacogenomics could help to understand why people respond differently to weight loss treatments or, conversely, why some prescription pharmaceuticals cause weight gain. For example, the insulin receptor (INSR) and the glucagon-like peptide receptor (GLP-1R) gene variants, which are targeted by topiramate and liraglutide, have been linked to treatment response differences [179, 180]. As a result, three silent variants in the gene encoding pancreatic lipase were discovered to alter orlistat efficacy [178]. Also, 13 SNPs have been found to be strongly linked to weight or BMI changes as a result of antipsychotic drug use [181].

### 5.15.6 Nutrigenetics and Nutrigenomics

Dietary nutrient intake, as an environmental factor, and its interaction with genes, plays a significant role in the management of health and the prevention of obesity and obesity-related disorders. Although obese individuals with the same eating pattern show significant diversity, distinct genetic variants may describe this heterogeneity which leading to the development of nutrigenetics theory [6].

This is critical to improve understanding of obesity as a result of an individual's genetic profile and to develop a concept of "personalized nutrition" in order for successful obesity prevention and treatment.

Nutrigenetics or nutrigenomics (in broadest sense) as a key part of personalized medicine is the science that defines how genetic variations influence on differential response to specific nutrients and interaction between this variation and different disorder phenotypes like obesity [182]. Personalized nutrition recommendations based on a person's genetic background may improve the outcomes of a specific dietary intervention and provide a novel dietary method to prevent obesity.

## 5.16 Challenges

Precision medicine-based studies are paving the way for personalized treatment of obesity. A better understanding of biological variations has aided the development of therapeutic strategies that may have a great impact on obesity treatment outcomes in the future.

Several of these techniques are still under development, and their therapeutic efficacy has yet to be determined. While multi-omics information has provided a more in-depth understanding of obesity progression, more trials are needed before some of these biomarkers for obesity may be considered or used as part of predictive models.

The overarching goal is to integrate multi-omics data into obesity precision medicine to improve therapeutics outcomes.

Healthcare providers may be hesitant to order specialized tests that could help them better understand a patient's condition. Furthermore, there are few subjects on the personalized approach in health professions curriculum. Obesity precision medicine will be possible only if all parties concerned work together.

Despite the abundance of data that will aid in the development of obesity precision medicine, a knowledge gap remains between the relation of biological markers and obesity and the mechanisms that underpin these associations. To develop particular strategies that facilitate customized medicine for obesity, this gap must be closed. Large cohorts and biobanks are critical to precision medicine research. These data are extremely sensitive, yet they might be widely disseminated in order to answer research questions that have yet to be created. Therefore, protecting privacy while supporting the appropriate and secure storage, transmission, and use of data should be prioritized. Recent breakthroughs in systems biology may enable us to gain unique mechanistic insights and discover possible treatment targets by adding new degrees of understanding.

However, integrating the intricacies of an incredible quantity of systems biology evidences into relevant biological interpretations is currently a major issue for scientific communities.

## 5.17 Future Perspective of Precision Medicine in Endocrinology

In the future, genetic information from patients with type 2 diabetes could be integrated with other clinical markers to help guide a stratified prescription of the most effective glucose-lowering treatment for a specific person. Single SNPs and polygenic risk scores, as well as nongenetic variables, may be helpful in this regard [183].

Precision diabetes medicine has experienced significant growth into the diagnosis and management of monogenic diabetes as it is also known in pharmacogenomics [9].

The discovery, measurement, and global implementation of agents for diagnosis and therapy will determine progress in translating breakthroughs in biology and technology; therefore, wide stakeholder engagement is required.

Rapid progress in science and technology (such as artificial intelligence) over the last few decades has motivated the development of novel approaches for solving scientific medicine problems at a speed well above the scientific community's capabilities. Artificial intelligence has also been studied in a variety of fields of health and medicine, including precision medicine [184].

Currently, clinical decision-making is based on evidence-based medicine (EBM) and clinical trials, but in the near future the use of clinical data and "omics" (big data) may influence the use of precision medicine for precise decisions in the practice of medicine [157, 185].

Many achievements in precision medicine have still yet to reach the clinic due to common challenges in the area. There is a lack of standard outcomes to define clinical value for ideas now in the translational phase [117].

# References

1. Ginsburg GS, Phillips KA. Precision medicine: from science to value. Health Aff (Millwood). 2018;37(5):694–701.
2. National Library of Medicine (NIH). What is precision medicine? https://ghrnlmnihgov/primer/precisionmedicine/definition.
3. Meybodi HRA, Hasanzad M, Larijani B. Path to personalized medicine for type 2 diabetes mellitus: reality and hope. Acta Med Iran. 2017;55:166–74.
4. Mirnezami R, Nicholson J, Darzi A. Preparing for precision medicine. N Engl J Med. 2012;366(6):489–91.
5. Tuomi T, Santoro N, Caprio S, Cai M, Weng J, Groop L. The many faces of diabetes: a disease with increasing heterogeneity. Lancet (London, England). 2014;383(9922):1084–94.
6. Barrea L, Annunziata G, Bordoni L, Muscogiuri G, Colao A, Savastano S. Nutrigenetics—personalized nutrition in obesity and cardiovascular diseases. Int J Obes Suppl. 2020;10(1):1–13.
7. Merino J. Florez JCJAotNYAoS. Precision medicine in diabetes: an opportunity for clinical translation. Ann N Y Acad Sci. 2018;1411(1):140–52.
8. Shepherd M, Shields B, Ellard S, Rubio-Cabezas O, Hattersley AT. A genetic diagnosis of HNF1A diabetes alters treatment and improves glycaemic control in the majority of insulin-treated patients. Diabet Med. 2009;26(4):437–41.
9. Chung WK, Erion K, Florez JC, Hattersley AT, Hivert MF, Lee CG, et al. Precision medicine in diabetes: a consensus report from the American Diabetes Association (ADA) and the European Association for the Study of Diabetes (EASD). Diabetes Care. 2020;43(7):1617–35.
10. Linsley PS, Greenbaum CJ, Nepom GT. Uncovering pathways to personalized therapies in type 1 diabetes. Diabetes. 2021;70(4):831–41.
11. Dufort MJ, Greenbaum CJ, Speake C, Linsley PS. Cell type-specific immune phenotypes predict loss of insulin secretion in new-onset type 1 diabetes. JCI Insight. 2019;4(4):e125556.
12. Maurano MT, Humbert R, Rynes E, Thurman RE, Haugen E, Wang H, et al. Systematic localization of common disease-associated variation in regulatory DNA. Science (New York, NY). 2012;337(6099):1190–5.
13. Mahajan A, Taliun D, Thurner M, Robertson NR, Torres JM, Rayner NW, et al. Fine-mapping type 2 diabetes loci to single-variant resolution using high-density imputation and islet-specific epigenome maps. Nat Genet. 2018;50(11):1505–13.
14. Scott RA, Freitag DF, Li L, Chu AY, Surendran P, Young R, et al. A genomic approach to therapeutic target validation identifies a glucose-lowering GLP1R variant protective for coronary heart disease Sci Transl Med. 2016;8(341):341ra76.
15. Reddy MA, Zhang E, Natarajan R. Epigenetic mechanisms in diabetic complications and metabolic memory. Diabetologia. 2015;58(3):443–55.
16. Pearson ER, Flechtner I, Njølstad PR, Malecki MT, Flanagan SE, Larkin B, et al. Switching from insulin to oral sulfonylureas in patients with diabetes due to Kir6.2 mutations. N Engl J Med. 2006;355(5):467–77.
17. Thomas NJ, Jones SE, Weedon MN, Shields BM, Oram RA, Hattersley AT. Frequency and phenotype of type 1 diabetes in the first six decades of life: a cross-sectional, genetically stratified survival analysis from UK Biobank. Lancet Diabetes Endocrinol. 2018;6(2):122–9.
18. Dimas AS, Lagou V, Barker A, Knowles JW, Mägi R, Hivert MF, et al. Impact of type 2 diabetes susceptibility variants on quantitative glycemic traits reveals mechanistic heterogeneity. Diabetes. 2014;63(6):2158–71.
19. Gooding JR, Jensen MV, Newgard CB. Metabolomics applied to the pancreatic islet. Arch Biochem Biophys. 2016;589:120–30.
20. Bain JR, Stevens RD, Wenner BR, Ilkayeva O, Muoio DM, Newgard CB. Metabolomics applied to diabetes research: moving from information to knowledge. Diabetes. 2009;58(11):2429–43.
21. Del Prato S. Heterogeneity of diabetes: heralding the era of precision medicine. Lancet Diabetes Endocrinol. 2019;7(9):659–61.
22. Ahlqvist E, Storm P, Käräjämäki A, Martinell M, Dorkhan M, Carlsson A, et al. Novel subgroups of adult-onset diabetes and their association with outcomes: a data-driven cluster analysis of six variables. Lancet Diabetes Endocrinol. 2018;6(5):361–9.
23. Dennis JM, Shields BM, Henley WE, Jones AG, Hattersley AT. Disease progression and treatment response in data-driven subgroups of type 2 diabetes compared with models based on simple clinical features: an analysis using clinical trial data. Lancet Diabetes Endocrinol. 2019;7(6):442–51.
24. Insel RA, Dunne JL, Atkinson MA, Chiang JL, Dabelea D, Gottlieb PA, et al. Staging presymptomatic type 1 diabetes: a scientific statement of JDRF, the Endocrine Society, and the American Diabetes Association. Diabetes Care. 2015;38(10):1964–74.
25. Krischer JP. The use of intermediate endpoints in the design of type 1 diabetes prevention trials. Diabetologia. 2013;56(9):1919–24.
26. Ziegler AG, Rewers M, Simell O, Simell T, Lempainen J, Steck A, et al. Seroconversion to multiple islet autoantibodies and risk of progression to diabetes in children. JAMA. 2013;309(23):2473–9.
27. Skyler JS, Krischer JP, Becker DJ, Rewers M. Prevention of Type 1 Diabetes. Diabetes in America. 3rd ed. Bethesda, MD: National Institute of Diabetes and Digestive and Kidney Diseases (US); 2018. CHAPTER 37
28. DECODE Study Group, the European Diabetes Epidemiology Group. Glucose tolerance and car-

diovascular mortality: comparison of fasting and 2-hour diagnostic criteria. Arch Intern Med. 2001;161(3):397–405.

29. Diabetes Prevention Program Research Group. Within-trial cost-effectiveness of lifestyle intervention or metformin for the primary prevention of type 2 diabetes. Diabetes Care. 2003;26(9):2518–23.

30. Langenberg C, Sharp SJ, Franks PW, Scott RA, Deloukas P, Forouhi NG, et al. Gene-lifestyle interaction and type 2 diabetes: the EPIC interact case-cohort study. PLoS Med. 2014;11(5):e1001647.

31. Hivert MF, Christophi CA, Franks PW, Jablonski KA, Ehrmann DA, Kahn SE, et al. Lifestyle and metformin ameliorate insulin sensitivity independently of the genetic burden of established insulin resistance variants in diabetes prevention program participants. Diabetes. 2016;65(2):520–6.

32. Godino JG, van Sluijs EM, Marteau TM, Sutton S, Sharp SJ, Griffin SJ. Lifestyle advice combined with personalized estimates of genetic or phenotypic risk of type 2 diabetes, and objectively measured physical activity: a randomized controlled trial. PLoS Med. 2016;13(11):e1002185.

33. Committee ADAPP, Care ADAPPCJD 9. Pharmacologic approaches to glycemic treatment: Standards of Medical Care in Diabetes—2022. Diabetes Care. 2022;45(Supplement_1):S125–43.

34. DiStefano JK, Watanabe RMJP. Pharmacogenetics of anti-diabetes drugs. Pharmaceuticals (Basel). 2010;3(8):2610–46.

35. Ahmed S, Zhou Z, Zhou J, Chen SQ. Pharmacogenomics of drug metabolizing enzymes and transporters: relevance to precision medicine. Genomics Proteomics Bioinformatics. 2016;14(5):298–313.

36. Pearson ER. Pharmacogenetics and target identification in diabetes. Curr Opin Genet Dev. 2018;50:68–73.

37. Ordelheide AM, Hrabe de Angelis M, Haring HU, Staiger H. Pharmacogenetics of oral antidiabetic therapy. Pharmacogenomics. 2018;19(6):577–87.

38. Lancia P, Adam de Beaumais T, Jacqz-Aigrain E. Pharmacogenetics of posttransplant diabetes mellitus. Pharmacogenomics J. 2017;17(3):209–21.

39. Karras SN, Rapti E, Koufakis T, Kyriazou A, Goulis DG, Kotsa K. Pharmacogenetics of glucagon-like Peptide-1 agonists for the treatment of type 2 diabetes mellitus. Curr Clin Pharmacol. 2017;12(4):202–9.

40. Florez JC. The pharmacogenetics of metformin. Diabetologia. 2017;60(9):1648–55.

41. Florez JC. Pharmacogenetics in type 2 diabetes: precision medicine or discovery tool? Diabetologia. 2017;60(5):800–7.

42. Zhou K, Pedersen HK, Dawed AY, Pearson ER. Pharmacogenomics in diabetes mellitus: insights into drug action and drug discovery. Nat Rev Endocrinol. 2016;12(6):337–46.

43. Singh S, Usman K, Banerjee M. Pharmacogenetic studies update in type 2 diabetes mellitus. World J Diabetes. 2016;7(15):302–15.

44. Scheen AJ. Precision medicine: The future in diabetes care? Diabetes Res Clin Pract. 2016;117: 12–21.

45. Pearson ER. Personalized medicine in diabetes: the role of 'omics' and biomarkers. Diabet Med. 2016;33(6):712–7.

46. Dawed AY, Zhou K, Pearson ER. Pharmacogenetics in type 2 diabetes: influence on response to oral hypoglycemic agents. Pharmgenomics Pers Med. 2016;9:17–29.

47. Pirmohamed M. Pharmacogenetics and pharmacogenomics. Br J Clin Pharmacol. 2001;52(4):345–7.

48. Pearson ER. Diabetes: is there a future for pharmacogenomics guided treatment? Clin Pharmacol Ther. 2019;106(2):329–37.

49. Zhou K, Donnelly L, Burch L, Tavendale R, Doney AS, Leese G, et al. Loss-of-function CYP2C9 variants improve therapeutic response to sulfonylureas in type 2 diabetes: a Go-DARTS study. Clin Pharmacol Ther. 2010;87(1):52–6.

50. Shu Y, Sheardown SA, Brown C, Owen RP, Zhang S, Castro RA, et al. Effect of genetic variation in the organic cation transporter 1 (OCT1) on metformin action. J Clin Invest. 2007;117(5):1422–31.

51. Shah HS, Gao H, Morieri ML, Skupien J, Marvel S, Paré G, et al. Genetic predictors of cardiovascular mortality during intensive glycemic control in type 2 diabetes: findings from the ACCORD clinical trial. Diabetes Care. 2016;39(11):1915–24.

52. Nasykhova YA, Tonyan ZN, Mikhailova AA, Danilova MM, Glotov AS. Pharmacogenetics of type 2 diabetes-Progress and prospects. Int J Mol Sci. 2020;21(18):6842.

53. Daniels MA, Kan C, Willmes DM, Ismail K, Pistrosch F, Hopkins D, et al. Pharmacogenomics in type 2 diabetes: oral antidiabetic drugs. Pharmacogenomics J. 2016;16(5):399–410.

54. Mannino GC, Andreozzi F, Sesti G. Pharmacogenetics of type 2 diabetes mellitus, the route toward tailored medicine. Diabetes Metab Res Rev. 2019;35(3):e3109.

55. Bedene A, Mencej Bedrač S, Ješe L, Marc J, Vrtačnik P, Prezelj J, et al. MiR-148a the epigenetic regulator of bone homeostasis is increased in plasma of osteoporotic postmenopausal women. Wien Klin Wochenschr. 2016;128(7):519–26.

56. Johnell O, Kanis J. An estimate of the worldwide prevalence and disability associated with osteoporotic fractures. Osteoporos Int. 2006;17(12):1726–33.

57. Ralston SH, De Crombrugghe BJG. Development. Genetic regulation of bone mass and susceptibility to osteoporosis. Genes Dev. 2006;20(18):2492–506.

58. US Food and Drug Adminstration. Drug development tools (DDT) qualification programs. https://www. fda.gov/drugs/development-approval-process-drugs/drug-development-tool-ddt-qualification-programs

59. O'Brien J, Hayder H, Zayed Y, Peng C. Overview of microRNA biogenesis, mechanisms of actions, and circulation. Front Endocrinol (Lausanne). 2018;9:402.

60. Van Wijnen AJ, Van De Peppel J, Van Leeuwen JP, Lian JB, Stein GS, Westendorf JJ, et al. MicroRNA functions in osteogenesis and dysfunctions in osteoporosis. Curr Osteoporos Rep. 2013;11(2):72–82.

61. Yahata T, Quan J, Tamura N, Nagata H, Kurabayashi T, Tanaka K. Association between single nucleotide polymorphisms of estrogen receptor α gene and efficacy of HRT on bone mineral density in post-menopausal Japanese women. Hum Reprod. 2005;20(7):1860–6.

62. Kurabayashi T, Matsushita H, Tomita M, Kato N, Kikuchi M, Nagata H, et al. Association of vitamin D and estrogen receptor gene polymorphism with the effects of longterm hormone replacement therapy on bone mineral density. J Bone Miner Metab. 2004;22(3):241–7.

63. Li W-F, Hou S-X, Yu B, Li M-M, Férec C, Chen J-M. Genetics of osteoporosis: accelerating pace in gene identification and validation. Hum Genet. 2010;127(3):249–85.

64. Li W-F, Hou S-X, Yu B, Jin D, Férec C, Chen J-M. Genetics of osteoporosis: perspectives for personalized medicine. Pers Med. 2010;7(6):655–68.

65. Marini F, Brandi ML. Pharmacogenetics of osteoporosis. Best Pract Res Clin Endocrinol Metab. 2014;28(6):783–93.

66. Giguère Y, Dodin S, Blanchet C, Morgan K, Rousseau F. The association between heel ultrasound and hormone replacement therapy is modulated by a two-locus vitamin D and estrogen receptor genotype. J Bone Miner Res. 2000;15(6):1076–84.

67. Simsek M, Cetin Z, Bilgen T, Taskin O, Luleci G, Keser I. Effects of hormone replacement therapy on bone mineral density in Turkish patients with or without COL1A1 Sp1 binding site polymorphism. J Obstet Gynaecol Res. 2008;34(1):73–7.

68. Heilberg IP, Hernandez E, Alonzo E, Valera R, Ferreira LG, Gomes SA, et al. Estrogen receptor (ER) gene polymorphism may predict the bone mineral density response to raloxifene in postmenopausal women on chronic hemodialysis. Ren Fail. 2005;27(2):155–61.

69. Diamanti-Kandarakis E, Dunaif A. Insulin resistance and the polycystic ovary syndrome revisited: an update on mechanisms and implications. Endocr Rev. 2012;33(6):981–1030.

70. Azziz R. Controversy in clinical endocrinology: diagnosis of polycystic ovarian syndrome: the Rotterdam criteria are premature. J Clin Endocrinol Metab. 2006;91(3):781–5.

71. Zawdaki J, Dunaif A. Diagnostic criteria for polycystic ovarian syndrome: towards a rational approach. In: Polycystic ovarian syndrome current issues in endocrinology and metabolism. Boston, MA: Blackwell Scientific; 1992.

72. Kahsar-Miller MD, Nixon C, Boots LR, Go RC, Azziz RJF. Sterility. Prevalence of polycystic ovary syndrome (PCOS) in first-degree relatives of patients with PCOS. Fertil Steril. 2001;75(1):53–8.

73. Hayes MG, Urbanek M, Ehrmann DA, Armstrong LL, Lee JY, Sisk R, et al. Genome-wide association of polycystic ovary syndrome implicates alterations in gonadotropin secretion in European ancestry populations. Nat Commun. 2015;6(1):1–13.

74. Day FR, Hinds DA, Tung JY, Stolk L, Styrkarsdottir U, Saxena R, et al. Causal mechanisms and balancing selection inferred from genetic associations with polycystic ovary syndrome. Nat Commun. 2015;6(1):1–7.

75. Chen Z-J, Zhao H, He L, Shi Y, Qin Y, Shi Y, et al. Genome-wide association study identifies susceptibility loci for polycystic ovary syndrome on chromosome 2p16. 3, 2p21 and 9q33. 3. Nat Genet. 2011;43(1):55–9.

76. Mutharasan P, Galdones E, Peñalver Bernabé B, Garcia OA, Jafari N, Shea LD, et al. Evidence for chromosome 2p16. 3 polycystic ovary syndrome susceptibility locus in affected women of European ancestry. J Clin Endocrinol Metab. 2013;98(1):E185–E90.

77. Shi Y, Zhao H, Shi Y, Cao Y, Yang D, Li Z, et al. Genome-wide association study identifies eight new risk loci for polycystic ovary syndrome. Nat Genet. 2012;44(9):1020–5.

78. Pau C, Saxena R, Welt CKJF. Sterility. Evaluating reported candidate gene associations with polycystic ovary syndrome. Fertil Steril. 2013;99(6):1774–8.

79. Lee H, Oh J-Y, Sung Y-A, Chung H, Kim H-L, Kim GS, et al. Genome-wide association study identified new susceptibility loci for polycystic ovary syndrome. Hum Reprod. 2015;30(3):723–31.

80. Hong S-H, Hong YS, Jeong K, Chung H, Lee H, Sung Y-A. Relationship between the characteristic traits of polycystic ovary syndrome and susceptibility genes. Sci Rep. 2020;10(1):10479.

81. Day F, Karaderi T, Jones MR, Meun C, He C, Drong A, et al. Large-scale genome-wide meta-analysis of polycystic ovary syndrome suggests shared genetic architecture for different diagnosis criteria. PLoS Genet. 2018;14(12):e1007813.

82. Sung H, Ferlay J, Siegel RL, Laversanne M, Soerjomataram I, Jemal A, et al. Global cancer statistics 2020: GLOBOCAN estimates of incidence and mortality worldwide for 36 cancers in 185 countries. CA Cancer J Clin. 2021;71(3):209–49.

83. Kitahara CM, Schneider AB, Brenner AV. Thyroid cancer. In: Thun M, Linet MS, Cerhan JR, Haiman CA, Schottenfeld D, eds. Cancer Epidemiology and Prevention. 4th ed. New York, NY: Oxford University Press; 2016:839–860.

84. Cooper DS, Doherty GM, Haugen BR, Kloos RT, Lee SL, Mandel SJ, et al. Management guidelines for patients with thyroid nodules and differentiated thyroid cancer: The American Thyroid Association Guidelines Taskforce. Thyroid. 2006;16(2):109–42.

85. Salter KD, Andersen PE, Cohen JI, Schuff KG, Lester L, Shindo ML, et al. Central nodal metastases in papillary thyroid carcinoma based on tumor histo-

logic type and focality. Arch Otolaryngol Head Neck Surg. 2010;136(7):692–6.

86. Ferrari SM, Fallahi P, La Motta C, Elia G, Ragusa F, Ruffilli I, et al. Recent advances in precision medicine for the treatment of anaplastic thyroid cancer. Expert Rev Precis Med Drug Dev. 2019;4(1):37–49.

87. Lloyd R, Osamura RY, Klöppel G, Rosai J. WHO classification of tumours of endocrine organs. Lyon: IARC Press; 2017.

88. Prete A, Borges de Souza P, Censi S, Muzza M, Nucci N, Sponziello M. Update on fundamental mechanisms of thyroid cancer. Front Endocrinol (Lausanne). 2020;11:102.

89. Antonelli A, Ferrari SM, Fallahi P, Berti P, Materazzi G, Minuto M, et al. Thiazolidinediones and antiblastics in primary human anaplastic thyroid cancer cells. Clin Endocrinol. 2009;70(6):946–53.

90. Ferrari SM, Ruffilli I, Centanni M, Virili C, Materazzi G, Alexopoulou M, et al. Lenvatinib in the therapy of aggressive thyroid cancer: state of the art and new perspectives with patents recently applied. R Recent Pat Anticancer Drug Discov. 2018;13(2):201–8.

91. Miccoli P, Antonelli A, Spinelli C, Ferdeghini M, Fallahi P, Baschieri L. Completion total thyroidectomy in children with thyroid cancer secondary to the Chernobyl accident. Arch Surg. 1998;133(1):89–93.

92. Asa SL. The current histologic classification of thyroid cancer. Endocrinol Metab Clin N Am. 2019;48(1):1–22.

93. Xing M. Molecular pathogenesis and mechanisms of thyroid cancer. Nat Rev Cancer. 2013;13(3):184–99.

94. Cancer Genome Atlas Research Network. Integrated genomic characterization of papillary thyroid carcinoma. Cell. 2014;159(3):676–90.

95. Xing M. Genetic-guided risk assessment and management of thyroid cancer. Endocrinol Metab Clin N Am. 2019;48(1):109–24.

96. Mazzaferri EL. An overview of the management of papillary and follicular thyroid carcinoma. Thyroid. 1999;9(5):421–7.

97. Landa I, Ibrahimpasic T, Boucai L, Sinha R, Knauf JA, Shah RH, et al. Genomic and transcriptomic hallmarks of poorly differentiated and anaplastic thyroid cancers. J Clin Invest. 2016;126(3):1052–66.

98. Sheng Z, Sun W, Smith E, Cohen C, Sheng Z, Xu X-XJO. Restoration of positioning control following Disabled-2 expression in ovarian and breast tumor cells. Oncogene. 2000;19(42):4847–54.

99. Romei C, Ciampi R, Elisei RJNRE. A comprehensive overview of the role of the RET proto-oncogene in thyroid carcinoma. Nat Rev Endocrinol. 2016;12(4):192–202.

100. Aksoy A, Ally A, Arachchi H, Asa S, Auman J, Balasundaram MJC. Integrated genomic characterization of papillary thyroid. Cell. 2014;159:676–90.

101. Chua EL, Wu WM, Tran KT, McCarthy SW, Lauer CS, Dubourdieu D, et al. Prevalence and distribution of ret/ptc 1, 2, and 3 in papillary thyroid carcinoma in New Caledonia and Australia. J Clin Endocrinol Metab. 2000;85(8):2733–9.

102. Adeniran AJ, Zhu Z, Gandhi M, Steward DL, Fidler JP, Giordano TJ, et al. Correlation between genetic alterations and microscopic features, clinical manifestations, and prognostic characteristics of thyroid papillary carcinomas. Am J Surg Pathol. 2006;30(2):216–22.

103. Khan MS, Qadri Q, Makhdoomi MJ, Wani MA, Malik AA, Niyaz M, et al. RET/PTC gene rearrangements in thyroid carcinogenesis: assessment and clinico-pathological correlations. Pathol Oncol Res. 2020;26(1):507–13.

104. Paulson VA, Rudzinski ER, Hawkins DSJG. Thyroid cancer in the pediatric population. Genes (Basel). 2019;10(9):723.

105. Vuong HG, Altibi AM, Duong UN, Hassell L. Prognostic implication of BRAF and TERT promoter mutation combination in papillary thyroid carcinoma—a meta-analysis. Clin Endocrinol. 2017;87(5):411–7.

106. Yuan X, Liu T, Xu D. Telomerase reverse transcriptase promoter mutations in thyroid carcinomas: implications in precision oncology—a narrative review. Ann Transl Med. 2020;8(19):1244.

107. Boos LA, Dettmer M, Schmitt A, Rudolph T, Steinert H, Moch H, et al. Diagnostic and prognostic implications of the PAX 8–PPAR γ translocation in thyroid carcinomas—a TMA-based study of 226 cases. Histopathology. 2013;63(2):234–41.

108. Nikiforova MN, Biddinger PW, Caudill CM, Kroll TG, Nikiforov YE. PAX8-PPARγ rearrangement in thyroid tumors: RT-PCR and immunohistochemical analyses. Am J Surg Pathol. 2002;26(8):1016–23.

109. Yang J, Gong Y, Yan S, Chen H, Qin S, Gong R. Association between TERT promoter mutations and clinical behaviors in differentiated thyroid carcinoma: a systematic review and meta-analysis. Endocrine. 2020;67(1):44–57.

110. Romei C, Tacito A, Molinaro E, Piaggi P, Cappagli V, Pieruzzi L, et al. Clinical, pathological and genetic features of anaplastic and poorly differentiated thyroid cancer: a single institute experience. Oncol Lett. 2018;15(6):9174–82.

111. Smith N, Nucera C. Personalized therapy in patients with anaplastic thyroid cancer: targeting genetic and epigenetic alterations. J Clin Endocrinol Metab. 2015;100(1):35–42.

112. Quiros RM, Ding HG, Gattuso P, Prinz RA, Xu X. Evidence that one subset of anaplastic thyroid carcinomas are derived from papillary carcinomas due to BRAF and p53 mutations. Cancer. 2005;103(11):2261–8.

113. Fagin JA, Matsuo K, Karmakar A, Chen DL, Tang S-H, Koeffler HP. High prevalence of mutations of the p53 gene in poorly differentiated human thyroid carcinomas. J Clin Invest. 1993;91(1):179–84.

114. Pozdeyev N, Gay LM, Sokol ES, Hartmaier R, Deaver KE, Davis S, et al. Genetic analysis of 779 advanced differentiated and anaplastic thyroid cancers. Clin Cancer Res. 2018;24(13):3059–68.

115. Elisei R, Tacito A, Ramone T, Ciampi R, Bottici V, Cappagli V, et al. Twenty-five years experience on RET genetic screening on hereditary MTC: an update on the prevalence of germline RET mutations. Genes. 2019;10(9):698.

116. Tate John G, Sally B, Jubb Harry C, Zbyslaw S, Beare David M, Nidhi B, Ray S, Thompson Sam L, Shicai W, Sari W, Campbell Peter J, Forbes Simon A, et al. COSMIC: the catalogue of somatic mutations in cancer. Nucleic Acids Res. 2018;47(D1):D941–7.

117. The Lancet. 20 years of precision medicine in oncology. Lancet (London, England). 2021;397(10287):1781.

118. National Cancer Institute (NIH). Precision Medicine. https://www.cancergov/publications/dictionaries/cancer-terms/def/precision-medicine.

119. Xiao H, Liu R, Yu S. Towards precision medicine in thyroid cancer. Ann Transl Med. 2020;8(19):1212.

120. Khatami F, Larijani B, Nikfar S, Hasanzad M, Fendereski K, Tavangar SM. Personalized treatment options for thyroid cancer: current perspectives. Pharmgenomics Pers Med. 2019;12:235–45.

121. Miccoli P, Materazzi G, Antonelli A, Panicucci E, Frustaci G, Berti P. New trends in the treatment of undifferentiated carcinomas of the thyroid. Langenbeck's Arch Surg. 2007;392(4):397–404.

122. Smallridge RC, Ain KB, Asa SL, Bible KC, Brierley JD, Burman KD, et al. American Thyroid Association guidelines for management of patients with anaplastic thyroid cancer. Thyroid. 2012;22(11):1104–39.

123. Pinto N, Black M, Patel K, Yoo J, Mymryk JS, Barrett JW, et al. Genomically driven precision medicine to improve outcomes in anaplastic thyroid cancer. J Oncol. 2014;2014:936285.

124. Samimi H, Fallah P, Sohi AN, Tavakoli R, Naderi M, Soleimani M, et al. Precision medicine approach to anaplastic thyroid cancer: advances in targeted drug therapy based on specific signaling pathways. Acta Med Iran. 2017;55:200–8.

125. Collaborators GO. Health effects of overweight and obesity in 195 countries over 25 years. N Engl J Med. 2017;377(1):13–27.

126. Hruby A, Hu FB. The epidemiology of obesity: a big picture. PharmacoEconomics. 2015;33(7):673–89.

127. Beckmann JS, Lew D. Reconciling evidence-based medicine and precision medicine in the era of big data: challenges and opportunities. Genome Med. 2016;8(1):1–11.

128. Collins FS, Varmus H. A new initiative on precision medicine. N Engl J Med. 2015;372(9):793–5.

129. Cifuentes L, Eckel-Passow J, Acosta A. Precision medicine for Obesity. Dig Dis Interv. 2021;5:239–48.

130. Heart N, Lung, Institute B, Diabetes NIo, Digestive, Diseases K. Clinical guidelines on the identification, evaluation, and treatment of overweight and obesity in adults: the evidence report. National Heart, Lung, and Blood Institute; 1998.

131. Blüher M. The distinction of metabolically 'healthy' from 'unhealthy' obese individuals. Curr Opin Lipidol. 2010;21(1):38–43.

132. Goodarzi MO. Genetics of obesity: what genetic association studies have taught us about the biology of obesity and its complications. Lancet Diabetes Endocrinol. 2018;6(3):223–36.

133. Chami N, Preuss M, Walker RW, Moscati A, Loos RJ. The role of polygenic susceptibility to obesity among carriers of pathogenic mutations in MC4R in the UK biobank population. PLoS Med. 2020;17(7):e1003196.

134. Loos RJ, Yeo GS. The genetics of obesity: from discovery to biology. Nat Rev Genet. 2022;23(2):120–33.

135. Chen R, Mias GI, Li-Pook-Than J, Jiang L, Lam HY, Chen R, et al. Personal omics profiling reveals dynamic molecular and medical phenotypes. Cell. 2012;148(6):1293–307.

136. Hasin Y, Seldin M, Lusis A. Multi-omics approaches to disease. Genome Biol. 2017;18(1):1–15.

137. Aleksandrova K, Rodrigues CE, Floegel A, Ahrens W. Omics biomarkers in obesity: novel etiological insights and targets for precision prevention. Curr Obes Rep. 2020;9(3):219–30.

138. Bouchard C. Genetics of obesity: what we have learned over decades of research. Obesity. 2021;29(5):802–20.

139. Loos RJ, Lindgren CM, Li S, Wheeler E, Zhao JH, Prokopenko I, et al. Common variants near MC4R are associated with fat mass, weight and risk of obesity. Nat Genet. 2008;40(6):768–75.

140. Chambers JC, Elliott P, Zabaneh D, Zhang W, Li Y, Froguel P, et al. Common genetic variation near MC4R is associated with waist circumference and insulin resistance. Nat Genet. 2008;40(6):716–8.

141. Thorleifsson G, Walters GB, Gudbjartsson DF, Steinthorsdottir V, Sulem P, Helgadottir A, et al. Genome-wide association yields new sequence variants at seven loci that associate with measures of obesity. Nat Genet. 2009;41(1):18–24.

142. Speliotes EK, Willer CJ, Berndt SI, Monda KL, Thorleifsson G, Jackson AU, et al. Association analyses of 249,796 individuals reveal 18 new loci associated with body mass index. Nat Genet. 2010;42(11):937–48.

143. Kilpeläinen TO, Carli JFM, Skowronski AA, Sun Q, Kriebel J, Feitosa MF, et al. Genome-wide meta-analysis uncovers novel loci influencing circulating leptin levels. Nat Commun. 2016;7(1):1–14.

144. Yaghootkar H, Zhang Y, Spracklen CN, Karaderi T, Huang LO, Bradfield J, et al. Genetic studies of leptin concentrations implicate leptin in the regulation of early adiposity. Diabetes. 2020;69(12):2806–18.

145. Kichaev G, Bhatia G, Loh P-R, Gazal S, Burch K, Freund MK, et al. Leveraging polygenic functional enrichment to improve GWAS power. Am J Hum Genet. 2019;104(1):65–75.

146. Sun Q, Cornelis MC, Kraft P, Qi L, van Dam RM, Girman CJ, et al. Genome-wide association study identifies polymorphisms in LEPR as determinants of plasma soluble leptin receptor levels. Hum Mol Genet. 2010;19(9):1846–55.

147. Pulit SL, Stoneman C, Morris AP, Wood AR, Glastonbury CA, Tyrrell J, et al. Meta-analysis of genome-wide association studies for body fat distribution in 694 649 individuals of European ancestry. Hum Mol Genet. 2019;28(1):166–74.

148. Willer CJ, Speliotes EK, Loos RJ, Li S, Lindgren CM, Heid IM, et al. Six new loci associated with body mass index highlight a neuronal influence on body weight regulation. Nat Genet. 2009;41(1):25–34.

149. Akiyama M, Okada Y, Kanai M, Takahashi A, Momozawa Y, Ikeda M, et al. Genome-wide association study identifies 112 new loci for body mass index in the Japanese population. Nat Genet. 2017;49(10):1458–67.

150. Turcot V, Lu Y, Highland HM, Schurmann C, Justice AE, Fine RS, et al. Protein-altering variants associated with body mass index implicate pathways that control energy intake and expenditure in obesity. Nat Genet. 2018;50(1):26–41.

151. Farooqi IS, Matarese G, Lord GM, Keogh JM, Lawrence E, Agwu C, et al. Beneficial effects of leptin on obesity, T cell hyporesponsiveness, and neuroendocrine/metabolic dysfunction of human congenital leptin deficiency. J Clin Invest. 2002;110(8):1093–103.

152. Collet T-H, Dubern B, Mokrosinski J, Connors H, Keogh JM, de Oliveira EM, et al. Evaluation of a melanocortin-4 receptor (MC4R) agonist (Setmelanotide) in MC4R deficiency. Mol Metab. 2017;6(10):1321–9.

153. Roth JD, Roland BL, Cole RL, Trevaskis JL, Weyer C, Koda JE, et al. Leptin responsiveness restored by amylin agonism in diet-induced obesity: evidence from nonclinical and clinical studies. Proc Natl Acad Sci U S A. 2008;105(20):7257–62.

154. Brunkwall L, Chen Y, Hindy G, Rukh G, Ericson U, Barroso I, et al. Sugar-sweetened beverage consumption and genetic predisposition to obesity in 2 Swedish cohorts. Am J Clin Nutr. 2016;104(3):809–15.

155. Marigorta UM, Gibson G. A simulation study of gene-by-environment interactions in GWAS implies ample hidden effects. Front Genetics. 2014;5:225.

156. Wang Z, Gerstein M, Snyder M. RNA-Seq: a revolutionary tool for transcriptomics. Nat Rev Genet. 2009;10(1):57–63.

157. Hasanzad M, Sarhangi N, Ehsani Chimeh S, Ayati N, Afzali M, Khatami F, et al. Precision medicine journey through omics approach. J Diabetes Metab Disord. 2021;21:881–8.

158. Amri E-Z, Scheideler M. Small non coding RNAs in adipocyte biology and obesity. Mol Cell Endocrinol. 2017;456:87–94.

159. Deutsch A, Feng D, Pessin JE, Shinoda K. The impact of single-cell genomics on adipose tissue research. Int J Mol Sci. 2020;21(13):4773.

160. González-Plaza JJ, Gutiérrez-Repiso C, García-Serrano S, Rodriguez-Pacheco F, Garrido-Sánchez L, Santiago-Fernández C, et al. Effect of Roux-en-Y gastric bypass-induced weight loss on the transcriptomic profiling of subcutaneous adipose tissue. Surg Obes Relat Dis. 2016;12(2):257–63.

161. Armenise C, Lefebvre G, Carayol JM, Bonnel S, Bolton J, Di Cara A, et al. Transcriptome profiling from adipose tissue during a low-calorie diet reveals predictors of weight and glycemic outcomes in obese, nondiabetic subjects. Am J Clin Nutr. 2017;106(3):736–46.

162. Rangel-Huerta OD, Pastor-Villaescusa B, Gil A. Are we close to defining a metabolomic signature of human obesity? A systematic review of metabolomics studies. Metabolomics. 2019;15(6):1–31.

163. Suhre K, Shin S-Y, Petersen A-K, Mohney RP, Meredith D, Wägele B, et al. Human metabolic individuality in biomedical and pharmaceutical research. Nature. 2011;477(7362):54–60.

164. Bakar MHA, Sarmidi MR, Cheng K-K, Khan AA, Suan CL, Huri HZ, et al. Metabolomics–the complementary field in systems biology: a review on obesity and type 2 diabetes. Mol BioSyst. 2015;11(7):1742–74.

165. Xie B, Waters MJ, Schirra HJ. Investigating potential mechanisms of obesity by metabolomics. J Biomed Biotechnol. 2012;2012:805683.

166. Cirulli ET, Guo L, Swisher CL, Shah N, Huang L, Napier LA, et al. Profound perturbation of the metabolome in obesity is associated with health risk. Cell Metab. 2019;29(2):488–500.e2.

167. Geidenstam N, Magnusson M, Danielsson AP, Gerszten RE, Wang TJ, Reinius LE, et al. Amino acid signatures to evaluate the beneficial effects of weight loss. Int J Endocrinol. 2017;2017:6490473.

168. Zhao L. The gut microbiota and obesity: from correlation to causality. Nat Rev Microbiol. 2013;11(9):639–47.

169. John GK, Mullin GE. The gut microbiome and obesity. Curr Oncol Rep. 2016;18(7):45.

170. Ley RE, Peterson DA, Gordon JI. Ecological and evolutionary forces shaping microbial diversity in the human intestine. Cell. 2006;124(4):837–48.

171. Rad SS, Nikkhah A, Orvatinia M, Ejtahed H-S, Sarhangi N, Jamaldini SH, et al. Gut microbiota: a perspective of precision medicine in endocrine disorders. J Diabetes Metab Disord. 2020;19:1827–34.

172. Bäckhed F, Ding H, Wang T, Hooper LV, Koh GY, Nagy A, et al. The gut microbiota as an environmental factor that regulates fat storage. Proc Natl Acad Sci U S A. 2004;101(44):15718–23.

173. Bauer PV, Hamr SC, Duca FA. Regulation of energy balance by a gut–brain axis and involvement of the gut microbiota. Cell Mol Life Sci. 2016;73(4):737–55.

174. Ejtahed H-S, Soroush A-R, Angoorani P, Larijani B, Hasani-Ranjbar S. Gut microbiota as a target in the pathogenesis of metabolic disorders: a new approach to novel therapeutic agents. Horm Metab Res. 2016;48(06):349–58.

175. Solas M, Milagro FI, Martínez-Urbistondo D, Ramirez MJ, Martínez JA. Precision obesity treatments including pharmacogenetic and

nutrigenetic approaches. Trends Pharmacol Sci. 2016;37(7):575–93.

176. Goni L, Milagro FI, Cuervo M, Martínez JA. Single-nucleotide polymorphisms and DNA methylation markers associated with central obesity and regulation of body weight. Nutr Rev. 2014;72(11):673–90.

177. Guzman A, Ding M, Xie Y, Martin K. Pharmacogenetics of obesity drug therapy. Curr Mol Med. 2014;14(7):891–908.

178. O'Connor A, Swick AG. Interface between pharmacotherapy and genes in human obesity. Hum Hered. 2013;75(2–4):116–26.

179. Li QS, Lenhard JM, Zhan Y, Konvicka K, Athanasiou MC, Strauss RS, et al. A candidate-gene association study of topiramate-induced weight loss in obese patients with and without type 2 diabetes mellitus. Pharmacogenet Genomics. 2016;26(2):53–65.

180. de Luis DA, Soto GD, Izaola O, Romero E. Evaluation of weight loss and metabolic changes in diabetic patients treated with liraglutide, effect of RS 6923761 gene variant of glucagon-like peptide 1 receptor. J Diabetes Complicat. 2015;29(4):595–8.

181. Zhang J-P, Lencz T, Zhang RX, Nitta M, Maayan L, John M, et al. Pharmacogenetic associations of antipsychotic drug-related weight gain: a systematic review and meta-analysis. Schizophr Bull. 2016;42(6):1418–37.

182. Doo M, Kim Y. Obesity: interactions of genome and nutrients intake. Prev Nutr Food Sci. 2015;20(1):1–7.

183. Rathmann W, Bongaerts B. Pharmacogenetics of novel glucose-lowering drugs. Diabetologia. 2021;64(6):1201–12.

184. Hasanzad M, Aghaei Meybodi HR, Sarhangi N, Larijani B. Artificial intelligence perspective in the future of endocrine diseases. J Diabetes Metab Disord. 2022;21(1):971–78.

185. Hasanzad M, Sarhangi N, Naghavi A, Ghavimehr E, Khatami F, Ehsani Chimeh S, et al. Genomic medicine on the frontier of precision medicine. J Diabetes Metab Disord. 2021;21(1):853–61.

# Precision Medicine in Psychiatric Disorders

6

Xenia Gonda, Kinga Gecse, Zsofia Gal, and Gabriella Juhasz

**What Will You Learn in This Chapter?**

In this chapter, you will learn about the special characteristics of psychiatric illness, which impact the development and implementation of precision psychiatry. These include the multifactorial background, high heterogeneity, the blood-brain barrier impeding direct sampling of central nervous pathophysiology, the highly subjective nature of psychiatric symptoms, and the lack of biomarkers. We will also discuss the difficulties of understanding underlying pathophysiological processes and the problems with current classification systems in psychiatry and the need for precise and objective diagnostic and prognostic predictive models. We will overview the present methods for choosing treatment and the subjective evaluation-based trial-and-error method of matching medications to psychiatric patients and the need for developing more precise models for predicting treatment efficacy. Finally, psychiatry-specific challenges and obstacles will be discussed.

**Rationale and Importance**

Psychiatry significantly lags behind other medical specialties in employing a precision approach.

X. Gonda
NAP-2-SE New Antidepressant Target Research Group, Hungarian Brain Research Program, Semmelweis University, Budapest, Hungary

Department of Psychiatry and Psychotherapy, Faculty of Medicine, Semmelweis University, Budapest, Hungary

K. Gecse
Department of Pharmacodynamics, Faculty of Pharmacy, Semmelweis University, Budapest, Hungary

SE-NAP 2 Genetic Brain Imaging Migraine Research Group, Hungarian Brain Research Program, Semmelweis University, Budapest, Hungary

Z. Gal
Department of Pharmacodynamics, Faculty of Pharmacy, Semmelweis University, Budapest, Hungary

TRAJECTOME Research Group, Budapest, Hungary

G. Juhasz (✉)
Department of Pharmacodynamics, Faculty of Pharmacy, Semmelweis University, Budapest, Hungary

SE-NAP 2 Genetic Brain Imaging Migraine Research Group, Hungarian Brain Research Program, Semmelweis University, Budapest, Hungary

TRAJECTOME Research Group, Budapest, Hungary
e-mail: juhasz.gabriella@pharma.semmelweis-univ.hu

As a consequence, diagnosis and treatment are still based on subjective evaluations and decisions. Therefore, there is a significant delay in accurate diagnosis and effective treatment, causing not only suffering but also prolonged and often lasting and progressive dysfunction. This is at least in part related to our current lack of proper understanding of underlying pathophysiological processes and lack of objective markers. Identifying novel and objective data sources and processing this information with machine learning approaches may yield more accurate predictive models for establishing the diagnosis, prognosis, and treatment which can be implemented not only for better management of psychiatric illness but also for predicting risk in premorbid stages opening the possibility for prevention.

## 6.1 Introduction

The need for a new era in medicine, which radically transforms the traditional concepts and the way we understand illness and apply care, has recently gained priority [1, 2]. Precision medicine could be such a new paradigm for the prevention, diagnosis, and treatment of disorders, which is novel in looking beyond the illness and its clinical characteristics and also considers unique patient features including variability in genetic background, environmental exposures, as well as lifestyle, in order to tailor and personalize treatment and prioritize individualization of care [3–5]. Within this framework, precision psychiatry is expected to shift clinical mental healthcare paradigms from our traditional, currently employed "evidence-based" approach based on data gathered and synthetized from findings in a large population of patients to a practice building on individualized deep knowledge of phenotypical, clinical, and biological characteristics [6], with the major aim of developing a more precise diagnostic conceptualization and classification of individual patients and identifying effective prevention and treatment for specific patients according to the relevant etiopathophysiological background of their risk factors and symptoms [7].

## 6.2 A Special Case for Psychiatry

Mental disorders are estimated to account for 2.9% of years lived with disability [8–10] and contribute to the expected rise to 6 trillion USD of the global financial burden by 2030 [11]. This is due in part to the loss of productivity and quality of life and, more importantly, to both the lack of neurobiologically based disease classifications and the current trial-and-error-based clinical decisions, resulting in multiple changes in medications before reaching symptomatic remission, in those cases where it can be reached at all [4]. Several unique features of psychiatric disorders emphasize the urgent need for precision psychiatry including their chronic, lifelong, and in many cases progressive nature leading to functional decline often from younger ages, the significant response variability, and remarkable lack of efficacy of therapeutic interventions including both pharmacotherapy and psychotherapy, and the long time needed to achieve benefit from the majority of treatments [12].

Reducing burden and disability, as well as costs, and increasing efficacy would require a better understanding of etiopathophysiology, improved classification of diseases, more precise diagnoses, more precise identifications of individuals at risk aiding early recognition and intervention, and novel treatments more precisely targeting individual symptom constellations. However, psychiatry, from several important aspects, is different from other branches of medicine which must be considered in order to develop and implement precision psychiatry.

### 6.2.1 Psychiatric Disorders Are Multigenic, Multifactorial, and Highly Heterogeneous

One unique challenge for precision psychiatry is disentangling the strikingly huge and intricate complexity of psychiatric disorders, whose biological underpinnings are only partially known. Psychiatric disorders are multigenic and multifactorial, determined by an interaction of genetic and both early, distal adversities which also contribute toward pathophysiological differences and recent, proximal stressors. Thus psychiatric disorders are both shaped by and manifested as individual experiences, which contributes to a remarkably high symptomatic heterogeneity even within the same diagnostic categories reflecting very different etiologies involved both on the neurobiological and the environmental effect level. Considering the symptomatic heterogeneity, recently, in an analysis of the STAR*D study, it has been concluded that, for example, depression can be manifested as more than 1000 different symptomatic profiles [13], likely reflecting equally divergent pathophysiological processes. Furthermore, while symptoms greatly vary between patients within the same diagnostic category, they often greatly overlap between different diagnostic categories [14–16] as reflected in the increasing number of transdiagnostic symptoms and markers described. Thus, two patients falling into identical diagnostic categories may have entirely distinct symptomatic profiles underlined by massively divergent psychopathologies, requiring entirely different treatments [17], which also suggests that treatment decisions can be insufficiently built upon diagnostic categories. This shows not only the shortcomings of our current classification and diagnostic systems and explains why our unsophisticated approach of trying to assign medication groups to patients based on such unrefined diagnostic grouping fails but also reflects our remarkable lack of understanding of the neurobiological background of psychiatric symptoms and illness processes.

In order to tackle this extreme phenotypic complexity, our current psychiatric classification systems may need to be thoroughly revised if not abandoned in the favor of one which reflects biological mechanisms underlying symptoms and symptom constellations, allowing for decomposing complex and heterogeneous psychiatric disorders to individual features mapped to neurobiological processes in order to understand the involvement of complex patterns of biological pathways [18]. This aim has also been expressed in the development of the Research Domain Criteria (RDoC) developed by the NIMH

[19], a data-driven dimensional approach to reduce clinical heterogeneity across current diagnostic classifications, in order to identify homogeneous categories and create a neurobiologically valid framework for understanding and classifying psychiatric disorders and aid development of interventions specifically targeting neurobiological underpinnings [2, 20]. This would lead to shifting from diagnosing and grouping patients based on the "average" "theoretical" patient to reclassifying them along with relevant clinical features and endophenotypes, allowing for more precisely matching medications and predicting response [2].

It must also be noted that not only psychiatric disorders are highly heterogeneous but also psychiatric medications. Currently, for the same indications, we possess several psychiatric medications in various classes of different pharmacodynamic profiles, which are only partially known, and with important within-class differences in pharmacodynamic and pharmacokinetic profiles. Medications in the same class with presumably similar properties, targets, and action mechanisms will have divergent effects in patients belonging to the same diagnostic group, producing either response, partial response, therapy resistance, and different side effect profiles; thus, the effects of such medications, and especially in patients with diverse and unique neurobiological and pathophysiological illness backgrounds, will be difficult to predict [12].

## 6.2.2 From Evidence-Based Through Personalized Toward Precision Psychiatry

As psychiatric disorders involve alterations in subjective states and individual life experiences factor in as key etiological contributors, psychiatry has traditionally and inherently been "personalized" focusing on the individual analyses and conceptualizations of the given patient's pathological psychological processes [6]. However, several years ago, the evidence-based movement became the mainstream approach in medicine including psychiatry, which, instead of looking for individual characteristics, pushed toward the uniformization and categorization of patients, building on results from randomized controlled studies, summarizing vast knowledge from meta-analyses and systematic reviews and average effects in theoretically homogeneous patient sub-groups to define diagnosis and treatment (Fig. 6.1). Thus instead of the individual unique subject, "evidence-based" psychiatry focuses on idealized average theoretical patients who would only meet unrealistic inclusion and exclusion criteria [21]. It must also be noted that one big

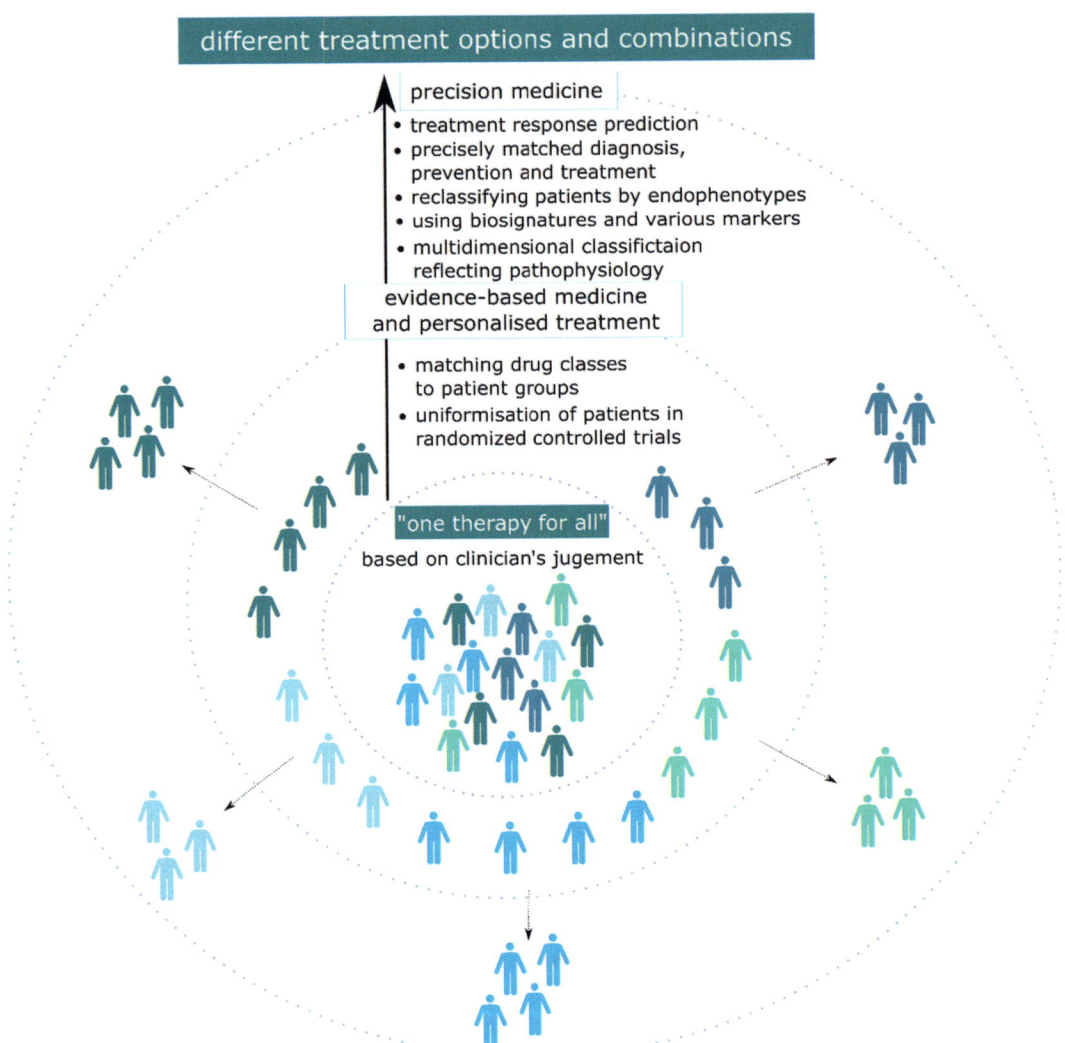

**Fig. 6.1** Aims of precision psychiatry. Precision psychiatry aims at employing the understanding of the multifactorial background of psychiatric disorders in both classification and treatment, rather than fitting patients into predetermined groups based on a subjective evaluation of symptoms. After abandoning the initial concept of "one therapy for all," the evidence-based medicine movement became mainstream in the treatment psychiatric disorders with the concept of establishing effective medications; however, the realization of the concept shifted toward uniformization of patients and generalization of treatment for subgroups. Precision psychiatry aims to take into account the unique characteristics of individual patients as well as the multidimensionality of psychiatric disorders in developing precision predictive models for diagnosis, prognosis, and treatment, also considering the longitudinal perspective and staging approaches in psychiatry

shortcoming of psychiatric clinical trials is involving subject populations who are in several important ways different from real-world patients; thus, key characteristics and idiosyncrasies are specifically excluded [21]. Such clinical trials and guidelines while allowing for broader generalizations about specific treatments in specific populations, and thus for those individuals who fit the inclusion and exclusion criteria for these study populations, at the same time fail to detect sophisticated individual characteristics and differences. As a consequence, findings are significant in large study populations and will not be able to provide the same benefit for several individual patients. A growing range of "evidence-based" treatments established based on randomized controlled studies are found to work only for a small portion of patients within a diagnostic group [21]. Guidelines developed based on the results of such studies focus on patient groups with similar average illness presentations and suggest matching drug classes to these groups, ignoring important differences between patients in each group and medications in each class and providing no information on how to choose between molecules within a drug class [12]. This has been increasingly pressing, for example, in the case of major depression where only about one-third of the patients respond to the first treatment trial, and especially as we increasingly understand that besides alleviating suffering and restoring function, the success of early treatment has an impact on illness trajectory in case of several psychiatric disorders [22]. Therefore, the need to identify in case of each treatment those subsets of people who possess a high probability of responding to that given treatment, based on their individual characteristics, has emerged [22], which led to the personalized psychiatry approach. However, while personalized psychiatry, as expressed in its name, intended to individualize treatment in the case of every single patient, it failed to fulfill this aim, as it assigns individual patients to groups based on their susceptibility toward a particular disease subtype and thus the likeliness to respond to a given drug, instead of creating individual medications fitting the unique patient [17]. The novel

concept of precision psychiatry [23] besides acknowledging that psychiatric treatments are not individually developed for the individual patient proposed that personalization could be achieved via introducing highly exact and accurate measurements and evaluations, contributing toward a very sophisticated and detailed and multidimensional classification system by the application of which true personalization of treatment could be achieved [2]. Eventually, while psychiatric patients won't get a custom-made, personally created medication based on their unique pathological features as it is now becoming possible in some other medical specialties, these patients would be matched to an existing but personally chosen treatment based on their own unique and very detailed classification [2].

### 6.2.3  The Issue of Psychiatric Measurement and Phenotyping: Objective Markers for Subjective Suffering

The majority of psychiatric symptoms show a subjective nature, and in the absence of objective markers, their evaluation takes place based on clinical features, evaluated based on the subjective account of the patient as assessed by the clinician, which in turn is also influenced by several subjective factors including knowledge, experience, personal opinion, and clinical judgment, which guide not only diagnosis but also treatment [12]. A possible approach to decomposing psychiatric disorders to biologically quantifiable information more closely related to etiological and underlying pathophysiological processes and establishing measurable biomarkers guiding diagnosis and treatment choice is the RDoC approach which has recently been modified to include, in addition to phenomenology-based observable clinical criteria, genomic or proteomic information [4, 17]. One of the major goals of precision psychiatry is to identify biomarkers that possess clinical utility and provide objective assessment, as indicators for risk, diagnosis, pathophysiology, or treatment outcomes.

These biomarkers allow longitudinal monitoring of changes within the patient, measured on their own or in interaction with environmental factors, which can prove transdiagnostic and reflect underlying pathological processes. It is also important to understand to what extent such biomarkers are either universal or specific by diagnoses [2, 24]. A major difficulty in finding biomarkers for psychiatric disorders is that the majority of etiopathological processes underlying psychiatric conditions are sealed inside the central nervous system, both physically and to a great extent chemically isolated by the blood-brain barrier, making sampling beyond our reach. Thus, we need to rely on measures estimated from peripheral blood or other accessible body fluids, which will not exactly reflect the actual conditions in the central nervous system and will not be able to provide topological information on processes.

Psychiatric biomarkers should be both of trait-like and trait-like nature and should be of several types beyond peripheral blood and genomic markers. Brain imaging technologies now possess both sufficient spatial and temporal resolution allowing quantification of neural connection and thus paving the way toward an in vivo conceptualization and understanding of mental illnesses reformulating them as brain functioning disorders [2, 25]. Psychosocial and psychological markers are also essential for a precision approach including specific neurocognitive features, which could be integrated with biological models and markers to predict not only diagnostic classifications but psychopharmacological treatment response [26, 27]. Novel types of behavioral phenotype markers also entered our arsenal, in part due to changes brought about by the COVID-19 pandemic which necessitated the expansion of telemedicinal approaches and paved the way for collecting data on pantomimic, voice, or social media behavior which provide another perspective on behavioral alterations characteristics of mental symptoms [4] (Fig. 6.2). The source of information is further expanding with the spreading of wearable sensors and devices providing a continuous data stream, which could open novel avenues of phenotyping as currently our information is based on either account of the state of the patient at visits or extrapolation between two visits based on retrospective information [21].

Another important aspect is to take into account the longitudinal perspective, considering that the majority of psychiatric disorders show characteristic developmental trajectories, often with childhood or early adolescence onset. The components and antecedents of most psychiatric disorders also involve state and trait-like features detectable well before the onset of the illness itself, thus offering the potential for prediction, possibly prevention, and fine phenotyping and subtyping. Thus, precision psychiatry also seeks to establish relevant risk and resilience factors, define their underlying neurobiological mechanisms and pathways, and understand and assess the effects of these factors on the brain and on behavior and their contribution not only to psychopathology but also to its targeted prevention and early intervention [28].

The final aim is not to identify singular biomarkers but rather biomarker profiles or signatures, mapping the heterogeneity of clinical manifestations seen in mental disorders, which would initially aid treatment choices and predict response in the case of our currently available medications and, in the long term, help understand pathophysiological illness processes and identify novel action mechanisms and treatment targets leading to the discovery of novel molecules [2].

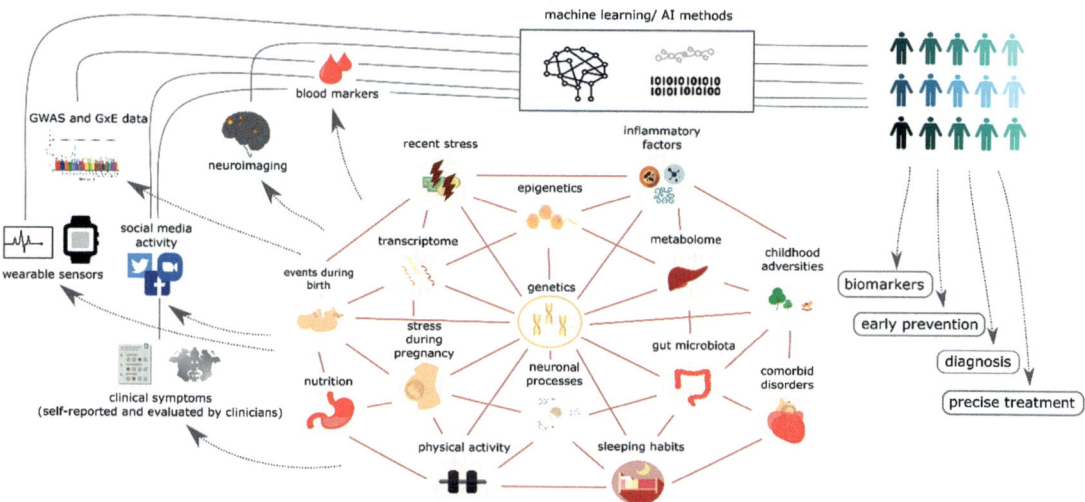

**Fig. 6.2** The multifactorial background of psychiatric disorders. The etiopathology of psychiatric disorders is highly multifactorial involving an interaction between genetic and various environmental factors with an important temporal dynamic pattern, including intrauterine effects, early proximal effects to the development of a diathesis, and current, proximal influences. The outcome of these direct and indirect effects influences a wide variety of disease characteristics including onset, prognosis and treatment, and impact illness trajectories and outcomes. Thus, precision psychiatry needs to collect detailed, objective, and often novel types of information on the above factors describing those processes which are involved in the development and progression of psychiatric symptoms and illness reflecting both state and trait-like characteristics and analyze the connection between them and interpret their impact on illness course and characteristics. Besides "well-known" clinical symptoms and the continuously growing amount of genetic data, other psychological, cognitive, and behavioral markers, data from wearable sensors, biomarkers, imaging markers, and the recording of environmental influences are needed to establish biosignatures for the precise prediction of diagnosis and treatment

## 6.3   Making Sense of the Information in Precision Psychiatry

### 6.3.1   Big Data Analytics and Machine Learning Toward a Systems Biological Approach in Understanding and Treating Neuropsychiatric Disorders

In current psychiatry, we focus on causality and aim to predict individual liability for a mental disorder in terms of linear associations between risk factors and clinical outcomes. However, this approach does not necessarily allow for clinically meaningful and accurate predictions concerning whether a patient will or will not develop a specific disorder, what will be the prognosis, and which medications will produce remission [21]. To reach such goals of precision psychiatry, we must go beyond restricting our models to pre-specified relationships and employ hypothesis-free and data-driven approaches, building on a multi-omic or pan-omic approach. This would involve analyzing genomic, epigenomic, transcriptomic, metabolomic, proteomic, as well as neuroimaging, biomarker, clinical, and environmental information, embedded in a systems biology approach, to uncover biological pathways involved in the background of psychopathology. This would provide both an unprecedented amount and novel types of data, requiring different methods of establishing relationships between them [2]. Novel big data analytical methods allow moving beyond the currently employed evidence-based group-level approaches, toward the identification of interaction patterns between a large number of different types of variables, and make use of machine learning to interpret these patterns and make data-driven decisions [21]. In the case of machine learning, applying an iterative and simultaneous analysis of multiple interacting associations, computers learn and improve automatically from previous experience, rather than being programmed, creating algorithms that facilitate recognition of patterns, identification of principles, as well as classification and prediction based on models which are derived from existing data [6, 29]. The ultimate goal is the development of a unified model using systems biological and computational psychiatry approach yielding a biosignature composed of a wide variety of markers providing better classification and diagnoses and guidance for tailored treatment and intervention options [2].

### 6.3.2 Modelling and Predicting Disease and Treatment Characteristics in Precision Psychiatry

One major focus of precision psychiatry is developing prediction and trajectory models concerning mental disorders which also contributes to our understanding of how potential psychobiological mechanisms develop over time and aids the identification of risk and resilience factors to help the development of preventive measures, interventions, and therapies [18]. Predictive models are aimed at estimating and forecasting the probability that a certain condition is present (diagnostic models) and that certain outcomes will occur (prognostic models) or of therapeutic response to interventions (in case of treatment prediction models) at the level of the individual patient [22]. Currently, there are several prediction models available that transdiagnostically target various psychiatric conditions, including affective disorders, therapy-resistant depression, psychotic disorders, risk and outcomes in schizophrenia, anxiety disorders, neurodevelopmental disorders, substance use disorders, as well as a range of clinically relevant outcomes such as suicide [6, 22]. However, a great obstacle in the way of clinically employing predictive models is the current lack of proper validation. One recent large-scale systematic review study identified almost 600 prediction models in the psychiatric literature, of which only 15% were validated, among them 10.4% internally and only 4.6% externally [22]. This also suggests that precision psychiatry tends to prioritize the development of novel models over the validation of already developed ones which is especially worrying considering the recent replication crises involving multiple fields of research [22] and also emphasizes that in the future, more emphasis should be put on reproducibility and replication also in order to improve the efficacy of research [30]. The majority of validated psychiatric prediction models in the above study were prognostic, followed by diagnostic models and least frequently predictive models, with the majority of models focusing on psychosis [22]. However, of the identified models, only 0.2% were considered for clinical implementation [22].

### 6.4 Special Aspects in Precision Psychiatry

### 6.4.1 The Role of the Blood-Brain-Barrier

As the majority of pathophysiological processes underlying psychiatric symptoms happen in the brain which is isolated from the rest of the body by a biological barrier, treatment efficacy, and likely etiopathology and development, of psychiatric disorders is also influenced by the characteristics of the blood-brain barrier (BBB) and its compounds. The BBB has the largest surface area [31] among the barriers sealing the paracellular routes between the periphery and the central nervous system. Evidence suggests that the integrity of its structure, consisting of neurons, pericytes, endothelial, and glial cells, may be impaired in psychiatric disorders [32]. The permeability of the BBB against passive transport pathways directly depends on tight junction (TJ) molecules, which show altered expression in several psychiatric disorders. Changes in the integrity of the

BBB are often associated with inflammatory processes [32, 33], which could indicate a distinct patient group with different pathophysiology among individuals with psychiatric disorders. Furthermore considering that the integrity of the BBB can change over the life span [34], age may also play a role in the divergent psychiatric etiologies as well as their treatments.

The major component of TJ is claudin-5, which is expressed in other tissues as well beyond the BBB [35]. Research focusing on the role of this protein in psychiatric disorders pointed out that therapeutic substances can also change its expression; in vitro and in vivo studies of schizophrenia showed that the level of claudin-5 shows a dose-dependent change after administration of antipsychotics haloperidol and chlorpromazine as well as lithium [36].

Besides the alteration in the integrity of BBB, the regulated transport pathways are also the focus of understanding the therapeutic efficacy of different psychiatric medications. The BBB includes multidrug-resistant proteins, which are efflux transporters taking care of eliminating substances from the brain parenchyma [33]. Among them, P-glycoprotein is the most widely researched, with extensive although sometimes contradictory results on the MDR1 gene (multidrug-resistant mutation1), also called ATP Binding Cassette Subfamily B Member 1 (ABCB1), encoding for P-glycoprotein, and its involvement in the therapeutic efficacy and side effects of drugs used in psychiatric disorders. Single nucleotide polymorphisms (SNP) of the MDR1 gene have been shown to affect remission rates in patients with chronic depression treated with sertraline [37], citalopram, paroxetine, amitriptyline, and venlafaxine [38]. In patients with schizophrenia, glucose intolerance as a side effect of olanzapine [39] and hyperprolactinemia after risperidone/paliperidone treatment [40] were associated with MDR1 gene variation. Further studies are warranted to thoroughly describe the involvement of the BBB in the effects and side effects of psychiatric medication, as well as its role in the development and characteristics of mental disorders, and this should also be included in our predictive models.

## 6.4.2 Longitudinal Perspective and Staging Model of Neuropsychiatric Disorders

Aiming at prevention or early intervention to improve disease trajectories and outcomes, we must also take into account the temporal dynamics and longitudinal course of mental disorders [6]. Several psychiatric disorders are currently conceptualized as having a neuroprogressive nature, proceeding in stages, characterized by different neurobiological alterations in each stage thus manifesting distinct biomarker profiles or signatures throughout their different neurodevelopmental or neuroprogressive stages in different ages [18] (Fig. 6.3).

There is usually a several years' gap between the first manifestations of signs or symptoms and first psychiatric contact [41], and by the time patients actually get into contact with mental healthcare, symptoms are usually more severe and have a more serious impact on function. The early signs of later more severe mental disorders are usually unspecific; however, certain mental disorders could hopefully be preventable or reversible or their neuroprogression halted or diverted toward a more benign trajectory. It would be key to study and understand relevant quantitative psychological and behavioral traits during neurodevelopment and to identify and develop early biological risk and resilience markers as well as biomarker signatures for early and even non-symptomatic illness stages that would predict development or course of mental disorders and would also allow for targeted prevention or intervention in younger ages [18]. This is relevant not only with respect to the development of mental illness but also in order to understand long-term brain changes associated with pharmacological treatment and its efficacy [6], or longitudinal interaction between different biological components and their impact on the recurrence of illness [6, 42].

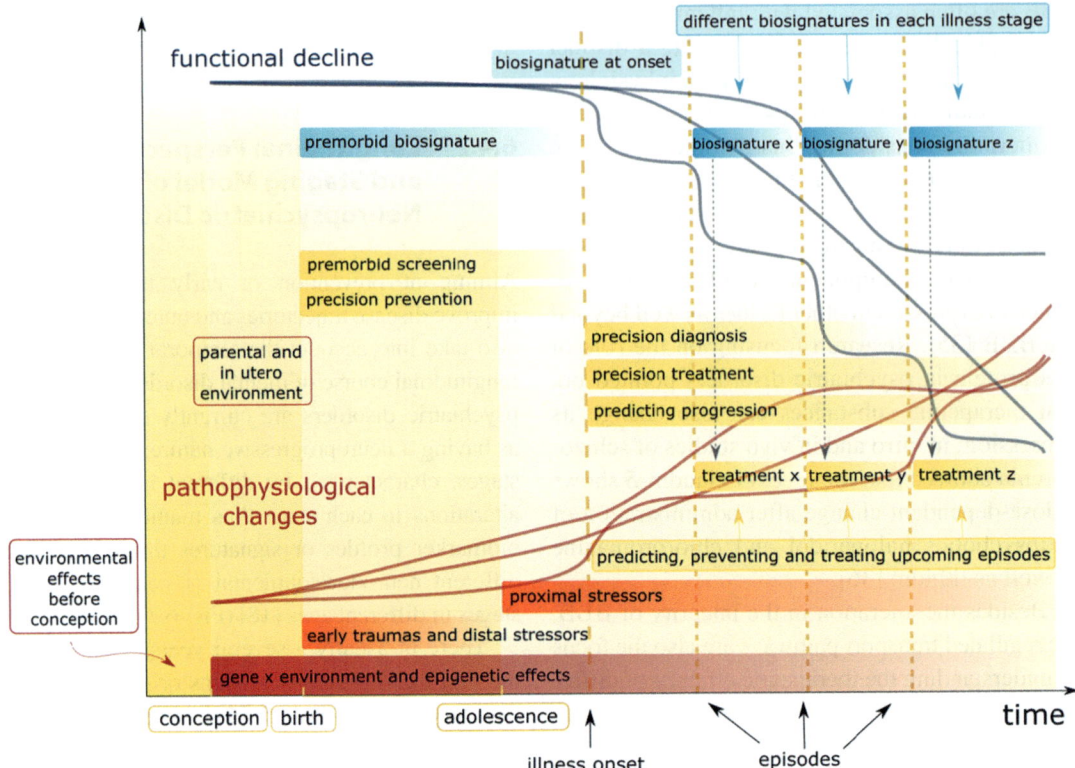

**Fig. 6.3** The staging model of psychiatric disorders. The current conceptualization of psychiatric disorders highlights the role of neuroprogression, postulating that a continuous pathophysiological progression underlies the emergence and course of psychiatric disorders, contributing to different alterations at different illness stages. This approach suggests the importance of taking into consideration the temporal dynamics in precision psychiatry, as in different stages, different biosignatures may reflect the actual pathophysiological stages, predict onset of episodes and progression, or determine which processes should be targeted and thus which treatment would be effective. This approach also postulates that a distinct biomarker profile characterizes not only each illness stage but also the premorbid stages, allowing for prediction and prevention measures

### 6.4.3 Precision Psychotherapy: Predicting the Efficacy of Psychological Treatment and Identifying Patients Who Would Benefit

Another unique aspect of psychiatry is psychotherapy as a non-molecular and non-biological treatment option unutilized in other medical specialties. As mentioned before, not only is the alteration of subjective experience a core feature of psychiatric disorders, but previous experiences, environmental influences, and the trace they leave on psychological function both via neurobiological alterations in case of early adversities and via learning and neuroplasticity in later stressors are important etiological processes, which can be modified by corrective emotional experiences, that is, psychotherapy. There are several issues psychotherapy must face especially in the era of precision medicine. First of all, psychotherapy is a lengthy and highly costly intervention, with in general a relative lack of randomized control studies to show its effectiveness. While there are several schools and branches of psychotherapy, in the case of the majority, individual skills, capacities, and experience of the therapist may significantly influence the effectiveness of treatment. Several patient characteristics also influence the efficacy of psychological interventions. While psychotherapy is not universally effective, currently used clinical and demographic markers cannot sufficiently predict if it will be in the case of a given patient. Recently there have been a few attempts to identify psychological as well as biological markers that would predict liability and response toward psychotherapy in order to be able to identify those patients who could actually benefit from this cost- and time-consuming and low-throughput method. A meta-analysis of 22 studies focusing on candidate genes supposedly playing a role in determining susceptibility toward environmental influences including *5-HTTLPR*, *DRD4*, *MAOA*, *DAT*, and *NR3C1* and looking at gene-environment interaction effects both in the development and psychotherapy of childhood mental disorders found that psychological interventions were effective only in those carrying the "susceptibility variants" [43]. Other studies employing a combined analysis of a small number of susceptibility alleles similarly found that genetic variation may predict response to psychotherapy. One study reported that in the case of mindfulness-based cognitive therapy (MBCT), the efficacy of treatment was mediated by a gene set including *BDNF*, *AchRMusc2*, *DRD4*, and *OPRM1* [44], while another one reported that a risk index based on variation in *5-HTTLPR* and *NGF* rs6330 combined with clinical and demographic data predicted response to cognitive behavior therapy (CBT) [45]. A few GWAS studies only found suggestive significance in association with the effectiveness of different types of CBT [46], but a polygenic risk score (PRS) analysis similarly found that a higher genetic predisposition toward environmental sensitivity predicted increased response to high-intensity psychotherapy [47]. While the largest psychotherapy-GWAS meta-analysis so far found no overall significant effect, it also reported suggestive associations of PRS of "subjective well-being" with fewer post-treatment symptoms and autism spectrum disorder PRS to predict worse treatment outcomes [48]. In a twin study, PRS calculated for environmental sensitivity significantly moderated both the effect of upbringing on emotional problems and the effects of CBT treatment on the alleviation of symptoms [47]. This model actually could predict both risk of psychiatric problems and response to treatment, indicating that those with the highest sensitivity react best to individual therapy and those with the lowest do not benefit more from high-intensity individual interventions compared to group therapy or parent-led therapy. Thus this model is also useful in selecting who is the appropriate candidate for individual psychotherapy [47].

### 6.4.4 Focusing on Comorbidities and Environmental Effects in Precision Psychiatry: The Diseasome Approach

In precision medicine, stratification of disorders within the same diagnostic category into different disease subgroups receives increasing emphasis. One aspect of stratification is to characterize the given psychiatric disorder together with its comorbid diseases and their common symptoms. In the case of major depressive disorder, there is a wide variety of direct (e.g., obesity) and indirect (e.g., cardiovascular disorders) comorbid disorders based on molecular mechanisms [49], which may be exploited to mark different strata within the etiology of this heterogenic disorder in order to identify more effective treatments more accurately targeting the involved pathophysiological processes. The symptoms of such comorbid somatic disorders are often detected as side effects (e.g., metabolic disturbances, cardiovascular or genitourinary alterations) of psychiatric pharmacological treatment [50], which could also support the pleiotropic biological and genetic connection between the disorders in a comorbidity group. Examining the shared genetic background and the presence of various polymorphisms could help identify the causal effects as well as improve the efficacy of different substances used in the treatment of psychiatric disorders.

Besides comorbid disorders, environmental factors such as stress may also contribute to the appearance of psychiatric disorders, and the differences in stress responses are also potential stratifying factors, for example, in depression research and also choosing between the treatment options [51]. As additive effects of all SNPs across the genome explain 44–54% of total brain volume [52], great space is left for the impact of environmental effects to significantly impact neural development, brain structure, and function [53, 54] as well as the emergence of psychiatric disorders [55]. Early life conditions and adverse experiences, such as neighborhood deprivation, air pollution, as well as common stressful life events, may have a lasting impact on mental health, although the neurobiological underpinnings of these relationships are not always known [18]. Childhood trauma is an established etiological factor in the development of psychiatric disorders with long-term consequences on brain structure and function, but its role is less investigated and understood in stratifying response to treatment. A meta-analysis of responses to ten different antidepressive medications found that those exposed to childhood traumas showed worse responses to pharmacotherapy [56], and specifically variation in the noradrenalin transporter [57] and *TPH2* [58] genes is associated with worse responses in those experiencing early adversities. Stress experienced in the year preceding the onset of the first mood episode has been shown to lead to a worse response to fluvoxamine only in those carrying the short allele of *5HTTLPR* [59].

Recent or proximal stress is similarly important both in the development and treatment response in affective disorders. Depressive episodes appearing following stress exposure, previously termed "reactive" depression, were shown to respond better to psychotherapy or placebo than pharmacotherapy [60], while depressive episodes developing independently of environmental effects respond better to pharmacological treatment [60] and to tricyclic agents than to SSRIs [60]. The presence of the *5HTTLPR* short allele was also shown to have a deleterious effect on citalopram response only in those exposed to stress before the onset of an episode, while no such effect of stress was observed in the case of nortriptyline [61]. Recent stress was also reported to influence treatment efficacy in interaction with variations in *FKBP5*, *CHRH1* [60], and *HTR1B* [58], just to cite a few preliminary results.

Understanding the disease and the complex effect of early and recent environment, even by stratifying further the given diagnosis, could provide a strong basis for precision medicine in the field of pharmaceutical treatment of psychiatric disorders.

## 6.5 Clinical Application of Precision Psychiatry

Precision psychiatry aims at developing clinical tools from machine learning models, such as the establishment of calculators, possibly integrated with electronic medical records, yielding precise calculated diagnostic and prognostic estimations and calculated treatment plans instead of the current subjective evaluation of clinical signs and trial-and-error search for effective medication [21]. At this point, some treatment response calculators have already been developed, for example, in the case of antidepressants, antipsychotics, or psychotherapy, although they have not been shown to be more effective than clinicians in making treatment recommendations [21]. While the application of artificial intelligence in psychiatry is slowly entering clinical practice, unfortunately, there are no guidelines on application. The International Network for Digital Mental Health has been established in order to ensure the implementation of digital innovations in psychiatric clinical practice with the aim of personalization, including the development of i-PROACH (Intelligent Platform for Research, Outcome, Assessment, and Care in Mental Health) which includes algorithms on digital phenotyping, genetic data, artificial intelligence tools, and a clinical decision support system [4, 62].

### 6.5.1 The Current State of Precision Psychiatry in Mood Disorders

Among psychiatric disorders currently, the largest burden both in terms of well-being and economic costs is associated with major depressive disorder [63]. In spite of this, mood disorders are frequently misdiagnosed or underdiagnosed and suboptimally treated, leading to self-medication with substances and to meagre outcomes including suicide, which is paralleled by the lack of objective tests and markers [24]. Although we have multimodal intervention options with pharmacological, psychotherapeutic, and biological or neuromodulation treatments including more than 40 different antidepressants, only about one-third of patients achieve remission with the first antidepressive medication; the majority require several trials before an effective antidepressant can be found, and one-third of patients is resistant to treatment with any currently available antidepressant medication [64, 65], the use of which is also limited by undesirable side effects leading to non-adherence to and discontinuation of treatment [4]. A longer duration of time spent in depression leads to an increased likelihood of residual functional impairments [20, 66].

Another difficulty hindering early intervention is determining the diagnosis and predicting the illness course. What cross-sectionally and initially appears as a depressive episode may be part of a bipolar disorder, with more than two-thirds of bipolar I patients initially misdiagnosed as unipolar. This leads to a significant delay between illness onset and diagnosis and also implies that in many cases, in the crucial earlier phases of the illness, bipolar patients will be treated with antidepressants without mood stabilizers with potentially harmful consequences [67, 68]. Thus, it is of key importance to develop models that predict not only the onset of depressive illness but also if the bipolar conversion can be expected, in addition to aiding in finding an antidepressant that will be effective in the case of the given patient.

Choice of antidepressant and diagnosis of mood disorders is currently based mostly on individual medical expertise and clinical skills, relying on sociodemographic and clinical characteristics such as individual symptom profiles, illness onset, previous treatment response, pharmacological history in the family, possible pharmacokinetic and pharmacodynamics features and interactions, and comorbid mental and physical disorders, but do not include objective criteria reflecting pathophysiological and neuro-

biological characteristics [4]. Individual genetic profile influences not only liability to mood disorders and symptomatic manifestations but also 42%–50% of antidepressant response and tolerability variance [4, 12, 69], although the effects are small and not consistently reproducible in part due to the complexity of mechanisms involved [70]. While in other fields of medicine identifying the most appropriate treatment based on the genetic profile is already happening, in psychiatry, especially as downstream effects of antidepressant molecular actions are not sufficiently understood, only a few individual genetic markers are known, and none of them is utilized in treatment decisions. From the pharmacodynamics perspective, the first studies focusing on predicting antidepressant response based on genetic data found that 5-HTTLPR and a TPH promoter polymorphism predicted fluvoxamine response [71]. These and several other polymorphisms implicated in later studies, such as variants HTR2A, BDNF, and COMT, are represented in commercial pharmacogenetic testing kits although their contribution to both antidepressant effects and side effects is not clearly supported [4]. So far rather candidate genes influencing pharmacokinetics, mostly restricted to the CYP450 superfamily, were considered and endorsed in prescribing guidelines, indicated in the labeling of the medication, and included in commercial pharmacogenetics testing kits [4, 72], with most attention focused on CYP2D6 and CYP2C19. Metabolizing groups and especially poor metabolizers (PM) and ultrarapid metabolizers (UM) may show evidence in association with both antidepressant response and side effects, with some evidence on the association between PM and UM status and clinical outcomes in the case of seven tricyclic antidepressants and moderate evidence due to the nonlinear relationship between plasma levels and clinical outcomes in case of certain SSRIs including citalopram, escitalopram, paroxetine, sertraline, and fluvoxamine [4]. While psychiatric genetics including antidepressant pharmacogenetics has moved from the candidate gene approach to GWAS, no clinical applications have come out of these studies so far due to limited power, difficulties in identifying the causal variants, and difficulty in identifying methods which can capture the poligenicity of antidepressant outcomes [4]. Novel approaches combining effects of functionally related genes in pathways, as well as combining risk alleles in polygenic risks scores, are needed to exploit pharmacogenetics in antidepressant choice [4].

In fact, considering the currently used information in antidepressant choice including consideration of symptomatic profile, compound profile, previous efficacy profile, family efficacy profile, medical and psychiatric comorbidity profile, metabolic profile, tolerability profile, family history, and individual preferences in itself would require complex algorithms to make choice more accurate and less subjective [12]. It is a question of whether this information should be only supplemented or completely replaced by biomarkers, as it appears that not only biological but also phenomenological features should be better characterized in order to predict both illness trajectory and therapeutic recovery [20, 73]. Using the STAR*D study sample and statistical approaches including machine learning, it has been established that several clinical variables such as ethnicity, recurrent illness, education, depression score, and PTSD predicted a therapy-resistant illness course and, more importantly, that the highest risk of treatment-resistant depression could be predicted based only on patient-reported data or measures [6, 74, 75]. A recent meta-analysis evaluating 20 previous studies employing machine learning models to predict therapeutic outcomes in unipolar and bipolar depression found a clinically relevant accuracy of 0.82 and concluded that those models which performed the best included multiple data types with genetic, neuroimaging, and such clinical predictors as the number of previous episodes, global functioning, symptom severity, anhedonia, anxiety, and sociodemographic variables such as education, employment, and income [20] and emphasized the importance of multiple data sources to increase the predictive accuracy of models [6].

There have been several attempts to identify biomarkers related to depression, and some of the earlier identified blood gene expression biomarkers were able to reflect changes in mood states including tracking response to psychotherapy [24, 76, 77] or suicidality [78–80]. Following up on such findings, a recent longitudinal study [24] focused on blood gene expression changes in association with self-reported mood states and in addition identified biomarkers based on previous animal and human studies; validated the top markers in an independent patient cohort with severe depression and mania, followed up by an analysis of the biological networks and pathways the identified biomarkers are involved in; and identified those biomarkers which are targeted by existing drug molecules and can thus be exploited for pharmacogenomic population stratification and measurement of treatment response. In the final step, the identified biomarker gene expression signatures were fed to connectivity map databases to identify non-psychiatric medications and nutraceutical compounds that can be repurposed for the prevention and treatment of depression [24]. The generated personalized patient report included an objective assessment of the depressive state, the future risk of severe depression, and diagnostic conversion to bipolarity, matching with existing psychiatric and non-psychiatric repurposable medications, and monitoring response. Notably, this study identified several nominally significant biomarkers which reproduced genes also implicated in related studies [81], and the majority of the identified top biomarkers were also reported in prior postmortem brain datasets as relevant to mood disorders as well as other psychiatric disorders including suicide (92%), stress (91%), aging (83%), and Alzheimer's disease (75%) [24]. Altogether this study provides a major step toward precision diagnosis and treatment of mood disorders [24] and toward our understanding that more sophisticated stratification of patients even within mood disorders is necessary to yield better clinical outcomes through the understanding that specific medications can be targeted only at smaller groups rather than the whole depressed patient population [17].

## 6.5.2 The Current State of Precision Psychiatry in Schizophrenia

Up to this point, the greatest portion (36.4%) of research in precision psychiatric modeling focused on psychoses with individual prediction models most intensively and extensively developed and investigated in the case of psychotic and especially schizophrenia spectrum disorders. This focused interest is a consequence of the lifelong and progressive nature of these disorders which, following an early onset, often lead to significant functional decline and thus are associated with a high personal, clinical, and societal burden, in parallel with a still limited neurobiological understanding of pathophysiological processes [22]. The onset of schizophrenia spectrum disorders is usually not sharp and clear-cut but in the majority of cases is preceded by gradual premorbid personality changes, followed by a prodromal phase or at-risk state with attenuated psychotic manifestations, before the full-blown disorder manifests. Therefore, it would be key to develop algorithms and identify markers to predict schizophrenia already at these stages. In the earlier phases, it would also be important to be able to differentiate other psychotic manifestations, including delusional disorders, drug-induced psychotic states, affective psychoses, or psychotic decompensations related to other psychotic conditions, from the early phases of schizophrenia, as most of these disorders do not have a progressive nature, require a different treatment approach, and are less stigmatizing. Predicting schizophrenia in presymptomatic people, and identifying medications, which, unlike currently used antipsychotics, could halt the neuroprogression of the disorder, would be the ultimate aim.

The appearance of a clinical staging model for psychosis, together with the emergence of the concept of clinical high-risk state for psychosis (CHR-O), with the associated need to stratify or personalize early intervention or preventive treatment for psychosis [82] prompted research into prognostic prediction models [22]. Currently, the major focus of precision psychiatry research in schizophrenia is improving the predictive accuracy of diagnostic conversion in at-risk subjects, accurately predicting symptomatic recurrences, and identifying subgroups of schizophrenic patients with more homogeneous characteristics that could be targeted by specific interventions [6].

## 6.6 Challenges, Obstacles, and Limitations in Precision Psychiatry

Moving toward precision psychiatry both in thinking and clinical application will pose multiple challenges of technical, ethical, and practical nature, and a number of issues, specific to psychiatry, will need to be addressed [4, 6].

Psychiatry has been traditionally subjective both in the conditions it deals with and, accordingly, in the methods employed to evaluate and treat those conditions, with a greater emphasis on the personal nature of the treatment relationship. Thus, a complete paradigm shift will be required not only from scientists but also from clinicians and patients, toward employing more objective measurements including neuroimaging, genetic and blood biomarkers, as well as trusting the decision to computers and algorithms which is highly alien both to the psychiatrist and psychiatric patients [2]. Implementing precision medicine tools would thus require giving up what several mental health professionals consider the essence of their work, switching from understanding the subjective state of patients to using technical data instead, and acquiring a new set of skills and knowledge and a novel approach to the therapeutic relationship will be necessary [6]. Equally important is to what extent patients are willing, instead of engaging in a supporting personal therapeutic relationship with clinicians, to provide data and information toward computers and calculators. It is also to be seen to what extent both clinicians and patients would be willing to accept and adhere to diagnostic and prognostic decisions and treatment recommendations made by models and calculators [22]. An important technical obstacle is that currently available imaging, pharmacogenetics, and other omics measures, at least at this point, are hard to translate into psychiatric treatment decisions [6].

Ethical issues already well-known, for example, in genetic testing will become more highlighted with the ability to artificially model liability to mental illness and prognosis even before the onset of symptoms, and determining access to treatment options based on prognosis is an unimaginable psychological burden. Self-determination would be another important question, especially pressing in case of disorders of the mind and psychological functions, concerning whether a psychiatric patient informed about novel possibilities of diagnostic and prognostic approaches will have the ability and possibility to freely exert their right to choose or refuse such options. Furthermore, active participation from the side of the patient is necessary for a complete evaluation involving novel instruments such as wearable continuous monitoring tools, or undergoing neuroimaging or genotyping procedures, which may be problematic in case of certain psychiatric disorders and symptoms, meaning some patients will have unfairly limited access [6]. Finally, psychiatric disorders are associated with heightened stigma, including both public and self-stigmatization, which may also impede implementation and exploit the full potential of precision psychiatry, both by limiting public health interest in precision psychiatry and by patients trying to avoid involvement [6, 83].

Cost-effectiveness is going to be a great obstacle in implementing precision psychiatry, as the majority of novel evaluation, diagnostic, and possible treatment methods will be fairly expensive and thus may be available to only a limited number of patients; thus, specific criteria will need to be established to determine access, which in general raises the question concerning how fair and sustainable precision medicine is [6]. At the same time, prediction models will hopefully shorten the time for appropriate diagnosis and decrease the number of treatment trials and also side effects which will in general increase the cost-effectiveness of treatment. Most importantly, earlier treatment will mean less functional loss and disability, as well as increased productivity for patients which will in the long term decrease both medical and societal costs.

We must also be aware of bias and inequalities related to the fact that the development of precision medicine is overrepresented in high-income countries. This runs several risks, beyond ignoring the needs of those living in less developed regions where people actually experience the greatest burden of mental disorders. The greatest danger is ignoring such important contributors to psychopathology in our prediction models which stem from more unfortunate conditions. This may significantly widen the already existing gap in mental health treatment [28].

Still, one of the greatest obstacles to precision psychiatry is the lack of replication and validation and implementation of already developed predictive models in real-world clinical practice [22].

## 6.7 Conclusion

In our current era, novel advances in genomics and proteomics as well as various technical and analytical developments finally allow us to build more detailed and precise models of neurobiological and neurochemical abnormalities underlying mental health and brain functional disorders. While precision psychiatry does not at this point promise a solution to the highly variable and unpredictable treatment response in psychiatry, and we still lack prediction models aiding sophisticated diagnosis, projecting illness course and prognosis, in the longer term, the novel approach and paradigm of precision psychiatry will increasingly contribute toward improving the outlooks and prognosis of mental health patients [17]. Building on such drivers as deep phenotyping and the integration of psychosocial, demographic, clinical and molecular data with the means offered by machine learning, earlier detection, accurate diagnosis, higher treatment efficacy, reduced adverse effects and costs, as well as reduced suffering and improved disease course and long-term function become attainable goals [6]. Currently, precision psychiatry approaches are on the rise, and while implementation at this point seems challenging, it will ultimately redefine and improve the standard of clinical care for mental patients. While psychiatry is significantly delayed compared to other medical disciplines in adopting and implementing precision medicine methods, this also carries several potential. We can proceed faster compared to other medical specialties as we can readily employ already established and tried technologies, and we can learn from past failures and successes of precision medicine [2].

However, even when focusing on novel scientific and technological advances revolutionizing psychiatry, we must keep in mind that one of the biggest challenges is that the psychiatric patient, and their psychological distress and suffering, will have to remain in the center and focus of care [17].

**Acknowledgments** This work was supported by the Hungarian Brain Research Program (Grants: 2017-1.2.1-NKP-2017-00002; KTIA_13_NAPA-II/14); the National Development Agency (Grant: KTIA_NAP_13-1-2013-0001); the Hungarian Academy of Sciences, Hungarian National Development Agency, Semmelweis University, and the Hungarian Brain Research Program (Grant: KTIA_NAP_13-2-2015-0001) (MTA-SE-NAP B Genetic Brain Imaging Migraine Research Group); the Thematic Excellence Program (Tématerületi Kiválósági Program, 2020-4.1.1.-TKP2020) of the Ministry for Innovation and Technology in Hungary, within the framework of the Neurology and Translational Biotechnology thematic programs of the Semmelweis University; the National Research, Development and Innovation Office, Hungary (2019-2.1.7-ERA-NET-2020-00005), under the frame of ERA PerMed (ERAPERMED2019-108); and the TKP2021-EGA-25 project (Ministry of Innovation and Technology of Hungary, National Research, Development and Innovation Fund, financed under the TKP2021-EGA funding scheme).

# References

1. Collins FS, Varmus H. A new initiative on precision medicine. N Engl J Med. 2015;372(9):793–5.
2. Fernandes BS, Williams LM, Steiner J, Leboyer M, Carvalho AF, Berk M. The new field of 'precision psychiatry'. BMC Med. 2017;15(1):80. https://doi.org/10.1186/s12916-017-0849-x.
3. Council NR. Toward precision medicine: building a knowledge network for biomedical research and a new taxonomy of disease; 2011.
4. Zanardi R, Prestifilippo D, Fabbri C, Colombo C, Maron E, Serretti A. Precision psychiatry in clinical practice. Int J Psychiatry Clin Pract. 2021;25(1):19–27.
5. Terry SF. Obama's precision medicine initiative. Genet Test Mol Biomarkers. 2015;19(3):113–4.
6. Manchia M, Pisanu C, Squassina A, Carpiniello B. Challenges and future prospects of precision medicine in psychiatry. Pharmgenomics Pers Med. 2020;13:127–40. https://doi.org/10.2147/PGPM.S198225.
7. Schumann G, Binder EB, Holte A, de Kloet ER, Oedegaard KJ, Robbins TW, et al. Stratified medicine for mental disorders. Eur Neuropsychopharmacol. 2014;24(1):5–50.
8. Whiteford HA, Degenhardt L, Rehm J, Baxter AJ, Ferrari AJ, Erskine HE, et al. Global burden of disease attributable to mental and substance use disorders: findings from the global burden of disease study 2010. Lancet. 2013;382(9904):1575–86.
9. Wittchen H-U, Jacobi F, Rehm J, Gustavsson A, Svensson M, Jönsson B, et al. The size and burden of mental disorders and other disorders of the brain in Europe 2010. Eur Neuropsychopharmacol. 2011;21(9):655–79.
10. Trautmann S, Rehm J, Wittchen HU. The economic costs of mental disorders: do our societies react appropriately to the burden of mental disorders? EMBO Rep. 2016;17(9):1245–9.
11. Raber J, O'Shea RD, Bloom FE, Campbell IL. Modulation of hypothalamic-pituitary-adrenal function by transgenic expression of interleukin-6 in the CNS of mice. J Neurosci. 1997;17(24):9473–80. https://doi.org/10.1523/jneurosci.17-24-09473.1997.
12. Serretti A. The present and future of precision medicine in psychiatry: focus on clinical psychopharmacology of antidepressants. Clin Psychopharmacol Neurosci. 2018;16(1):1–6. https://doi.org/10.9758/cpn.2018.16.1.1.
13. Fried EI, Nesse RM. Depression is not a consistent syndrome: an investigation of unique symptom patterns in the STAR* D study. J Affect Disord. 2015;172:96–102.
14. Stephan KE, Bach DR, Fletcher PC, Flint J, Frank MJ, Friston KJ, et al. Charting the landscape of priority problems in psychiatry, part 1: classification and diagnosis. Lancet Psychiatry. 2016;3(1):77–83.
15. Tamminga CA. Approaching human neuroscience for disease understanding. World Psychiatry. 2014;13(1):41.
16. Kendler KS. The nature of psychiatric disorders. World Psychiatry. 2016;15(1):5–12.
17. Roche D, Russell V. Can precision medicine advance psychiatry? Ir J Psychol Med. 2021;38(3):163–8. https://doi.org/10.1017/ipm.2020.79.
18. Quinlan EB, Banaschewski T, Barker GJ, Bokde ALW, Bromberg U, Buchel C, et al. Identifying biological markers for improved precision medicine in psychiatry. Mol Psychiatry. 2020;25(2):243–53. https://doi.org/10.1038/s41380-019-0555-5.
19. Insel TR. The NIMH research domain criteria (RDoC) project: precision medicine for psychiatry. Am J Psychiatr. 2014;171(4):395–7.

20. Lee Y, Ragguett RM, Mansur RB, Boutilier JJ, Rosenblat JD, Trevizol A, et al. Applications of machine learning algorithms to predict therapeutic outcomes in depression: a meta-analysis and systematic review. J Affect Disord. 2018;241:519–32. https://doi.org/10.1016/j.jad.2018.08.073.

21. Passos IC, Ballester P, Rabelo-da-Ponte FD, Kapczinski F. Precision psychiatry: the future is now. Can J Psychiatr. 2021;67 https://doi.org/10.1177/0706743721998044.

22. Salazar de Pablo G, Studerus E, Vaquerizo-Serrano J, Irving J, Catalan A, Oliver D, et al. Implementing precision psychiatry: a systematic review of individualized prediction models for clinical practice. Schizophr Bull. 2021;47(2):284–97. https://doi.org/10.1093/schbul/sbaa120.

23. Vieta E. Personalised medicine applied to mental health: precision psychiatry; 2015.

24. Le-Niculescu H, Roseberry K, Gill SS, Levey DF, Phalen PL, Mullen J, et al. Precision medicine for mood disorders: objective assessment, risk prediction, pharmacogenomics, and repurposed drugs. Mol Psychiatry. 2021;26(7):2776–804. https://doi.org/10.1038/s41380-021-01061-w.

25. Tretter F, Gebicke-Haerter PJ. Systems biology in psychiatric research: from complex data sets over wiring diagrams to computer simulations. In: Psychiatric disorders. Cham: Springer; 2012. p. 567–92.

26. Godlewska B, Browning M, Norbury R, Cowen PJ, Harmer CJ. Early changes in emotional processing as a marker of clinical response to SSRI treatment in depression. Transl Psychiatry. 2016;6(11):e957.

27. Shiroma PR, Thuras P, Wels J, Erbes C, Kehle-Forbes S, Polusny M. A proof-of-concept study of subanesthetic intravenous ketamine combined with prolonged exposure therapy among veterans with posttraumatic stress disorder. J Clin Psychiatry. 2020;81(6) https://doi.org/10.4088/JCP.20l13406.

28. Schumann G, Benegal V, Yu C, Tao S, Jernigan T, Heinz A, et al. Precision medicine and global mental health. Lancet Glob Health. 2019;7(1) https://doi.org/10.1016/s2214-109x(18)30406-6.

29. Bzdok D, Meyer-Lindenberg A. Machine learning for precision psychiatry: opportunities and challenges. Biol Psychiatry Cogn Neurosci Neuroimaging. 2018;3(3):223–30. https://doi.org/10.1016/j.bpsc.2017.11.007.

30. Fusar-Poli P, Hijazi Z, Stahl D, Steyerberg EW. The science of prognosis in psychiatry: a review. JAMA Psychiat. 2018;75(12):1289–97.

31. Abbott NJ, Patabendige AA, Dolman DE, Yusof SR, Begley DJ. Structure and function of the blood-brain barrier. Neurobiol Dis. 2010;37(1):13–25. https://doi.org/10.1016/j.nbd.2009.07.030.

32. Kealy J, Greene C, Campbell M. Blood-brain barrier regulation in psychiatric disorders. Neurosci Lett. 2020;726:133664. https://doi.org/10.1016/j.neulet.2018.06.033.

33. Pollak TA, Drndarski S, Stone JM, David AS, McGuire P, Abbott NJ. The blood-brain barrier in psy-

chosis. Lancet Psychiatry. 2018;5(1):79–92. https://doi.org/10.1016/s2215-0366(17)30293-6.

34. Erdő F, Denes L, de Lange E. Age-associated physiological and pathological changes at the blood–brain barrier: a review. J Cereb Blood Flow Metab. 2016;37(1):4–24. https://doi.org/10.1177/0271678X16679420.

35. Greene C, Hanley N, Campbell M. Claudin-5: gatekeeper of neurological function. Fluids Barriers CNS. 2019;16(1):3. https://doi.org/10.1186/s12987-019-0123-z.

36. Greene C, Kealy J, Humphries M, Gong Y, Hou J, Hudson N, et al. Dose-dependent expression of claudin-5 is a modifying factor in schizophrenia. Mol Psychiatry. 2018;23(11):2156–66.

37. Ray A, Tennakoon L, Keller J, Sarginson JE, Ryan HS, Murphy GM, et al. ABCB1 (MDR1) predicts remission on P-gp substrates in chronic depression. Pharmacogenomics J. 2015;15(4):332–9. https://doi.org/10.1038/tpj.2014.72.

38. Uhr M, Tontsch A, Namendorf C, Ripke S, Lucae S, Ising M, et al. Polymorphisms in the drug transporter gene ABCB1 predict antidepressant treatment response in depression. Neuron. 2008;57(2):203–9. https://doi.org/10.1016/j.neuron.2007.11.017.

39. Kuzman MR, Medved V, Bozina N, Grubišin J, Jovanovic N, Sertic J. Association study of MDR1 and 5-HT2C genetic polymorphisms and antipsychotic-induced metabolic disturbances in female patients with schizophrenia. Pharmacogenomics J. 2011;11(1):35–44. https://doi.org/10.1038/tpj.2010.7.

40. Geers LM, Pozhidaev IV, Ivanova SA, Freidin MB, Schmidt AF, Cohen D, et al. Association between 8 P-glycoprotein (MDR1/ABCB1) gene polymorphisms and antipsychotic drug-induced hyperprolactinaemia. Br J Clin Pharmacol. 2020;86(9):1827–35. https://doi.org/10.1111/bcp.14288.

41. Pedersen CB, Mors O, Bertelsen A, Waltoft BL, Agerbo E, McGrath JJ, et al. A comprehensive nationwide study of the incidence rate and lifetime risk for treated mental disorders. JAMA Psychiat. 2014;71(5):573–81.

42. Manchia M, Squassina A, Pisanu C, Congiu D, Garzilli M, Guiso B, et al. Investigating the relationship between melatonin levels, melatonin system, microbiota composition and bipolar disorder psychopathology across the different phases of the disease. Int J Bipolar Disord. 2019;7(1):1–7.

43. Van Ijzendoorn MH, Bakermans-Kranenburg MJ. Genetic differential susceptibility on trial: meta-analytic support from randomized controlled experiments. Dev Psychopathol. 2015;27(1):151–62.

44. Bakker J, Lieverse R, Menne-Lothmann C, Viechtbauer W, Pishva E, Kenis G, et al. Therapygenetics in mindfulness-based cognitive therapy: do genes have an impact on therapy-induced change in real-life positive affective experiences? Translational Psychiatry. 2014;4(4):e384.

45. Hudson JL, Lester KJ, Lewis CM, Tropeano M, Creswell C, Collier DA, et al. Predicting outcomes following cognitive behaviour therapy in child anxiety disorders: the influence of genetic, demographic and clinical information. J Child Psychol Psychiatry. 2013;54(10):1086–94.

46. Coleman JR, Lester KJ, Keers R, Roberts S, Curtis C, Arendt K, et al. Genome-wide association study of response to cognitive–behavioural therapy in children with anxiety disorders. Br J Psychiatry. 2016;209(3):236–43.

47. Keers R, Coleman JR, Lester KJ, Roberts S, Breen G, Thastum M, et al. A genome-wide test of the differential susceptibility hypothesis reveals a genetic predictor of differential response to psychological treatments for child anxiety disorders. Psychother Psychosom. 2016;85(3):146–58.

48. Rayner C, Coleman JR, Purves KL, Hodsoll J, Goldsmith K, Alpers GW, et al. A genome-wide association meta-analysis of prognostic outcomes following cognitive behavioural therapy in individuals with anxiety and depressive disorders. Transl Psychiatry. 2019;9(1):1–13.

49. Marx P, Antal P, Bolgar B, Bagdy G, Deakin B, Juhasz G. Comorbidities in the diseasome are more apparent than real: what Bayesian filtering reveals about the comorbidities of depression. PLoS Comput Biol. 2017;13(6):e1005487. https://doi.org/10.1371/journal.pcbi.1005487.

50. Carvalho AF, Sharma MS, Brunoni AR, Vieta E, Fava GA. The safety, tolerability and risks associated with the use of newer generation antidepressant drugs: a critical review of the literature. Psychother Psychosom. 2016;85(5):270–88. https://doi.org/10.1159/000447034.

51. Juruena MF, Bocharova M, Agustini B, Young AH. Atypical depression and non-atypical depression: is HPA axis function a biomarker? A systematic review. J Affect Disord. 2018;233:45–67. https://doi.org/10.1016/j.jad.2017.09.052.

52. Toro R, Poline J-B, Huguet G, Loth E, Frouin V, Banaschewski T, et al. Genomic architecture of human neuroanatomical diversity. Mol Psychiatry. 2015;20(8):1011–6.

53. Quinlan E, Barker E, Luo Q, Banaschewski T, Bokde A, Bromberg U, Consortium, I.et al. Peer victimization and its impact on adolescent brain development and psychopathology. Mol Psychiatry. 2018;25(11):366–76.

54. Quinlan EB, Cattrell A, Jia T, Artiges E, Banaschewski T, Barker G, et al. Psychosocial stress and brain function in adolescent psychopathology. Am J Psychiatry. 2017;174(8):785–94.

55. Hullam G, Antal P, Petschner P, Gonda X, Bagdy G, Deakin B, et al. The UKB envirome of depression: from interactions to synergistic effects. Sci Rep. 2019;9(1):1–19.

56. Nanni V, Uher R, Danese A. Childhood maltreatment predicts unfavorable course of illness and treatment outcome in depression: a meta-analysis. Am J Psychiatr. 2012;169(2):141–51.

57. Xu Z, Zhang Z, Shi Y, Pu M, Yuan Y, Zhang X, et al. Influence and interaction of genetic polymorphisms in catecholamine neurotransmitter systems and early life stress on antidepressant drug response. J Affect Disord. 2011;133(1–2):165–73.

58. Xu Z, Zhang Z, Shi Y, Pu M, Yuan Y, Zhang X, et al. Influence and interaction of genetic polymorphisms in the serotonin system and life stress on antidepressant drug response. J Psychopharmacol. 2012;26(3):349–59.

59. Mandelli L, Marino E, Pirovano A, Calati R, Zanardi R, Colombo C, et al. Interaction between SERTPR and stressful life events on response to antidepressant treatment. Eur Neuropsychopharmacol. 2009;19(1):64–7.

60. Keers R, Uher R. Gene–environment interaction in major depression and antidepressant treatment response. Curr Psychiatry Rep. 2012;14(2):129–37.

61. Keers R, Uher R, Huezo-Diaz P, Smith R, Jaffee S, Rietschel M, et al. Interaction between serotonin transporter gene variants and life events predicts response to antidepressants in the GENDEP project. Pharmacogenomics J. 2011;11(2):138–45.

62. Maron E, Baldwin DS, Balõtšev R, Fabbri C, Gaur V, Hidalgo-Mazzei D, et al. Manifesto for an international digital mental health network. Digital Psychiatry. 2019;2(1):14–24.

63. Vos T, Barber RM, Bell B, Bertozzi-Villa A, Biryukov S, Bolliger I, et al. Global, regional, and national incidence, prevalence, and years lived with disability for 301 acute and chronic diseases and injuries in 188 countries, 1990–2013: a systematic analysis for the global burden of disease study 2013. Lancet. 2015;386(9995):743–800.

64. Trivedi MH, Rush AJ, Wisniewski SR, Nierenberg AA, Warden D, Ritz L, et al. Evaluation of outcomes with citalopram for depression using measurement-based care in STAR* D: implications for clinical practice. Am J Psychiatr. 2006;163(1):28–40.

65. Souery D, Serretti A, Calati R, Oswald P, Massat I, Konstantinidis A, et al. Switching antidepressant class does not improve response or remission in treatment-resistant depression. J Clin Psychopharmacol. 2011;31(4):512–6.

66. Trivedi MH, Morris DW, Wisniewski SR, Lesser I, Nierenberg AA, Daly E, et al. Increase in work productivity of depressed individuals with improvement in depressive symptom severity. Am J Psychiatr. 2013;170(6):633–41.

67. Hirschfeld RM, Vornik LA. Perceptions and impact of bipolar disorder: how far have we really come? Results of the national depressive and manic-depressive association 2000 survey of individuals with bipolar disorder. J Clin Psychiatry. 2003;64(2):161–74.

68. Berk M, Dodd S, Callaly P, Berk L, Fitzgerald P, De Castella A, et al. History of illness prior to a diagnosis of bipolar disorder or schizoaffective disorder. J Affect Disord. 2007;103(1–3):181–6.

69. Tansey K, Rucker JJ, Kavanagh D, Guipponi M, Perroud N, Bondolfi G, et al. Copy number variants and therapeutic response to antidepressant medication in major depressive disorder. Pharmacogenomics J. 2014;14(4):395–9.

70. Border R, Johnson EC, Evans LM, Smolen A, Berley N, Sullivan PF, et al. No support for historical candidate gene or candidate gene-by-interaction hypotheses for major depression across multiple large samples. Am J Psychiatr. 2019;176(5):376–87.

71. Serretti A, Smeraldi E. Neural network analysis in pharmacogenetics of mood disorders. BMC Med Genet. 2004;5(1):1–6.

72. Fabbri C, Zohar J, Serretti A. Pharmacogenetic tests to guide drug treatment in depression: comparison of the available testing kits and clinical trials. Prog Neuro-Psychopharmacol Biol Psychiatry. 2018;86:36–44.

73. Rosenblat JD, Lee Y, McIntyre RS. Does pharmacogenomic testing improve clinical outcomes for major depressive disorder? A systematic review of clinical trials and cost-effectiveness studies. J Clin Psychiatry. 2017;78(6):720–9.

74. Nie Z, Vairavan S, Narayan VA, Ye J, Li QS. Predictive modeling of treatment resistant depression using data from STAR* D and an independent clinical study. PLoS One. 2018;13(6):e0197268.

75. Perlis RH. A clinical risk stratification tool for predicting treatment resistance in major depressive disorder. Biol Psychiatry. 2013;74(1):7–14. https://doi.org/10.1016/j.biopsych.2012.12.007.

76. Le-Niculescu H, Kurian S, Yehyawi N, Dike C, Patel S, Edenberg H, et al. Identifying blood biomarkers for mood disorders using convergent functional genomics. Mol Psychiatry. 2009;14(2):156–74.

77. Kéri S, Szabó C, Kelemen O. Blood biomarkers of depression track clinical changes during cognitive-behavioral therapy. J Affect Disord. 2014;164:118–22.

78. Niculescu A, Levey D, Phalen P, Le-Niculescu H, Dainton H, Jain N, et al. Understanding and predicting suicidality using a combined genomic and clinical risk assessment approach. Mol Psychiatry. 2015;20(11):1266–85.

79. Levey DF, Niculescu EM, Le-Niculescu H, Dainton H, Phalen PL, Ladd TB, et al. Towards understanding and predicting suicidality in women: biomarkers and clinical risk assessment. Mol Psychiatry. 2016;21(6):768–85.

80. Niculescu AB, Le-Niculescu H. Dissecting suicidality using a combined genomic and clinical approach; 2017.

81. Howard DM, Adams MJ, Clarke TK, Hafferty JD, Gibson J, Shirali M, et al. Genome-wide meta-analysis of depression identifies 102 independent variants and highlights the importance of the prefrontal brain regions. Nat Neurosci. 2019;22(3):343–52. https://doi.org/10.1038/s41593-018-0326-7.

82. Davies C, Cipriani A, Ioannidis JP, Radua J, Stahl D, Provenzani U, et al. Lack of evidence to favor specific preventive interventions in psychosis: a network meta-analysis. World Psychiatry. 2018;17(2):196–209.

83. Carpiniello B, Pinna F. The reciprocal relationship between suicidality and stigma. Front Psych. 2017;8:35.

# Precision Public Health Perspectives

7

Maria Josefina Ruiz Alvarez ⓘ

M. J. R. Alvarez (✉)
Research Coordination and Support Service,
Istituto Superiore di Sanità, Rome, Italy
e-mail: mariajose.ruizalvarez@iss.it

**What Will You Learn in This Chapter?**

By 2030, the primary focus of healthcare wants to shift to the optimization of healthcare systems with an integration of the societal perspective. In this way, the economic sustainability and societal benefits that PM can offer can help integrate new risk-sharing processes and PM approaches for the entire life cycle. From this perspective, this chapter illustrates the definition of public health, population health, and their link with personalized medicine. The objective is to describe the most significant points of this ecosystem such as risk definition, patient stratification, health promotion, and disease prevention strategies of particular value for aging societies. The chapter also describes the equity impact and citizen empowerment from a personalized (precision) perspective.

**Rational and Importance**

Personalized medicine looks to incorporate new technologies into healthcare, supported by data collection, integrating clinical phenotypes, and biological information from imaging to laboratory tests and health records. The analysis of data identifies, prevents, and treats better individual patient diseases. Indeed, the application of this approach to public health can improve the management and prevention of both communicable, or infectious diseases, and non-communicable or chronic diseases, starting from each person and reaching the whole community. In this chapter, the emerging field of "personalized" public health is reflected through the polygenic studies, epigenetics, omics, and citizens' involvement. Obstacles in public policy include uncertain regulatory requirements, insufficient insurance reimbursement for diagnostic tests linked to preventive care, incomplete legal protections to prevent genetic discrimination, and the lack of a comprehensive technology system.

## 7.1 Population Health Versus Public Health

Currently, there is some debate about the difference between population health and public health. For the scope of this chapter, both terms are considered in the same folder (gaps and challenges), with a previous clarification of both definitions.

*Health system definition*: The key structural arms of the Public Health system include the configuration and the design of health services influencing the way in which services are delivered (including health workers and mechanisms of governance and administrative decision making); the aspects of the behaviour (workers and population), performance, health facility and the nature of participation on the appropiate section; and, the last arm is the effect of the interventions provided by the health system, including facilities and personnel [1].[1]

Components:

1. Structure: design of health services that influence the way in which services are delivered, from the health workers to mechanisms of governance and administrative decision-making
2. Processes: aspects of the behaviour (workers and population) or performance or health facility and the nature of participation on the part of people it serves
3. Outcomes: result from the interventions provided by the health system, the facilities, personnel, and the actions of the targets of the interventions [1]

*Primary care*: It guarantees person-focused care over time to a defined population, accessibility to facilitate first care when needed, comprehensiveness, and coordination, such that it integrates all care facets (wherever received) [1].

*Population health definition*: This term refers to the health outcomes of a group of individuals, including the distribution of such outcomes within the group, including health outcomes, patterns of health determinants, and policies and interventions that link these two [2]. In other words, it focuses on understanding the factors influencing population health over lifetimes and measures occurrences of certain problems.

*Public health definition*: it is defined as the art and science of preventing disease, prolonging life, and promoting health through the organized efforts of society[2]. Both population and public

health definitions work to improve health in the public itself [3].

## 7.2 Introduction to Public Health (PH) and Personalized Medicine (PM)

One of the initial definitions of public health (PH) was the science and art of preventing disease, prolonging life, and promoting physical health and efficiency. They will ensure every individual in the community has a standard of living adequate for health maintenance. This term includes organized community efforts for the sanitation of the environment, the control of community infections, the individual education in principles of personal hygiene, the organization of medical and nursing services for the early diagnosis and preventive treatment of disease, and the development of social machinery [4].

Later on, concepts such as "protect and improve" were added to the definition: "the science and art of preventing disease, prolonging life and promoting, protecting and improving health through the organized efforts of society." Indeed, the link between health and mortality was reflected, emphasizing the need to eliminate health inequalities [5].

Both definitions focused on improving health through society-wide measures like vaccinations, the fluoridation of consuming water, or policies such as the mandatory seatbelt and non-smoking laws. In this way, these terms are linked to the wider definition of health, found in the preamble of the constitution of the World Health Organization (1948), where health is referred to as "a state of complete physical, mental and social well-being and not merely the absence of disease" [6].

Nevertheless, the concept has changed over the years due to changes in the health status of the population and health-determining situations. Currently, PH is the science of protecting and improving the health of people and their commu-

---

[1]WHO adopted definition https://www.euro.who.int/en/health-topics/Health-systems

[2]WHO adopted definition https://www.euro.who.int/en/health-topics/Health-systems/public-health-services/public-health-services

nities; the status of health is achieved by promoting healthy lifestyles, researching disease and injury prevention, and detecting, preventing, and responding to infectious diseases.

The introduction of the discussion on population health term follows the understanding that policies and higher-level interventions are crucial in determining health together with genetics, levels of activity, nutritional intake, and other individual behaviors.

There are several definitions of personalized medicine and precision medicine already described in this book. Before 1990, patient management followed sociological, educational, and psychological bases. Then, the approach changed following the personalized medicine (interchangeably in this book with precision medicine) implementation, which has become more and more common. Currently, it includes mainly the term "genetic" and new biomarkers and their application in pharmacotherapy, molecularly targeted therapies in oncology, and the application of novel therapeutic agents. In 2015, in his State of the Union Address, President Barack Obama opened the door to the PM initiative that included "delivering the right treatments, at the right time, every time to the right person."[3]

In the context of this chapter, the definition of Horizon 2020 Advisory Group is applied: Personalized medicine is "a medical model using the characterization of individuals' phenotypes and genotypes (e.g. molecular profiling, medical imaging, lifestyle data) for tailoring the right therapeutic strategy for the right person at the right time, and/or to determine the predisposition to disease and/or to deliver timely and targeted prevention."[4]

Nowadays, public health has to find a balance among individualized approaches that focus on diseased individuals and on population-based preventive programs and health promotion that consider the behavioral, environmental, and social determinants of health.

## 7.3 Impact of Personalized Medicine in Public Health

Nowadays, medicine is moving from a reactive to a proactive discipline. Adapting to these changes, WHO developed a framework to promote an understanding of the attributes and objectives to strengthen a health system. It is useful to identify gaps for appropriate health interventions focused on an experienced health workforce; essential medicines, vaccines, health products; health information; and service delivery including health facilities, centers, clinics, and hospitals. In this way, a health system is effective if it has the ability to provide the ten essential public health functions defined by the WHO: surveillance; response to emergencies; health protection; health promotion; disease prevention; governance; workforce; finance; communication and social mobilization; and research.

For this reason, the P4 medicine, which includes predictive, preventive, personalized, and participatory medicine, needs to integrate the population/public perspective (5P) into each of the other four components [7].

Population perspective merges predictive medicine into the ecological model of health, applies population screening to preventive medicine, uses evidence-based practice (best examples) to personalized medicine, and supports participatory medicine with the three core functions of public health: assessment, policy development, and assurance.

Key elements of the personalized public integration are the right balance between "premature translation," leading to increased healthcare costs and potential for harm, and "lost in translation," leading to exacerbation of social, economic, and health disparities [7]. There is a clear need to evaluate the benefits, harms, and costs of person-

[3]Background and links related to personalized medicine initiative in the USA available from URL: https://www.genome.gov/about-genomics

[4]Background, conference reports, publications and links related to personalise medicine in EU available from URL: https://ec.europa.eu/research/health/index.cfm?pg=policy&policyname=personalised

alized interventions compared to the already existing ones.

One described example is the screening for prostate cancer [8]: the prostate-specific antigen (PSA) detects many cases of asymptomatic prostate cancer. The issue is that most asymptomatic cancers detected by PSA screening seem not to be able to progress or affect life span. However, this diagnostic involves serious treatment with surgery, radiation, or other therapy. The consequence is a loss of quality of life and higher societal investment.

There are also available commercially genomic risk tests (multiple single nucleotide polymorphism (SNP) profiles). These tests, sold directly to consumers, do not have a full clinical validation or formal assessments of benefits and harms, or the involvement of healthcare providers, with the evident issue [9].

Finally, the promise of molecular biomarkers, which can offer data to estimate the transition from health to disease, is frequently not used based on incomplete evidence. They need to be strictly evaluated for their potential benefits and harms at both the individual and population levels [10].

Subsequently, applying a personalized public health perspective, tests should be prioritized for validation based on principles of population screening, such as disease burden and the effectiveness and acceptability of interventions.

Noteworthy, nutrition is another example. It is now evident that nutrition participates in human development to ensure life expectancy and well-being, and it is not only related to food transformation into energy.

Moreover, according to the emerging health outlines, food intake should be assessed in relationship with social, safety, and sustainable dimensions. In this context, the global public health perspective and the precision-personalized nutrition paradigm are complementary and should be harmonized.

Indeed, precision nutrition considers factors involved in global quality of life and metabolic well-being depending not only on the genotype but also on the dietary intake and associated healthy lifestyles as well as environmental factors [11].

The current challenge for the healthcare system is to shift from a reactive healthcare system to a personalized health approach and from episodic and acute care models (where individuals presenting some symptoms receive a similar treatment) to the use of more individual care, predictive, and preventive tools for stratification of at-risk individuals. This stratification can facilitate the intervention before the onset of symptoms or identify the risk before symptoms appear.

Currently, the health system faces up to the coordination for implementing this new vision and the effective initial economic investments.

The recent pandemic is overwhelming the health system around the world, already with an ongoing increasing burden of health assistance and social needs mainly due to the ageing of the population, the health workforce shortages[5, 6] as well as other neurodegenerative or rare diseases. This complexity of elder patients is mainly due to chronic conditions of multi-morbidity associated to the rising burden of preventable, caused by risk factors such as tobacco, alcohol, and obesity. As example, in EU Member States Public, spending on health and long-term care is gradually rising and continuous in this direction. In 2014, €1.39 trillion has been the EU-28's total healthcare expenditure (10% of the EU's GDP). This is expected to increase to 30% by 2060. These trends are a hard problem for the sustainability of worldwide healthcare systems.[7]

Subsequently, and for easy comprehension, the impact of PM on public health can be grouped into two diverse impacts, direct and indirect.

- Direct impact: It includes the direct effects of new technologies on the reorganization of the public health system: mHealth, Internet of Things (IoT), artificial intelligence (AI),

[5]https://www.axios.com/2022/05/24/the-health-care-workforce-shortage-problem

[6]https://www.who.int/health-topics/health-workforce#tab=tab_1

[7]https://ec.europa.eu/eurostat/statistics-explained/index.php?title=Healthcare_expenditure_statistics

imaging, data sharing, new -omics technologies, and the health technology assessment.

- Indirect impact: It includes all indirect effects of genetic information on preventive applications, diagnostic diseases, and targeted therapies.

## 7.4 Direct Impact

**mHealth** The use of applications and/or mobile-connected devices for supporting medical and public health practices is defined as mobile health (mHealth). Mobile health can play a positive role in various domains of health (well-being, prevention, care, monitoring or surveillance of diagnosed diseases, etc.) and in the healthcare system as a whole. To face the current financial difficulties of our health system, treating the patient at home (de-hospitalization) can be a solution. In this sense, the pandemic has accelerated the use of telemedicine and teleconsultation services.

mHealth has two kinds of software programs. One works as a medical device (SaMD), performing medical functions through software installation on generic devices such as tablets or smartphones. The second is a software program included in a medical device (SiMD) and a mHealth application as interface that interacts with a material medical device (MD). Both SaMDs and SiMDs must comply with the regulatory frameworks established by national and international authorities for marketing, quality, the safety of use, usability, and data security. Two international organizations, the World Health Organization (WHO) and the International Medical Device Regulators Forum (IMDRF)[8], have been developing regulations for MDs in collaboration with a group of Member States. The IMDRF proposes strategies, policies, and orientations for the deployment of MDs under different involved stakeholders' visions [12]. The 2018 World Health Assembly [13] adopted a resolution to develop a global strategy on mHealth to support efforts toward universal health coverage (2020–2025).

Material MDs have long been required to undergo certification by national and international regulatory systems. By contrast, the requirement for certification of SaMDs, particularly mHealth applications, was implemented only recently. Regulation (EU) 2017/745 on medical devices becomes applicable in the European Union in May 2021.[9] This MDR 2017/745 regulation includes a detailed list of safety and performance requirements that all medical devices placed on the market in Europe must comply with in order to guarantee a high level of quality, and it has been extended to a device for non-medical purposes. One of the objectives of the new EU regulation is the creation of a complete European bank of Medici devices – EUDAMED (European Databank for Medical Devices) – which is not only used for cataloging and medical devices but also for monitoring the life cycle. The new bank of EUDAMED will increase the transparency and the coordination of the information on its devices and represents an important openness to citizen engagement. If it becomes data interoperable, harmonized, adapted to international regulation, and under quality evaluation, the mHealth can become an important keystone of a successful personalized public health system.

*Artificial intelligence*: It has had an increasing role in the healthcare revolution and has shown great potential in developing effective prevention intervention strategies such as the prevention of HIV infection (see Chap. 13, Personalized Medicine in Infectious Diseases). With the aim to benefit from AI, the transformation of the market including new legal and ethical frameworks must be considered by public policy. In this direction, the European Alliance for Industrial Data, Edge, and Cloud was launched in 2021 by the European

---

[8]IMDRF. Work items. IMDRF, 2020, available from URL: http://www.imdrf.org/workitems/work.asp

[9]Background and links related to medical device regulation, available from URL: https://www.ema.europa.eu/en/news/medical-device-regulation-comes-application

Commission[10] with the aim to ensure Europe's competitiveness in the research and deployment of AI and to follow the associated social, economic, ethical, and legal. This Alliance has two main objectives, strengthening the position of the EU industry on cloud and edge technologies and meeting the needs of EU businesses and public administrations that process sensitive categories of data.

*Imaging*: The innovation in imaging has improved the diagnosis of disease and at the same time has been useful in different population-based screening such as for breast, lung, and prostate cancer. For example, on breast screening guidelines, the recommendations were based on age (following the age incidence) rather than on genetic factors. However, there are predisposing genetic variants with a polygenic risk score of breast cancer. New guidelines combine molecular testing with the current approach, with no clear consensus on how to lead with the different tiers of cancer screening risk and the knowledge and response of the public.

*Data production and data sharing*: Data from multiples sources and disciplines that need integrated solutions to enable cross-border data exchange; standardization for data sharing and analysis; and promote legal, organizational, semantic, and technical interoperability

*Machine learning*: The explosion of new concepts such as machine learning (e.g., transfer learning, distance metric learning, semi-supervised learning, structured ML meta-learning, multiview learning, and generative models), data processing techniques (e.g., new dimension reduction approaches, outlier removal methods, data augmentation techniques), and model validation methods (e.g., bootstrapping), if standardized, may facilitate follow-up research and subsequently public health implementation. Then, guidelines on the best approach to do available public knowledge and data sources can per-

mit to take advantage of previous and new biomarkers and create robust and interpretable biomarker models, always with the integration of specific expertise in data analysis and regulatory and legal fields.

*Electronic medical records (EMRs)*, based on *big data technology*, are improving predictive modeling, clinical decision support, and safety surveillance. EMR system starts as a small data repository of a patient receiving care under a given healthcare system (i.e., hospital, clinic, etc.) and offers clinicians a clear vision of patient care. The creation of a "big data" repository with the combination of other data sources facilitates data sharing with others. In this way, EMRs open all available information while determining diagnosis and patient prognosis. It would be possible to ensure the best and most timely treatment decisions on an individual basis. Indeed, the parallel development of training strategies is requested to support health workers and to ensure the implementation of innovation will be sustainable.

These technological developments involved directly the health industry. In this way, the PM development is also modifying the approach to the health industry as it operates to the benefit of the patient. Patient engagement is required for the health system decisions, so both the health system and health industry focus [14] on a user-centered approach "human-centered design (HCD)." This approach is based on principles such as the inclusion of the entire user experience. Users are involved during design and development, design that is guided by a user-centered assessment. Multidisciplinary skills and perspectives (e.g., doctor, nurse, citizen, designer) are involved interactively in the design process until satisfactory data results.

*Technological innovation* can help de-hospitalization, which improves the quality of care, thanks to medical devices. Thanks to these technologies, the doctor has an updated "image" of the patient's state and the actions already carried out, and inconvenience due to the movement of sick and disabled people, the social cost both for the patient and for the family members accompanying him, and the public and private cost of health care are reduced.

---

[10] Background and links related to the European Alliance for Industrial Data, Edge and Cloud 2021, available from URL: https://ec.europa.eu/growth/industry/strategy/industrial-alliances/european-alliance-industrial-data-edge-and-cloud_en

Indeed, new digital technologies such as wearable devices and interconnected products according to the Internet of Things (IoT) paradigm help maintain certain independence of patients, ensuring a valid and less expensive alternative to institutionalized care.

Each device can store and process the information on the network independently but also communicate with other devices belonging to the network and facilitate remote monitoring.

All are supported by robotics, artificial intelligence, and multidisciplinary teams composed of researchers in health, architecture, design, psychology, environment, and geological sciences and, of course, patients. Patients and general practitioners need to be involved as it has been demonstrated to have huge influence over daily decisions So, collaboration among all levels is needed to achieve changes in clinical practice, changes aimed at optimizing care and treatments for patients and their caregivers and prevention for all citizens.

*Internet of medical things (IoMT)* includes medical devices connected to a facility or healthcare provider via the Internet. Devices are able to generate, collect, analyze, and transmit health data such as smartphones and health apps, simple wearable devices, tools for remote patient monitoring, infusion pumps, drug tracking systems, specialized tools for monitoring, and medical equipment. Their impact on patient health management is increasing including diagnostics, bioinformatics collection, data sharing, rapid analysis, and timely therapy decisions. The impact of IoMT is higher and higher. Since IoMT started [15], it is growing exponentially to 10 billion connected IoT devices at present with a predicted increase to about 25 billion by 2025 [16]. In the current pandemic, intercommunication difficulties have been resolved using remote monitoring, telemedicine, robotics, sensors, etc. However, mass adoption seems challenging due to factors like privacy and security of data, management of a large amount of data, scalability, upgrade, etc., considering the start of economic involvement in the healthcare system.

*Health technology assessment*: An evidence-based process that enables competent authorities to determine the relative effectiveness of new or existing technologies, which in turn empowers national health authorities to make pricing or reimbursement decisions within health insurance. New rules for all these innovative health technologies and prevention and treatment methods are needed. So, in December 2021, the Regulation on Health Technology Assessment (HTA)[11] has been adopted. The Regulation will also ensure the efficient use of resources, strengthen the quality of HTA across the EU, save national HTA bodies and industry from duplicating their efforts, reassure businesses, and ensure the long-term sustainability of EU HTA cooperation. The new regulation, fully effective in January 2025, introduces a permanent framework. It will make it possible to unify procedures, work on joint and centralized clinical evaluations, promote unified scientific consultations, improve the identification of emerging health technologies, and favor voluntary cooperation mechanisms. This regulation includes the recommendation of patient input inclusion in all the regulatory decisions on the use of health technologies. One important point is the intention to facilitate faster authorizations, with centralized and more compatible criteria between countries, market unity, and, finally, more equitable access for all Europeans to innovation. So, Article 13 describes a joint clinical evaluation model through the collaboration of all countries and establishes that the Member States "shall pay due attention to the joint clinical evaluation report when carrying out a national HTA on health technology." In addition, for each national evaluation (undergone joint clinical evaluation), Member States will provide information on this development in their respective national processes. Indeed, each country will be able to complement the joint clinical evaluation with additional clinical analyses that may be necessary for their national regulations.

---

[11] https://ec.europa.eu/commission/presscorner/detail/en/ip_21_6771

## 7.5    Indirect Impact: Genetic Information on Preventive Applications, Diagnostic Disease, and Targeted Therapies

The appropriate function of health systems is in the pursuit of a standardized and rapid flow of digital information, including genomic, clinical outcome, and requested data. It will be feasible to drive treatments tailored to individuals' genetic structures, prescriptions could be analyzed in advance for likely effectiveness, and researchers will be able to study clinical data in real time to determine success.

Regarding this needed genetic information, important human genome map initiatives (Personalized Medicine Initiative in the USA,[12] 100000 Genomes Initiative in the UK[13], the Million European Genomes Alliance in Europe, and the Beyond 1 Million Genomes (B1MG) project[14]) have been launched worldwide. These initiatives include correlated populations' genetic information, environment, lifestyle, and clinical data. These combinations of information will help find lines of prevention and prediction and target better the treatment, and the health system can economize using therapies that will not be effective for a specific patient.

We need to consider that the advances in the field of genomics have led to substantial reductions in the cost of genome sequencing. The National Human Genome Research Institute (NHGRI) has carefully estimated the cost of whole exosome sequencing, and it is now less than $1.000. It is decreasing and is more accurate

than the first human genome sequenced in 2001, which costs $95 million.[15]

Another important genetic data application is the surveillance and identification of genetic relatedness in outbreaks. For example, in a study of extensively drug-resistant tuberculosis (XDRTB), investigators used targeted and whole-genome sequencing to account for the geographic distribution of XDR-TB strains [17]. This study shows that the combination of PM tools with previous epidemiologic methods may rise to disease mapping and lead health policy decisions.

The other example is the actual advance of genomic surveillance in Africa. The continent is getting close to sequencing up to 50,000 genomes in 2021, thanks to investment and capacity building since the beginning of the pandemic, with successful implementation in South Africa, Angola, Nigeria, and Kenya and starting in Botswana. KRISP and CERI (working with WHO and Africa Centres for Disease Control and Prevention (Africa CDC)) are able to do genomic sequencing for many African countries. The most important is the share protocols with countries, as well as train technicians to boost their capacity.[16]

Personalized approach is also useful for improving the classification of diseases: Usually, cancer was divided into its histological subtypes and clinical phenotypes. Nowadays, molecular testing reclassifies subtypes providing a more precise classification of diseases, for example, with the abnormalities on the surface of the cancer cell such as the presence of the epidermal growth factor receptor (EGFR) in lung cancer, and compares its phenotype or histological group and the new targeted therapies [18]. Rare diseases can use the PM to analyze the hereditary condition and relate specific mutations with clinical phenotype. In this direction, molecular profiling can provide prevention strategies and prognosis information and drive treatment strategies. However, in some diseases without treatment options, a patient can receive early symptomatic and supportive care. The challenge

---

[12]Background and links related to personalised medicine initiative in the USA available from URL: to https://www.genome.gov/about-genomics

[13]Background and links related to 100000 Genomes Initiative in the UK available from URL: https://www.genomicsengland.co.uk/about-genomics-england/the-100000-genomes-project/

[14]Background and links related to Million European Genomes Alliance in Europe available from URL https://digital-strategy.ec.europa.eu/en/policies/1-million-genomes; Beyond 1 Million Genomes (B1MG) project https://b1mg-project.eu/

[15]https://www.genome.gov/

[16]https://www.afro.who.int/news/why-genomic-sequencing-crucial-covid-19-response

for public health is that a more precise stratification of disease can increase the financial burden with little clinical benefit for overall population health. This is the case when the molecular classification of one disease fragments the management of this disease with more health services involved and more difficulty to apply a specific treatment [19].

*Molecular profiling and biomarkers* can identify persons at high risk of developing a disease, mainly among family members, and save an unaffected member from unnecessary routine screening procedures. One biomarker is more reliable than a clinical marker in predicting disease, for example, on family history of cancer, diabetes, or heart disease. However, it is not clear whether a person positive for genetic predisposition is able to decrease the risk through behaviors or lifestyle changes. In the same way, people can feel like not being able to decrease the risk for a disease or for addiction once the risk has been detected. On the other hand, biomarkers are sometimes difficult to validate in clinical settings, and it remains a lack generalizability. A recently systematic review regarding biomarkers for patient stratification illustrates how successful clinical biomarker translation is really providing applicable information for the design of new health public programs. However, it is far needed to guide clinicians involved in biomarker discovery with standard guidelines on methodologies for omics biomarker discovery [20].

*The polygenic risk score* (PGS) is the process by which people can learn about their risk of developing a specific disease. This score is based on the total number of changes in either one or many of their genes related to the disease across different populations, regularly coupled with environmental factors[17]. In general, the GWAS (genome-wide association study) estimates the polygenic risk as the result of the sum of risk alleles that an individual has, weighted by the risk-allele effect sizes on the phenotype. This estimation is considered a "relative risk" for a disease because it is independent of the baseline or timeframe for the progression of a disease. In addition, the clinical practice shows that the PGS needs to be integrated with other risk algorithms using environmental factors. Accordingly, people with polygenic high-risk percentage scores should discuss this risk with their medical doctor and or genetic counsellors for further health assessments.

**Epigenetics Biomarkers** Epigenetics is the study of altered gene expression without change in base pairs. The PGS with other omics data and environmental data for predicting programs will facilitate the implementation of PPH. Epigenetics biomarkers are any mark or altered epigenetic mechanism that:

- Can be measured in the body fluids or tissues
- Defines a disease (detection)
- Predicts the outcome of disease (prognostic)
- Responds to therapy (predictive)
- Monitors responses to therapy or medication (therapy monitoring)
- Predicts risk of future disease development (risk)

One example is the D-methylation changes described in neurodegenerative and neuropsychiatric diseases [21]. However, the adaptation of new technologies and methods will increase the adoption of epigenetic biomarkers in the diagnostic process.

Lastly, PM modern technology and therapeutics have the risk to create disparity among the population due to the cost of these services and the differences in healthcare system approach and the insurance that cannot cover genetic sequencing. In addition, genomic-wide association studies are doing an effort to include participants worldwide, but currently, there is a low representation of minorities that can drive a misclassification of disease in these groups, and for this, prevention strategies can be far from being effective in this groups.

In conclusion, the genetic risk approach cannot be predicted by genetic information without the combination with the clinical and familiar history and with an analysis of the environmental factors.

---

[17] https://www.genome.gov/Health/Genomics-and-Medicine/Polygenic-risk-scores

**Gene Therapy and Gene Editing Techniques** Targeting the mutated genes is becoming a therapeutic option. An example is a study for hemophilia B [22] using gene editing tools such as CRISPR/Cas9 or ZFNs (zinc finger nucleases), which correct the defective genes responsible for the disease. After that, adeno-associated viral (AAVs) or lentiviral (LVs) vectors carried by the "therapeutic gene" are administered systemically to the patient.

Another alternative therapy is done with stem cells, or already differentiated cells, transfected or otherwise, to correct a deficiency in the patient's physiological function.

Impact of personalized medicine in public health will follow the several genetic tests already available, medical device and biological date, mobile health application and data of behaviors, electronic health records, and clinical history. However, we need to consider onboard social

inequities, poverty, and racism. In this way, the incorporation of data at multiple levels, including environmental data, can support the public health PM approach.

Lastly, a combined approach can improve the integration of PM in public health by extending the training and education of healthcare providers and citizens, opening access to genetic testing infrastructure, and increasing access to novel drugs.

Following this address, personal health suggests involving multidisciplinary stakeholders and listening to peoples' needs; the multiplicity of data sources available can drive the PM implementation to develop and foster the uptake of those technologies that meet peoples' needs.

As a result, integrating these technologies in the healthcare systems requires monitoring data use over time and the data outcomes. This continuous monitoring will promote risk minimization and assess the scientific validity of new technologies (Fig. 7.1).

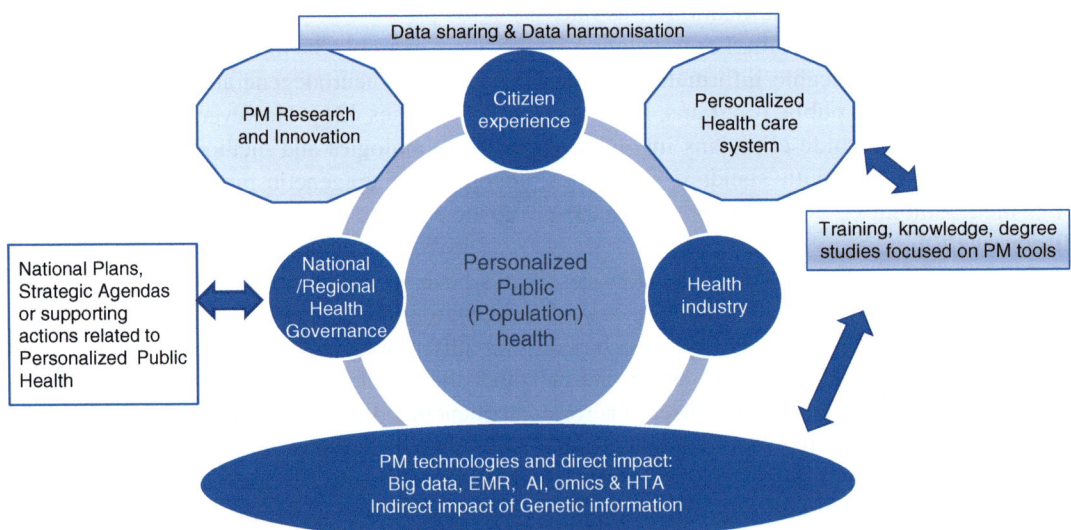

**Fig. 7.1** Cooperation and coordination for personalized public health. Three main stones have been indicated for the PM interaction with public health and implementation aspects. Following the results of the research and the implementation of the healthcare system. Scientific and technological needs to enable personalized medicine implementation. All based on the successful PM technolo-gies (starting from the center-left in the figure and proceeding clockwise); linked to the existence of national strategic plans, programs, and actions supporting PM-related basic, translational, and clinical research; infrastructures for PM research (i.e., biobanks, large-scale genomic databases, DNA sequencing facilities, etc.); data sharing and data harmonization and academia-industry relationships

## 7.6   Public Involvement in PM

It is needed to include here some already known definitions:

- Expertise: Convey a combination of specific education, training, or professional/personal experience
- Experience: Convey practical disease knowledge obtained from direct experience with the disease (affected person or close contact with affected person, e.g., family, carrier) or its treatment (e.g., healthcare professional)
- Advocacy: Act on behalf of the affected patients in defense of their rights; provide a patient-oriented public health/healthcare policy perspective
- Empowerment or engagement: Participate in the decision-making process within the committee; having access to information and process on behalf of patients and healthcare professionals

In the year 2000, The Council of Europe declared that the right of the public to be involved in the decision-making processes affecting healthcare is a basic and essential part of any democratic society [23]. The European Medicines Agency (EMA) considers that transparency and trust justify the participation of patient and citizen on their scientific committees, which improves the quality of the given opinion. In this way, patients are included as members in four of the six human EMA scientific committees: the Committee for Orphan Medicinal Products (COMP), the Paediatric Committee (PDCO), the Committee for Advanced Therapies (CAT), and the Pharmacovigilance and Risk Assessment Committee (PRAC). Their participation (i.e., members, alternates, experts, observers, or representatives) has been structured in four different features as indicated textually in the EMA report (which are not mutually exclusive)[18]:

In the same direction, the personalised medicine approach promotes citizen's (patient and public in general) active participation in research and healthcare. Although there is a clear interconnection between PM research and healthcare, the role of public involvement is different in them. Citizen involvement in public health follows the concepts of consumer, service user, community, and the general public. The main scope of this involvement is to engage them actively in decision-making about large-scale health changes.

Consequently, the necessity of citizen involvement has been slowly increasing due to the difficulties to find the best model of implementation. Above all, citizens can be involved in the evaluation of public health programs trying to better identify needs and for improving the quality of these programs; also, the participation in the process of development of these public health programs adjusted to social and geographical aspects; and the participation in future health planning to increase their voice on the healthcare. There are diverse involvement methods that need to match the purpose of involvement, the social context, and the population. So, the strategy to involve individuals with experience of cardiovascular diseases in designing health education prevention programs is diverse from involving citizens in a public conference around the health danger of gambling dependence.

Additionally, new resources are needed to increase the knowledge and the involvement of citizens in the process of access to medical data. This involvement will improve the trust in its quality. One example of the impact of omics on citizens' health is the direct-to-consumer genetic test (DTC-GTc) and also the high request rapid test for the infection of SARS-COVID during the pandemic [24].

*Community engagement*: A systematic review [25] evaluated the community engagement effectiveness of public health interventions across diverse health issues and with a high positive impact on health outcomes without a definitive conclusion on the best effective model of intervention. The two related health interventions with positive impacts were divided into health

---

[18]The role of patients as members of the EMA Human Scientific Committees, European Medicines Agency, 2011. https://www.ema.europa.eu/en/documents/other/role-members-representing-patients-healthcare-professionals-organisations-ema-scientific-committees_en.pdf

behaviors. Outcomes extracted were alcohol abuse, antenatal (prenatal) care, breastfeeding, cardiovascular disease, child illness, ill health, drug abuse, healthy eating, immunization, injury/safety, parenting, physical activity, smoking cessation, smoking/tobacco prevention, and health consequences.

Some of the significant recommendations for planning intervention are:

1. Involving the same community members in the delivery of the intervention. This is more effective if participants are classified as disadvantaged due to socioeconomic position (compared with those targeted to people based on their ethnicity, place of residence, or being at/high risk).
2. Single component interventions, both universal and targeted.
3. Consider personal skill development or training strategies or offer incentives.
4. Most effective in adult populations and less effective in general populations.

**For a final reflection on the implementation of Personalised Medicine on Public Health:** Patients' groups alone are not likely to change the prevailing pattern of public health, nor experts' groups at mechanisms for prioritization, neither would public funding alone solve the PM implementation. Suh as the Policies developed in the preapproval phase of drug development are needed, in the same way, the implementation of Personalised/ Precision Public health needs also a collaboration with industry and with input from the regulatory body [13].

In conclusion, the integration of every kind of PM innovation into Public health needs multidisciplinary public private engagement and the appropriate training of healthcare workers and of citizens.

Lastly, among the several ongoing projects, the International Hundred K+ Cohorts Consortium

IHCC[19] has as its main objective to increase the biological and genetic basis of disease and improve clinical care and population health. This consortium is currently working "to create a global platform for translational research, informing the biological and genetic basis for disease and improving clinical care and population health." In this way, this is an international-level project example of sharing of data, information, and resources linked to Population Health.

In conclusion, the integration of every kind of innovation into the health system needs engagement and appropriate training of healthcare workers and also of citizens.

## 7.7 Personalized Public Health and Equity

The CDC[20] defines health equity as "when every person has the opportunity to 'attain his or her full health potential' and no one is 'disadvantaged from achieving this potential because of social position or other socially determined circumstances'," describing their *goal to develop a set of priorities and actions that can help ensure that everyone has the opportunity to reap the health benefits of advances in genomics and precision medicine.* One of the primary goals of CDC's National Center for Chronic Disease Prevention and Health Promotion (NCCDPHP)[21] is to achieve health equity by eliminating health disparities and achieving optimal health for all.

Personalized public health wants to ensure equitable access to healthcare opportunities regardless of a person's age, gender, geographi-

---

[19] https://ihccglobal.org/

[20] Centers for Disease Control and Prevention Foundation. Atlanta, GA: Centers for Disease Control and Prevention. What is Heath equity

[21] https://www.cdc.gov/chronicdisease/healthequity/index.htm"https://www.cdc.gov/chronicdisease/healthequity/index.htm"/

cal origin, cultural, linguistic, religious background, communication, and accessibility needs. As a result of their application to public health, it can also provide a pathway to improve health equity across traditional barriers such as socioeconomic status, race/ethnicity, sex/gender, and geographical location. Considering that health disparities have often been linked to disparities in education, race, and income, several papers are concerned that the benefits of personalized medicine will go more to patients who have higher income and who are often not people of racial and/or ethnic minority groups, thus increasing the health disparities healthcare [26]. Based on the strategies proposed to optimize the benefits of personalized medicine [1] and lessen the potential worsening of health disparities, the priorities can be list as anticipated by Ward (Box 7.1) [27].

---

**Box 7.1 Priorities to Optimize the Benefits of PPM Approach**

1. Data must be collected in a systematic way, and it must be secured that health data flow in different socioeconomic statuses.
2. Dissemination models for developed therapies should include strategies that include outreach to patients who are not likely to have access to medications and healthcare, in other words an equitable access to public health service for all citizens.
3. Policies that foster the equitable development and access of prevention programs, medications, therapies, and devices that would be used in personalized medicine. Public health services are optimized in terms.

---

In this way, personalized public health should generate more specific and cost-effective prevention programs, enhance the impact of prevention and risk reduction campaigns, enhance coordinator among public health workers and the commu-

nity, and request equity of access and service for all, including marginalized sectors and underserved citizens [27].

**Healthcare and Personalized Medicine**  The process of the public health transformation needs to adopt the PM approach. In this way, the line of value based-health[22] needs to focus the personalized value and needs to understand how to measure this value. In this line, the value for patients cannot be assessed at the level of the hospital, a site of care, a medical specialty, a procedure, a primary care practice, or an entire population [28]. The value could look at the medical condition (unit of value) a patient has over the entire cycle of care for that condition. In primary and preventative care, value is created for segments of the population with similar needs. The most important practices in moving toward a value-based healthcare system were tried to be identified, and experts reached a consensus on the importance of outcome measurements, a focus on medical conditions, and full cycles of care. No consensus was reached on the importance of benchmarking with the purpose to improve efficiency, quality of care, patient safety, and patient satisfaction. More research studies through the value-based health concept are needed to assess the impact on the main quality of healthcare [29].

To sum up, the precision public health perspective is based mainly on two pillars, digital technology and genetic studies. The knowledge of the genetic diversity of the population will offer the needed background to guide decision-making concerning policies such as drug adoption, vaccination strategies, etc. However, crucial investments are needed for the appropriate implementation and sustainability [30], taking into account that the level of adoption will be variable and depending on regional geographic diversity, lifestyle population, and policies.

---

[22] https://hbr.org/2015/09/better-value-in-health-care-requires-focusing-on-outcomes

# References

1. StarfieldJ B. Basic concepts in population health and health care. Epidemiol Community Health. 2001;55:452–4. https://doi.org/10.1136/jech.55.7.452.

2. Kindig D, Stoddart G. What is population health? Am J Public Health. 2003;93(3):380–3. https://doi.org/10.2105/ajph.93.3.380.

3. Roux AV. On the distinction--or lack of distinction--between population health and public health. Am J Public Health. 2016;106(4):619–20. https://doi.org/10.2105/AJPH.2016.303097.

4. Winslow CEA. The untilled fields of public health. Science. 1920;51:23–33. https://doi.org/10.1126/science.51.1306.23.

5. Acheson D. Independent inquiry into inequalities in health: report. HMSO; 1998. Available from URL. https://assets.publishing.service.gov.uk/government/uploads/system/uploads/attachment_data/file/265503/ih.pdf

6. Grad FP. The Preamble of the Constitution of the World Health Organization. Bull World Health Org. 2002;80(12):984–1. https://apps.who.int/iris/handle/10665/268691. (World Health Organization)

7. Khoury MJ, Gwinn ML, Glasgow RE, Kramer BS. A population approach to precision medicine. Am J Prev Med. 2012;42(6):639–45. https://doi.org/10.1016/j.amepre.2012.02.012.

8. Eisinger F, Morère JF, Touboul C, et al. Prostate cancer screening: contrasting trends. Cancer Causes Control. 2015;26:949–52. https://doi.org/10.1007/s10552-015-0573-9.

9. Bloss CS, Darst BF, Topol EJ, Schork NJ. Direct-to-consumer personalized genomic testing. Hum Mol Genet. 2011;20(R2):R132–41. https://doi.org/10.1093/hmg/ddr349.

10. Gulcher J, Stefansson K. Genetic risk information for common diseases may indeed be useful for prevention and early detection. Eur J Clin Investig. 2010;40(1):56–63. https://doi.org/10.1111/j.1365-2362.2009.02233.x.

11. Martínez-González MA, Kim HS, Prakash V, Ramos-Lopez O, Zotor F, Martinez JA. Personalised, population and planetary nutrition for precision health. BMJ Nutr Prev Health. 2021;4(1):355–8. https://doi.org/10.1136/bmjnph-2021-000235. Published 2021 Jun 2

12. Hassanaly P, Dufour JC. Analysis of the Regulatory, Legal, and Medical Conditions for the Prescription of Mobile Health Applications in the United States, The European Union, and France. Med Dev (Auckl). 2021;14:389–409. https://doi.org/10.2147/MDER.S328996. Published 2021 Nov 24

13. WHO OMS. 71e World Health Assembly - main debates of May 25; 2018. Available from: https://www.who.int/fr/news/item/25-05-2018-seventy-first-world-health-assembly-update-25-may.

14. Liberati A. Need to realign patient-oriented and commercial and academic research. Lancet. 2011;378(9805):1777–8. https://doi.org/10.1016/S0140-6736(11)61772-8.

15. Ashton K. That 'internet of things' thing. RFID J. 2009;22:97–114.

16. Dwivedi R, Mehrotra D, Chandra S. Potential of internet of medical things (IoMT) applications in building a smart healthcare system: a systematic review. J Oral Biol Craniofac Res. 2021; https://doi.org/10.1016/j.jobcr.2021.11.010.

17. Shah NS, Auld SC, Brust JCM, Mathema B, Ismail N, Moodley P, Mlisana K, Allana S, Campbell A, Mthiyane T, Morris N, Mpangase P, Van Der Meulen H, Omar SV, Brown TS, Narechania A, Shaskina E, Kapwata T, Kreiswirth B, Gandhi NR. Transmission of extensively drug-resistant tuberculosis in South Africa. N Engl J Med. 2017;376(3):243–53. https://doi.org/10.1056/NEJMoa1604544.

18. Zhou C, Wu YL, Chen G, et al. Erlotinib versus chemotherapy as first-line treatment for patients with advanced EGFR mutation-positive non-small-cell lung cancer (OPTIMAL, CTONG-0802): a multicentre, open-label, randomised, phase 3 study. Lancet Oncol. 2011;12(8):735–42. https://doi.org/10.1016/S1470-2045(11)70184-X.

19. Wang C, Baer HM, Gaya DR, Nibbs RJB, Milling S. Can molecular stratification improve the treatment of inflammatory bowel disease? Pharmacol Res. 2019;148:104442. https://doi.org/10.1016/j.phrs.2019.104442.

20. Glaab E, Rauschenberger A, Banzi R, The PERMIT Group, et al. Biomarker discovery studies for patient stratification using machine learning analysis of omics data: a scoping review. BMJ Open. 2021;11:e053674. https://doi.org/10.1136/bmjopen-2021-053674.

21. García-Giménez JL, Ushijima T, Tollefsbol TO. Epigenetic biomarkers: new findings, perspectives, and future directions in diagnostics. In: García-Giménez JL, editor. Epigenetic biomarkers and diagnostics. Amsterdam: Elsevier Inc.; 2016. http://ezaccess.libraries.psu.edu/login?url=http://www.sciencedirect.com/science/book/9780128018996.

22. Rodríguez-Merchán EC, De Pablo-Moreno JA, Liras A. Gene therapy in hemophilia: recent advances. Int J Mol Sci. 2021;22(14):7647. https://doi.org/10.3390/ijms22147647. Published 2021 Jul 17

23. Council of Europe. The development of structures for citizen and patient participation in the decision-making process affecting health care: Recommendation Rec(2000)5 adopted by the Committee of Ministers of the Council of Europe on 24 February 2000 and explanatory memorandum. Health and society Strasbourg, : Croton-on-Hudson, NY : Council of Europe Pub. ; Manhattan

Pub. Co., c2001 60 p. ; ISBN:9789287145659; 9287145652.

24. Traversi D, Pulliero A, Izzotti A, Franchitti E, Iacoviello L, Gianfagna F, et al. Precision medicine and public health: new challenges for effective and sustainable health. J Personalized Med. 2021;11(2) https://doi.org/10.3390/jpm11020135.

25. O'Mara-Eves A, Brunton G, Oliver S, et al. The effectiveness of community engagement in public health interventions for disadvantaged groups: a meta-analysis. BMC Public Health. 2015;15:129. https://doi.org/10.1186/s12889-015-1352-y.

26. Estape EA, Mays MH, Sternke EA. Translation in data mining to advance personalized medicine for health equity. Intell Inf Manag. 2016;8:9–16. https://doi.org/10.4236/iim.2016.81002-.

27. Ward MM. Personalized therapeutics: a potential threat to health equity. J Gen Intern Med. 2012;27:868–70. https://doi.org/10.1007/s11606-012-2002-z.

28. Porter ME, Teisberg EO. Redefining competition in health care. Harv Bus Rev. 2004;82(6):64–76. 136

29. Steinmann G, Delnoij D, van de Bovenkamp H, Groote R, Ahaus K. Expert consensus on moving towards a value-based healthcare system in the Netherlands: a Delphi study. BMJ Open. 2021;11(4):e043367. Published 2021 Apr 12. https://doi.org/10.1136/bmjopen-2020-043367.

30. Ramaswami R, Bayer R, Galea S. Precision medicine from a public health perspective. Annu Rev Public Health. 2018;39:153–68. https://doi.org/10.1146/annurev-publhealth-040617-014158.

# DNA Technologies in Precision Medicine and Pharmacogenetics

**8**

Seyedeh Sedigheh Abedini, Niloofar Bazazzadegan, and Mandana Hasanzad

**What Will You Learn in This Chapter?**
A wide range of biological fields, including medicine and precision medicine, can benefit from next-generation sequencing. DNA technologies and their applications to pharmacogenomics, and drug prescribing in the context of genome genotyping, are discussed in this chapter. This chapter discusses a variety of NGS technologies, including whole-genome sequencing (WGS), clinical exome sequencing (CES) and whole-exome sequencing (WES), whole transcriptome sequencing (WTS), targeted sequencing (TS), single-cell sequencing (SCS), and DNA microarrays and their clinical and medical applications.

**Rationale and Importance**
Due to the genetic diversity, some individuals might show unexpected side effects and even drug resistance. Therefore, the genetic profile of these patients must be analyzed to determine molecular biomarkers and genetic data for prescription medicine. DNA technologies are

enabled to elucidate the profile of the human genome, which could result in improved drug treatments. Pharmacogenomics (PGx) is a critical component of personalized medicine since genomic information enables the development of safer, more effective, and more affordable drugs. Because of their low cost and accuracy, genotyping technologies like next-generation sequencing (NGS), microarrays, and bead arrays are expected to make their way into clinical application. These techniques provide a lot more information than other types of genetic testing, which could be extremely useful when attempting to figure out what is wrong with a patient.

Researchers and medical professionals will be able to use the NGS on precision medicine routinely due to developing chemistries, lowering cost, and the newest tools available to facilitate the analysis of genotyping data based on PGx. As a result, we expect genotyping technologies to have strong capabilities and more guidelines for personalized medicine and pharmacogenetics in the next years, leading to improved healthcare.

S. S. Abedini (✉) · N. Bazazzadegan
Genetics Research Center, University of Social Welfare and Rehabilitation Sciences, Tehran, Iran

M. Hasanzad
Personalized Medicine Research Center, Endocrinology and Metabolism Clinical Sciences Institute, Tehran University of Medical Sciences, Tehran, Iran

Medical Genomics Research Center, Tehran Medical Sciences, Islamic Azad University, Tehran, Iran

## 8.1 Introduction

Many studies indicate that drug-related genes, also referred to as "pharmacogenes," in the human genome contain extensive functional genetic variations (FGVs). Different alleles are associated with diverse outcomes of drug treatments [1–3]. Genetic variations and level of expression in drug-targeted molecules, containing membrane and nuclear receptors, signal transduction components, and enzymes, moreover drug transporters and drug-metabolizing enzymes may affect the incidence of the individual variations in response to a drug [4]. Around 97–98% of people have at least one actionable FGV in their drug-related genes. In addition, the possibility of the presence of a genetic variant that could result in a loss of function (LOF) variant in pharmacogenes is 93% for every individual [5]. Hence, identifying the different genetic variants associated with the drug metabolism would affect medication prescription, allowing for selecting the right drug and dose, thereby reducing the potential adverse effects or therapeutic inefficacy.

Breakthrough of NGS platforms throughout this decade now provides affordable and reliable high-throughput sequencing for assessment of functional DNA variations in many diseases, including both monogenic and polygenic phenotypes such as diabetes, cancer, and cardiovascular and neurological disorders as well as in the regulation of physiological conditions such as height, blood pressure, and body mass index [6–11]. Thus, there is an increasing excitement to apply individual genome sequencing for predicting disease risk, lifelong well-being of individuals, medical care, and response to drugs in the era of personalized medicine. Currently, the area of PGx is shifting from reactive testing of a single gene toward scanning a whole panel of genes concerned with drug absorption, distribution, metabolism, and excretion (ADME) before prescribing (preemptive genotyping) by using different types of next-generation sequencing (NGS) platforms [12]. DNA technologies have been used to detect variants that affect the drug's toxicity and efficacy. The properties of NGS technolo-

gies make them an exciting approach to performing clinical PGx testing. Several investigators have recently explored approaches utilizing NGS platforms, namely, targeted sequencing, whole-exome sequencing (WES), and whole-genome sequencing (WGS) in pharmacogenomics. Microarrays enable gene expression profiling of thousands of genes in tens of samples by research. Also, different gene clusters showed a correlation with discrete phenotypes in tumors offering that tumor grades are associated with distinct gene expression [13].

## 8.2 NGS Technology Procedures: Its Important Applications and Related Different Databases

Figure 8.1 indicates NGS steps, improvement in biochemical steps, the kind of machine types, performance of each phase, mechanism of preparation, and how each step is carried out for complete huge parallel sequencing [14–16].

NGS has diverse biological applications; nevertheless, the applications in precision medicine have extended tremendously in the last few years. Figure 8.2 represents some DNA technology applications in the field of precision medicine [17].

A growing number of databases can be used for showing disease associations. Population databases supply information concerning the frequencies of variants in populations. These databases might contain healthy and diseased patients and must be used cautiously to rely on the purpose of the association study. On the other hand, the disease databases include variants in patients suffering from a particular disease and relevant information about its pathogenicity. An outline of the most used databases is provided in Table 8.1.

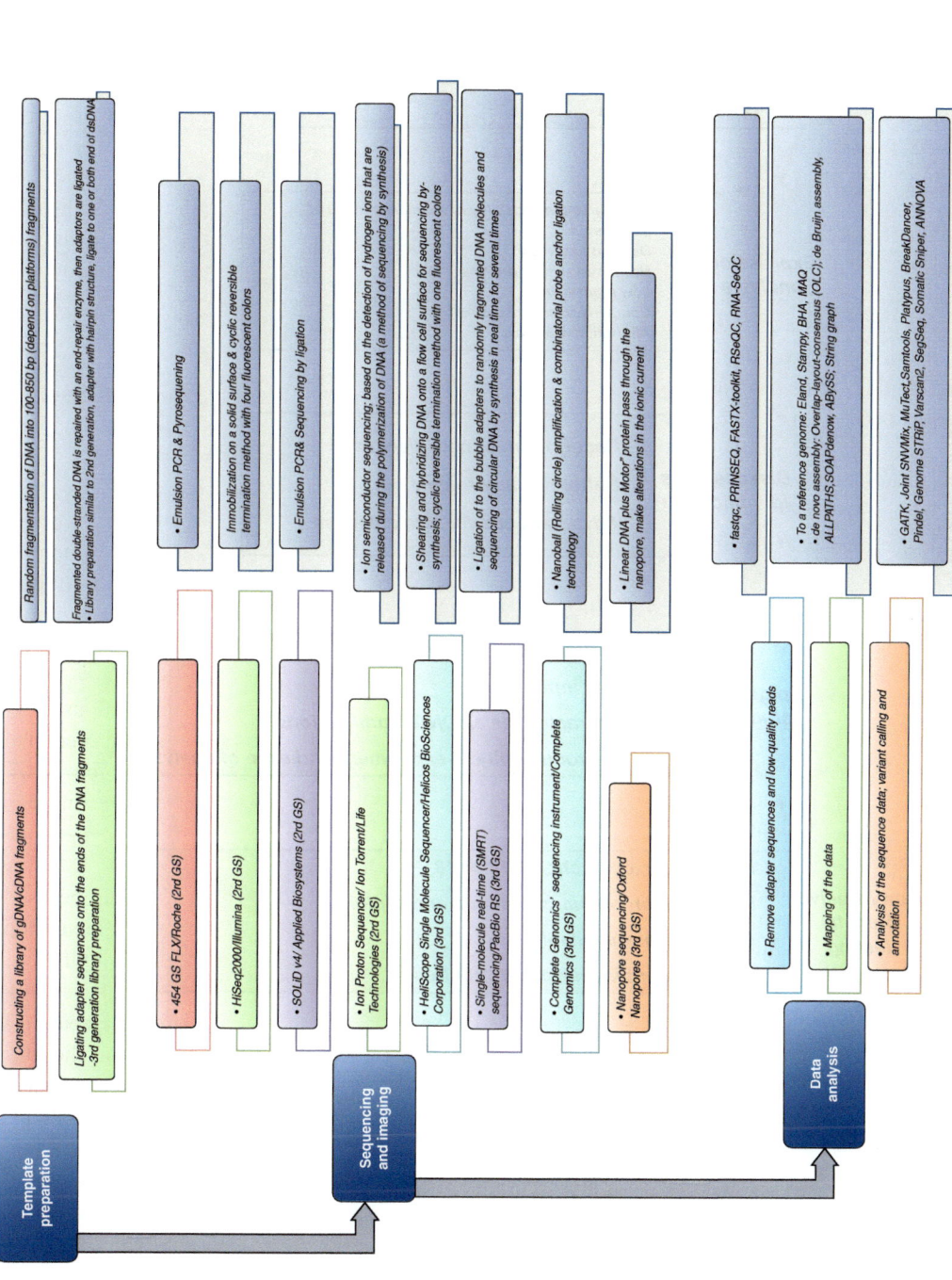

**Fig. 8.1** NGS step, development in biochemical steps, machine types, the performance of each step, and mechanism of preparation and how each step is performed for complete massive parallel sequencing is shown. "This figure is reprinted with permission from source Rabbani B, et al. Mol Biosyst. 2016. PMID: 27066891 Review"

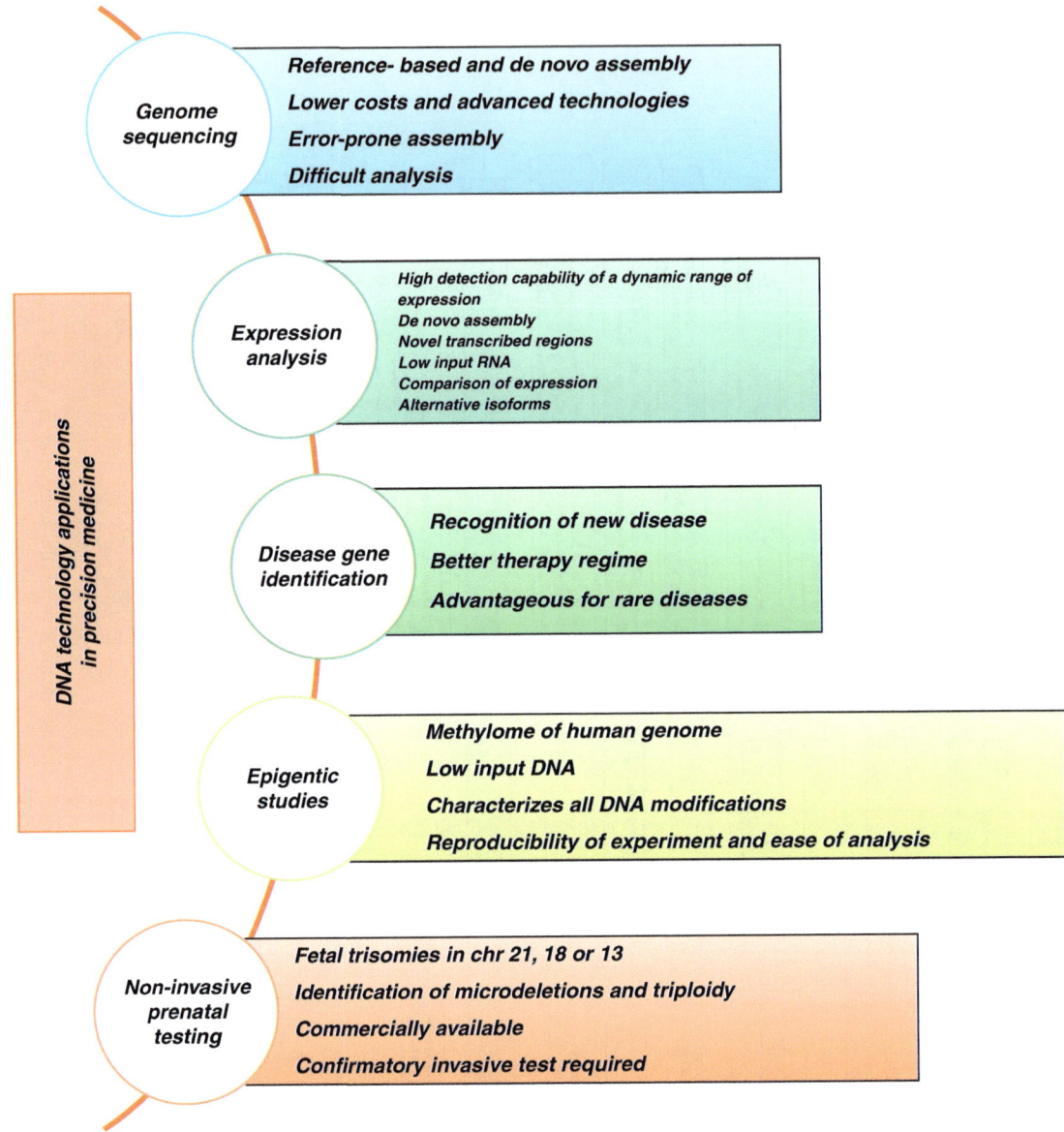

**Fig. 8.2** Overview of some major applications of next-generation sequencing

**Table 8.1** Summary of useful databases in medicine and pharmacogenomics

| Databases | Description |
|---|---|
| **Data interpretation and genomic test databases** | |
| Evidence aggregator *http://64.29.163.162:8080/GAPPKB/evidencerStartPage.do* | An application that makes finding evidence reports, systematic reviews, and guidelines for using genetic tests and other genomic applications easier |
| FDA Pharmacogenomic Biomarkers *http://www.fda.gov/drugs/scienceresearch/researchareas/pharmacogenetics/ucm083378.htm* | A list of FDA-approved medications that include pharmacogenomic data on the label |
| PharmGKB *http://www.pharmgkb.org* | Clinical information, such as dose guidelines and prescription labels, as well as potentially clinically actionable gene–drug connections and genotype-phenotype interactions, are all included in this pharmacogenomics knowledge repository |
| EGAPP *http://www.egappreviews.org/about.htm* | A collection of information on the validity and value of genetic testing in clinical practice, as well as advice from professional organizations and advisory committees on how to utilize genetic tests |
| GAPP Finder *http://64.29.163.162:8080/GAPPKB/topicStartPage.do* | A searchable collection of genetic testing and genomic applications that are in the process of transitioning from research to clinical and public health use. Disease, genes, drugs, tests, and other terms can be used in the search query |
| Genetic testing registry *http://www.ncbi.nlm.nih.gov/gtr/* | Information about test technique, validity, proof of the test's utility, and laboratory contacts and credentials are available in one central spot |
| **Locus/disease/ethnic/other-specific databases** | |
| DECIPHER *http://decipher.sanger.ac.uk* | DECIPHER provides information on over 19,000 patients (with more on the way) and allows for phenotype-genotype comparisons |
| Human Genome Variation Society (HGVS) *http://www.hgvs.org/locusspecific-mutation-databases* | HGVS gives data on variants that are unique to the locus of interest |
| Leiden Open Variation Database (LOVD) *http://www.lovd.nl* | LOVD is an open source for analyzing sequence variants and storing NGS data |
| **Sequence databases** | |
| Locus Reference Genomic (LRG) *http://www.lrg-sequence.org* | EBI and NCBI assemble and preserve LRG sequences, which have documented mutations with a permanent ID |
| NCBI Genome *http://www.ncbi.nlm.nih.gov/genome* | This NCBI database contains genome-related information like sequences, chromosomes, and annotations |
| RefSeqGene *http://www.ncbi.nlm.nih.gov/refseq/rsg* | By offering sequences that may be used as reference standards, RefSeqGene assists researchers in matching their data |
| MitoMap-Human Mito Seq *http://www.mitomap.org/MITOMAP* | The database contains data on human mitochondrial DNA variation from over 31,000 human mitochondrial DNA sequences, both published and unpublished (with more on the way) |

(continued)

**Table 8.1** (continued)

| Databases | Description |
|---|---|
| Disease databases | |
| Human Gene Mutation Database (HGMD)*http://www. hgmd.org* | HGMD is a licensed database that contains up-to-date information on mutations that cause human hereditary illnesses |
| ClinVar*http://www.ncbi.nlm.nih.gov/clinvar* | ClinVar is a free resource that contains information on genetic variants and phenotypes |
| Online Mendelian Inheritance in Man (OMIM)*http:// www.omim.org* | OMIM has up-to-date information on over 15,000 genes and their genotype-phenotype relationships |
| Population databases | |
| VarSome *https://varsome.com/* | A search engine, aggregator, and impact analysis tool for human genetic variation and a community-driven project aiming at sharing global expertise on human variants |
| dbSNP*http://www.ncbi.nlm.nih.gov/snp* | SNPs (single nucleotide polymorphisms) and other small-scale variants are stored in the dbSNP database |
| Exome Variant Server*http://evs.gs.washington.edu/ EVS* | Through next-generation sequencing data sets, this database provides information on new genes and pathways that contribute to heart, lung, and blood illnesses |
| Exome Aggregation Consortium (ExAC)*http://exac. broadinstitute.org/* | For the scientific community, this consortium provides exome sequencing data for over 60,000 individuals from large-scale sequencing efforts |
| dbVar*http://www.ncbi.nlm.nih.gov/dbvar* | dbVar collects data on structural variants such as translocations, duplications, deletions, insertions, and more |
| 1000 Genomes Project*http://browser.1000genomes.org* | This database is an extension of the Ensembl human genome browser, and it shows data on genetic variation in a large number of people |

## 8.3    Whole-Genome Sequencing

WGS is the sequencing of an entire organism's genome at a single time. WGS includes the sequencing of chromosomal DNA, mitochondrial DNA, and chloroplast DNA in plants [18]. Complete genomic variants (including PGx-related markers) for an individual would be available by utilizing the WGS approach. WGS was introduced to clinics in 2014 and has been chiefly used as a research tool [19–21]. Although the widespread data interpretation of such tests is still challenging, a reduction in sequencing costs alongside the comprehensiveness of WGS may also result in the method turning into a widespread platform for clinical PGx tests.

Using phase I WGS data from the 1000 Genome Project followed by annotations, the variant minor allele frequency was >1%, of which 8207 resulted in strong linkage disequilibrium (LD) (r2 > 0, 8) with known PGx variants. Differences were distributed in various genome components, introns, coding, and 5′-upstream and 3′-downstream regions. Finally, the authors identified putative functional variants within the known pharmacological genomics loci underlying the drug response phenotype and suggested direct testing instead of relying on LD, which will be different among populations [22].

Yang et al. conducted a three-way analysis with the Directorate of Medical Education and Training (DMET), WES, and WGS, to examine the concordance between PGx genotyping calls based on these various technologies. They showed a 94% concordance between the DMET and WES and a 96% concordance between the DMET and WGS [23]. The functional copy number variation (CNV) of the ADME genes was distributed in different populations at significantly different frequencies [24, 25]. NGS data can also be used for CNV calls with different ethnic backgrounds.

Researchers used integrated WGS and WES data from 1000 genomes and ExAC repositories for CNV identification of 208 pharmacogenes. Novel CNVs (deletion in 84% and duplications in 91% of genes) over six distinct populations of non-Finnish Europeans, Africans, Finns, East

Asians, South Asians, and mixed Americans were decoded effectively. The ultimate result highlighted the need for the comprehensive NGS-based genotyping of the pharmacogenes for the CNV distinguishing proof nearby their allele frequencies. The evaluation of the commitment of such CNVs to the medicate response results is additionally conceivable through a population-specific analysis of uncommon variants [26].

WGS has been performed on a patient with a family history of vascular disease and sudden early death in a study. However, no clinically significant medical records predict the potential risk of coronary artery disease and the cause of sudden cardiac death [27]. Rare variants of three genes, including *TMEM43* (MIM # 612048), *DSP* (MIM # 125647), and *MYBPC3* (MIM # 600958), have been found to be clinically associated with sudden cardiac death. This patient was heterozygous for a null mutation in the *CYP2C19* (MIM # 124020) gene, suggesting possible resistance to clopidogrel. The authors suggested that WGS could provide helpful and clinically relevant information for individual patients. Knowledge of pharmacogenetic variants may be essential for future personalized medicine of patients. Overall, personal genomic analysis is a field of genomics that ultimately provides experimental medical treatment to individuals based on genomic analysis. Predictive and preventive care results in advanced healthcare.

WES and WGS may give a promising approach to recognize low-frequency (1–5%) and uncommon (<1%) variations. The suitable medicate reaction loci with a genome-wide approach can be found instead of finding one gene [28]. Researchers at Washington University have utilized WGS to analyze a challenging leukemic disorder, appearing that this information can be assembled and analyzed within a time frame consistent with clinical decision-making [29].

Most WGS is generated using methods that the user can modify concerning laboratory standards [30]. Many scientists feel that their responsibility is to modify sequencing protocols to provide the best sequencing results possible. Just as microarray data depends on the method used to isolate mRNA and generate labeled cDNA [31], WGS results depend on changes made to the protocol developed, tested, and validated by the manufacturer [32, 33]. This variability leads to significant differences between WGS produced by different laboratories [32].

## 8.4   Whole-Exome Sequencing

Whole-exome sequencing (WES) is one of the primary applications of high-throughput DNA sequencing methodology for detecting different variations in the coding sequence, and additional relevant adjacent and untranslated regions of the genome. WES is a progressively critical technology and molecular diagnostic tool in rare diseases and drug response genetics [34, 35].

Since 2011, WES as a useful diagnostic tool has been routinely offered in clinical genetics laboratories [36]. Then, it has been consolidated into National Heart, Lung, and Blood Institute (NHLBI) "Grand Opportunity" Exome Sequencing Project (GOESP) (more than 6500 patients), DiscovEHR study (functional variants in 50,726 human genome), the 1000 Genome Project (variants in 1092 individuals) and the Exome Aggregation Consortium (ExAC) projects (pathogenic variants in 60,706 individuals) [37–41]. Distinct evaluation of the WES data in the different populations indicates the frequency and potential functional association of rare variants, almost novel SNVs, among many pharmacogenes [40, 42]. The recent studies especially phase I and II drug transporters and metabolic enzymes consisting of exome sequencing and SNVs data revealed that approximately 93% of all identified variants are rare (minor allele frequency [MAF] < 1%) or very rare (MAF < 0.01%) [42]. Another study, involving 14,002 subjects for investigation of a rare genetic variant by sequencing of 202 drug target genes, explored that almost variants have MAF below 0.5% and had not been previously identified. In addition, many of these variants are harmful which are associated with risk factors for developing a disease and drug response. Also, at least 5–10% of them have a critical role in the PGx panel [43].

Through the years, WES has dramatically made improvements in the robustness of research procedures and laboratory tests, dataset uniformity, and advances in filtering and interpretation of variants [44]. To date, greater than 80–85% of pathogenic mutations in Mendelian disorders and complex disorders have been detected within the exomes, and which WES technique offers a fair-minded approach to identify these variations and provides additional information about them in the era of personalized medicine and PGx profiling [14, 43–47]. This information leads to achieving maximum benefits for particular patients by adapting unique treatment based on genetic makeup. For example, affected individuals with pyridoxine-dependent epilepsy-*ALDH7A1* (OMIM ID, 266100) are commonly resistant to therapy with anticonvulsants; however, using massive doses of pyridoxine (vitamin B6) can be treated efficiently [48]. Prevention of futile drug use and consideration of the treatment strategy in these non-insensitive patients are done according to WES data as "game-changing technology" [48, 49].

WES technology maturity and process standardization exhibit impressive improvements in the different eras of personalized medicine including targeted therapeutic agents based on tumor biomarkers like PD-L1, larotrectinib (Vitrakvi) and olaparib, vemurafenib (BRAF-positive tumors), imatinib (KIT-positive tumors), and monoclonal antibody pembrolizumab in which all of them have been approved by the US Food and Drug Administration (FDA) [50–53]. It is anticipated that the digital genome market and the personalized medicine market will attain over $45 billion and $87.7 billion by 2024, respectively [35].

In recent years, the potential effectiveness of WES has been additionally studied for a wide range of pharmacogenes' profiling, drugs' pharmacokinetics (PK), and pharmacodynamics (PD) from absorption to excretion of certain diseases, such as nervous system, seizures, kidney transplantation, cancer, infectious diseases, and autoimmune disorders [40, 54, 55]. The first stage of xenobiotics metabolism involves cytochrome P450 (CYP450 from the CYP1 enzyme family) activities that may change by way of genetic variants positioned in their related genes. Therefore, identifying the genetic variation using drugs' PK and PD helps the clinician to choose suited therapeutic without toxicity [35, 56]. As a result, WES data has the ability to revolutionize the prevention and even therapy of human disease. In addition to the prediction of common drug reactions, genetic information and WES data are also used to select small molecular inhibitors and analyze different variants (somatic and germline) in various diseases.

According to WES analysis study performed by Van der Lee and his colleagues for providing the PGx panel with actionable Ubiquitous Pharmacogenomics (U-PGx; www.upgx.eu) panel, Clinical Pharmacogenomics Implementation Consortium (CPIC), and Dutch Pharmacogenetics Working Group (DPWG) guidelines, 39 out of 42 variants (86% of total) in 11 pharmacogenes represent linking genotype to drug response phenotypes. Recently, more than 21 important genes and 50 drugs have been recommended by a combination of CPIC and the DPWG. This data highlighted the ability of WES data to create a significant PGx panel based on critical pharmacogenes (7 out of 11 genes). However, this group did not identify any structural variations (SVs) in *CYP2C19*, *UGT1A1*, *CYP3A5*, and *CYP2D6* genes due to a limited sample size [57]. Proper coverage of variants of the *CYP2D6* gene is clinically important because the CYP2D6 enzyme accounts for 25–30% of commonly prescribed drugs. Mutations in the *VKOR* and *CYP2C9* genes lead to different metabolic capacities of the coding enzyme in the drug metabolization pathway [57].

The PGx panel is only informative when specific drugs are used and are expounded regarding each patient's genetic information and their medicine desires. Therefore, the results of PGx are beneficial when gene-drug interactions are considered and completely interpreted according to the obtained sequencing data of each patient. This genotype-phenotype correlation substantially decreases the risk of revealing unsolicited finding and accelerates the treatment processing.

Cousin and his colleagues revealed that a considerable percentage of patients had actionable PGx profiles based on current drug intake. They investigated 94 patients for PGx variants in the three important pharmacogenes (*CYP2C19*, *CYP2C9*, and *VKORC1* genes) using WES data and detected at least one actionable variant in 91% of all subjects. Twenty percent of total patients showed an immediate impact on current medicinal drug use (warfarin and clopidogrel) through the PGx finding [58]. The proper interpretation of PGx variants in this study was the key to inhibiting drug adverse effects and making individualizing prescribing decisions.

Several studies have been carried out to determine the accuracy and the concordance rate of WES technology in PGx and its application in precision medicine. Rennert et al. investigated 337 cancer patients with Exome Cancer Test v1.0 (EXaCT-1), and causative genetic mutation has been detected in 82% of all cases. The results suggest accurate cancer treatment and provide utilized information for precision medicine cancer care. The positive predictive value, specificity, and sensitivity were 99.2%, 99.9%, and 95.7%, respectively. This emphasizes the accuracy of WES for mutation detection and prescribed medications with improvement in saving the cost and time [59]. Yang et al. simultaneously examined three technologies, clinical genotyping (DMET array-based), WES, and WGS, for comparing PGx variants obtained from sequencing of 13 valuable pharmacogenes with ICI guidelines. The contradiction genotyping was observed between 4 out of 68 loci by DMET and WES and 3 out of 66 loci by DMET and WGS. They reported the concordance rate between WES and DMET and WGS and DMET is 94% and 96%, respectively. They confirmed that WES and WGS are capable of providing worthy and usability data for most pharmacogenes and prepare further validation of genomic sequences in clinical laboratories [23]. Another study for the assessment of the WES variant's integrity was performed by Chua et al. They used cross-comparison between the MiSeqR amplicon sequencing data and WES for two important pharmacogenes: *CYP2D6* and *CYP2C1*. They indicated the error rate is less than 1% and WES is a pioneer tool in providing PGx profiling, even if complex loci have been studied [45]. Other researchers have published similar results that the most useful outcomes are obtained from sequencing data compared to orthogonal tests [40, 52, 54, 57].

The improvement of WES accuracy and its cost make it as a usable molecular diagnostic tool for the evaluation of genetic disorders and pharmacogenetic tests. However, the obtained WES variants, read length, depth of coverage, and variant interpretation regarding the PGx panel for each patient to avoid any futile drugs should be considered in more detail.

## 8.5    Clinical Exome Sequencing

The clinical value of WES and WGS as a general test for mutation findings is now appropriate for the almost genetic diagnostic query. Although whole-exome sequencing and whole-genome sequencing are emerging, panel-based testing (based on clinical) is more practical for clinical annotation in the human genome and has a strong position in precision medicine. Clinical exome sequencing (CES) has become more viable – and possibly cost-effective – as a first-line diagnostic, rather than an alternative to exploring if other types of testing fail to offer a diagnosis, due to the rapid improvement of high-throughput sequencing technology in speed and cost. This approach concentrates on genes in which disease-causing mutations have been discovered and documented in the Human Mutation Database. Ambry Genetics (Aliso Viejo, CA, USA) was the first CLIA laboratory to use NGS technology for establishing a "Clinical Diagnostic Exome" in 2011 [34, 47, 60].

In comparison to WES and WGS, the CES dataset is substantially smaller, but it offers several advantages: firstly, not generating excessive numbers of uncertain significant variants, which simplifies genetic counseling; secondly, putting by the emphasis on clinically indicated genes, achieving trio analyses, and obtaining high-quality data (deep coverage) which is cost-effective; and finally, using an instrument such as the Illumina MiSeq, which can be used at a benchtop scale for data analyzing [17]. Therefore,

it helps to facilitate the identification of actionable variants for applicability in precision medicine and therapeutic decision-making. Many firms have offered different panels for some genetic disorders such as cancer, hearing loss, and cardiomyopathy that are used by researchers and clinicians. To date, different panels have been established from actionable gene panels, hotspot panels, and disease focus panels to comprehensive multigene panels. The panels will allow us to detect genetic variants responsible for diseases and predict treatment regimens that will be effective, leading to better and more prompt patient management. More recently, the CES application was mainly used for determining the risk of hereditary malignancies and drug decision-making for somatic cancers [60, 61].

Using a hotspot panel (comprised of common hotspot mutations), clinicians can identify mutations in regions of the genome relevant for treatment, diagnostics, or prognosis. The first commercially available hotspot panel was the AmpliSeq cancer panel V1 which covers 46 cancer genes (tumor suppressor and oncogenes) with 739 actionable mutations. The number of hotspot mutations is increased to 2855 from 50 cancer genes in the new version of this panel. In contrast with hotspot panels, actionable gene panels cover all exon and targeted genes to identify other harmful mutations outside of hotspot variants. The most common target genes of these panels are FDA-approved genes such as *BRAF*, *PIK3CA*, *KIT*, *ALK*, *NRAS*, *KRAS*, and *EGFR*. The TruSight Tumor panel was the first commercial actionable gene panel that covers 26 genes involved in melanoma and ovarian, gastric, lung, and colon cancers. The majority of these panels look into somatic mutations to help determine therapeutic options. Unlike the gene panels, the disease focus panels are mostly focused on identifying inherited diseases or detecting suspected genetic disorders based on germline mutations. The sensitivity, specificity, and depth of coverage can be increased via these panels consisting of a limited set of genes, while the cost is reduced [17, 34, 36, 47, 60–62].

While disease-specific panels have become increasingly popular, some potential barriers are propounded: (1) the limited quantity of samples for clinical testing, (2) the process of developing and validating the panels according to ACMG guidelines, and (3) the need to keep current panels up to date. These challenges have led investigators and clinicians to explore the new utilizing panel, comprehensive panels, which include different actionable genes associated with their related disorders. By using this panel, disease-specific testing would be simplified, while the medical significance of most variants could avoid interrogation. More than 60 valuable genes with 4813 causative genes for genetic disorders have been listed in Illumina's TruSight panels as a known and popular comprehensive multigene panel [60, 61].

Despite the potential advantages of CES for patients whose diseases are undiagnosed or whose results from disease-focused panels are inaccurate, we do not expect the full scale of this approach in clinical trial tests due to the limitation of the restricted number of specified genes and complex bioinformatics pipelines.

## 8.6    Whole Transcriptome Sequencing

Transcriptome sequencing, or gene expression arrays in general, has been a well-established diagnostic tool for characterizing and quantifying gene expression profiles and detecting fusion transcripts. Improvements in RNA sequencing (RNA-Seq), including polyA selection and WTS, can be used to develop analytical spectra that cover multiple transcriptional events (chimeric transcripts, isoform switching, expression, etc.) in a single approach. RNA-Seq provides single base-pair resolution and significantly less background noise, enabling distortion-free transcriptome evaluation compared to expression arrays [63]. Currently, the diagnosis of acute lymphoblastic leukemia patients requires several analyses encompassing morphology, immunophenotyping, molecular evaluation of gene fusions and mutations, and detection of numerical and structural abnormalities based totally on chromosomal banding analysis and

fluorescence in situ hybridization [64]. WTS parallel analysis of gene expression profiles allows for changes in fusion transcripts and copy numbers, leading to the specific characterization of patients' genetic profiles as the basis for disorder classification based on a single method dataset. Understanding the transcriptome is necessary to interpret the genome's functional elements and apprehend the underlying development and disease mechanisms.

Advancements in large-scale parallel DNA sequencing technology have enabled transcriptome sequencing (RNA-Seq) through cDNA sequencing. RNA-Seq quickly replaced microarray technology due to its high resolution and reproducibility. This method can be used to expand knowledge about alternative splicing events [65], new genes and transcripts [66], and fusion transcripts [67].

One difficulty involving the utility of RNA-Seq is estimating abundance at the gene level and differential expression at the transcriptional level under various conditions. RNA-Seq can determine the expression profile of normal and affected cells and tissues [68].

The etiology of Alzheimer's disease (AD) is complicated and remains challenging to research efforts worldwide. In the absence of a greater understanding of AD pathogenesis, cure strategies do not supply a treatment, however only deal with symptoms or decrease the price of onset. The transcriptome displays cellular activity within the tissue at a given time. Genome-wide expression studies, which are no longer influenced by deductive assumptions, provide an independent strategy for investigating the etiology of complicated ailments such as AD. Transcriptome analyses have been performed using transgenic animal models of AD and patient-derived cell lines [69, 70].

In contrast to these approaches, an autopsy of brain tissue is challenging to obtain, and some RNA quality issues can affect transcriptome studies [71, 72]. Nonetheless, the same postmortem brain tissue as the tissue affected by the disease remains the gold standard for evaluating against all other model systems. However, transcriptome studies of AD using brain tissue have yielded almost contradictory results. The latest improvement in next-generation sequencing offers a more complete and accurate tool for transcriptome analysis of this invaluable resource [73, 74].

Jinquan et al., in 2014, sought to identify differences in ATRX mRNA expression that extended the biological understanding of astrocytic tumors and supplied new possible markers of prognosis. They used RNA-Seq in 169 astrocyte tumor samples in which three levels of different ATRX mRNA expressions have been detected [75]. Their approach identified *ATRX* as a prognostic marker and highlighted the power of RNA-Seq technology in characterizing three subsets of astrocytic tumors [76, 77].

Information about environmental and other influences is partly captured in the transcriptome, which can be explored through RNA-Seq, part of the Large-scale Unbiased Sequencing (LUS) family of technologies. RNA-Seq signatures are currently under investigation (e.g., in some breast cancers) and may provide a previous opportunity to combine genomic and transcriptomic data [78]. RNA-Seq analysis of individual subsets of peripheral blood mononuclear cells in patients with autoimmune disease has potential future research.

The importance of RNA-Seq in drug development is becoming increasingly apparent to clinicians and drug developers. Divergence in the expression levels and splicing of drug-metabolizing enzymes, transporters, and targets, such as receptors and ion channels, have been associated with inter-individual differences in optimal drug dose, drug effectiveness, and adverse drug events [79, 80].

Therefore, a comprehensive study of variation in the transcriptome profiles of pharmacologically relevant tissues promises to yield significant insights into the molecular basis of variation in drug response. In pharmacogenomics, polymorphisms that influence the expression levels or effect in alternative splicing of drug-metabolizing enzymes significantly affect drug disposition and response. For instance, UGT1A1*28 (rs8175347), with seven thymine-adenine 13 repeats in the promoter region, leads to decreased transcription rates of this enzyme and substantial toxicity in patients obtaining the topoisomerase inhibitor, irinotecan [81, 82]. Also, alternative splicing of *CYP2D6* often arises in human populations and is liable for the reduced activity of the enzyme [83]. Given these significant and clinically meaningful effects in drug-metabolizing enzymes, a systematic study of the transcriptome focusing on pharmacogenes is needed. With the support of the NIH, the Pharmacogenomics Global Research Network (PGRN) has launched a transcriptome sequencing project to catalog differences in gene expression and splicing between individuals within tissues and pharmacologically significant genes. They used this approach to represent the expression of 389 genes of pharmacologic significance in some human tissue types and lymphoblastoid cell lines (LCLs). Different from many other transcriptome profiling studies using RNA-Seq, this study showed findings for numerous samples across tissues, authorizing the capture of inter-individual divergence in expression levels in addition to comparison of expression and splicing across various tissues [84]. It was possible that peripheral B-lymphocytes, the primary cells from which LCLs were derived, also displayed various expression patterns from the other four physiological tissues (human liver, heart, kidney, adipose tissue) included in the study. These results proposed that considering the phenotype and the gene of interest is essential when utilizing LCLs as a substitute for other tissues in pharmacogenetic consideration and when using tissues as proxies for each other. This study also showed significant variability in gene expression, particularly among drug transporters and drug-metabolizing enzymes. Several cytochrome P450 (CYP) enzymes revealed substantial variability in expression levels between individuals in the liver; such variability can cause differences in drug metabolism across individuals, directing to divergence in drug effectiveness and vulnerability to toxicity [85].

## 8.7    Targeted Sequencing

WES and WGS techniques can be combined in a targeted sequencing (TS) approach to preserve the accuracy and abundance of WGS data while lowering costs. Both coding and noncoding regions of interest genes are captured by this technology. The selected genes are sequenced at a high coverage level typically more than 30-fold, thus leading to improving genotype calling accuracy by reducing error rates and uncertainty in genotype analysis, which are commonly encountered in short-read sequences. Target-enrichment approaches provide rapid detection and analysis of common and rare genetic variations that affect response to therapeutic drugs or adverse effects. This information is critical and fundamental for tailoring personalized pharmacotherapy [40, 57, 61].

There are several custom pharmacogenetic panels including drug target genes and other pharmacogenes that are involved in ADMET (absorption, distribution, metabolism and excretion, and toxicity), such as the PGRNseq panel, xGen Pan-Cancer Panel v2.4, and CleanPlex NGS Panel which cover 84 pharmacogenes, 532 responsible genes for cancer, and 180 pharmacogenes, respectively [57, 61, 86]. The PGRNseq platform is utilized in different PGx profiling researches, and it has the ability to cover all complex variations in different regions of the most important pharmacogenes such as *CYP2A6*, *CYP2D6*, and *HLA*-B genes [40, 44, 86]. Similarly, valuable and comprehensive panels for studying PGx genes have been developed by other research groups. The accuracy and coverage in these panels for more than 100 PK/PD-related genes were higher than 99% [87, 88]. It has been demonstrated this approach is especially relevant in the area of pharmacogenetic research as well as to actionable clinical targets for individualized pharmacotherapy, since particular noncoding sequences of genes encoding phase I and II enzymes (*CYP2C19*, *CYP3A5*, *CYP3A4*, and *UGT1A1*) can be enriched and targeted [40, 87, 88]. Over 5000 patients have been sequenced by PGRNseq with collaboration between the Electronic Medical Records and Genomics (eMERGE) network, and most of identified variants are related to CPIC [40].

Target enrichment can be achieved through multiple strategies such as molecular inversion probe (MIP), polymerase chain reaction (PCR)-based, or hybridization capture-based, but the results can vary considerably among these approaches. Among these methods, the MIP-based approach has been used for large-scale sequencing in several versions with considerable advancements, and the modest quantities of input DNA can be captured with high specificity [89].

Han et al. developed an MPI strategy with capturing improvement based on the PharmaADME database and ADMET-PGx to detect relevant and rare variants of 114 PK-/PD-related genes in 375 Korean subjects. According to their finding, widespread profiling of pharmacogenes which are important for personalized medicine approaches can be easily accomplished using this method [88]. In addition to screening patients for functional variations, these panels aimed to be a diagnostic tool for rapid and reliable identification of rare, potentially clinically significant variations across a population.

The other research group designed an exome panel of capture probe (PGxseq panel) for 100 pharmacogenes including all SNVs and CNVs in 235 patients. They confirmed that a technique like the PGxseq panel can be used as a robust, fast, and accurate method to identify common as well as novel SNVs alongside CNVs in drug target genes, which will provide insights into the area of precision medicine [87]. In this study, the noncoding region has been not sequenced. Klein et al. performed a comprehensive study of 340 ADMET-related pharmacogenes using a targeted NGS-PGx panel with coverage at least 100-fold, and all SNVs, small Indel, and large structural variants were analyzed with MAF below 2%. Similar to other studies, they found that deleterious variations are more prevalent in less common variants, and also they demonstrated that this approach can provide a more accurate pharmacogenetic framework for the prediction of toxicity and adverse effects of drugs [57, 87, 88, 90].

Overall, targeted sequencing is the more cost-effective and higher level of coverage in comparison to other advanced techniques, providing valuable information for uncommon variants and unbiased PGx profiling. But despite this fact, there are some limitations and challenges with the TS technique. One of the major limitations is that only a tiny percentage of all medications' metabolism is regulated by a few genetic variations, while most of pharmaceuticals exhibit only minor effects of multiple variants (most of which are still unknown to date). The other is sometimes difficult to determine large Indels (more than 1 kb) with short-read sequencing since Indel length might exceed the length of the read. Besides this, developing a close collaboration network between clinicians and analysts can be a challenging task.

## 8.8    Single-Cell Sequencing

The current development of SCS techniques has led to a paradigm shift in genomics, away from bulk tissue analysis and toward distinctive and comprehensive research of individual cells. A significant milestone occurred in 2005 with the development of the first NGS technologies, which enabled genome-wide sequencing of DNA and RNA [91]. The highest point of these technologies led to the invention of the first genome-wide single-cell DNA [92] and RNA [93] techniques for mammalian cells. These preliminary studies led to the establishment of a new discipline of biology: single-cell sequencing.

The improvement of DNA SCS techniques has been established to be extra challenging than RNA. A single cell comprises only two copies of each DNA molecule but many copies of most RNA molecules. Due to the restrained amount of WGA input material, some technical errors such as coverage nonuniformity, allele dropout (ADO) events, false-positive (FP) errors, and false-negative (FN) errors occur [94, 95]. The first SCS method was developed for genomic DNA combined with degenerative oligonucleotide PCR with flow-sorting nuclei and NGS to create high-resolution copy number profiles for single mammalian cells [92, 96]. Since then, many SCS with higher coverage technologies have been used.

Single-cell RNA sequencing technology has shown remarkable progress in recent years. RNA must first be amplified by amplifying the entire transcriptome to sequence a single-cell transcriptome. This step is needed because a typical mammalian cell carries only 10 pg of total RNA and 0.1 pg of mRNA [97].

Epigenomic profiling of single cells remains one of the most significant technical challenges in the area. The difficulty is that standard epigenomic sequencing methods need a pool of DNA split into two separate fractions for treatment with bisulfite or methylation restriction enzymes before sequencing. The other technical barrier is that epigenetic DNA modifications cannot be amplified with DNA polymerases. Despite these technical hurdles, studies have made initial progress [98, 99]. SCS methods have impacted many broad fields of biology. They include microbiology, neurobiology, tissue mosaicism, germline transmission, organogenesis, immunology, cancer research, and clinical applications. Figure 8.3 represents the clinical applications of SCS in various fields [100]. Single-cell DNA and RNA sequencing methods supply a powerful new approach to unraveling microbial genomes and depicting intercellular diversity within different populations. However, bacteria and other microorganisms often have only femtograms of DNA

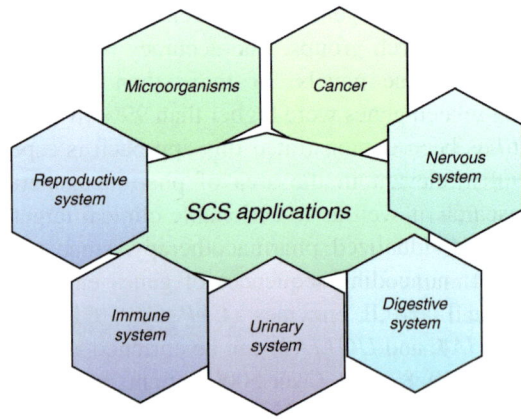

**Fig. 8.3** The clinical applications of SCS in different fields

and RNA, making it even more challenging to amplify than mammalian cells [95].

Single-cell RNA sequencing presents a practical and impartial technique for categorizing neurons based on transcriptional profiles. In a study by Qiu and his coworkers, the RNA sequencing of a single neuron was combined with electrophysiology to obtain transcriptional profiles of mouse embryonic hippocampus and neocortical neurons [101]. In a study done by Usoskin et al. using single-cell RNA sequencing, 622 sensory neurons in mice were profiled, revealing 11 novel expression classes of sensory neuron cell types [102].

A new method for studying the mechanisms that cause germline variation was presented as single-cell DNA sequencing. In a study on this topic, after single sperm cell sequencing, the results consisted of ~22.8 recombination events, 5–15 gene conversion events, and 25–36 de novo mutations in each sperm cell [103]. Copy number profile calculation showed that 7% of the single sperm cells had aneuploid genomes.

RNA SCS was used to analyze transcriptional reprogramming in vitro during the transition from the inner cell mass of blastocysts to pluripotent embryonic stem cells [104]. Also, it was used to study transcriptome dynamics from oocyte to morula development in human and mouse embryos, which delineated a stepwise advancement of pathways that regulate the cell cycle, gene regulation, translation, and metabolism [105].

RNA SCS methods provide a robust new fair approach to perform transcriptional profiling and determine groups of cells that share standard expression programs, representing specific cell types. In another research, RNA SCS was used to analyze lung epithelium development in the first study to apply this approach [106]. These data chased the development of lung progenitor cells that form the alveolar air sac that regulates gas exchange. Also, the authors recognized lots of novel markers for distinguishing the four essential cell types and used them to reconstruct the cell lineage throughout alveolar sac differentiation.

The primary immune cell types have been known for decades, but little is known about transcriptional heterogeneity within cell types that respond to antigens. One study used RNA-SCS to analyze bone marrow-derived dendritic cells from mice stimulated in vitro under various conditions, with individual cells showing different responses mediated by paracrine interferon signaling [107].

Most SCS studies of cancer research have centered on intra-tumor heterogeneity and clonal evolution in primary tumors. The first study used single-nucleus RNA sequencing (SNS) to observe the improvement of aneuploidy expansion in single cells of sufferers with triple-negative (ER/PR/HER2) breast cancer [92]. These data indicated that copy number aberrations developed in punctuated bursts of evolution, tracked by steady clonal expansions to form the tumor mass. In summary, SCS procedures have already substantially enhanced our fundamental understanding of intra-tumor heterogeneity, clonal evolution, and metastatic dissemination in human cancers [108, 109].

SCS techniques have direct translational applications in cancer therapy and prenatal genetic diagnosis (PGD) in clinical applications. In cancer research, intra-tumor heterogeneity shows a considerable challenge for clinical diagnostics because single samples may not represent the tumor as a whole. SCS supplies a potent tool for determining intra-tumor heterogeneity and steering targeted treatment toward the most malignant clones. SCS can also be utilized to estimate a diversity index for each cancer patient, which may have prognostic utility for predicting poor survival and unsatisfactory response to chemotherapy.

Single-cell sequencing translation applications in precision cancer treatment can improve cancer diagnosis, prognosis, targeted therapy, early detection, and noninvasive monitoring [110]. Single-cell sequencing enables sensitive detection of rare mutations and cell-specific gene expression profiles. This method can identify rare tumor tissue variants that may promote drug resistance or act as biomarkers for successful treatment, ultimately advancing cancer genomics [111]. Drug resistance dynamics have been formerly modeled in metastatic breast cancer cell lines using RNA-Seq technology [111].

Treatment of metastatic breast cancer cells with paclitaxel causes the stressed cells to stop and die, but those rare drug-resistant cells resume proliferation, and clones expand. The strength to profile the genome and transcriptome of the same cell can potentially unravel heterogeneity at the genomic, epigenomic, and transcriptomic levels. SCS in drug development grows on bulk genomic data by proposing a more complete and comprehensive picture of responders' underlying genetics, epigenetics, and transcriptomics versus nonresponders at an individual cell level. Applications of SCS in pharmaceutical development include identifying drug candidates and drug targets, drug resistance, and drug reactions and toxicities [112].

## 8.9    DNA Microarray

DNA microarray technology has the potential to be a swift, reliable, and affordable technique for pharmacologic research and clinical activities by allowing investigators to study the expression of the entire human genome simultaneously. DNA microarrays are commonly used to analyze changes in gene expression patterns across the genome to link genes or proteins to drug responses. Disease prevention, drug response prediction, personalized medicine, and the molecular fingerprints of different genetic diseases such as nervous system disorders and cancer could be as results of studying gene expression profiles. Gene expression changes are hierarchic, regulated, and compatible with the phenotypic and physiological responses to medication.

The most widely used platforms for measuring gene expression are Affymetrix and Illumina. The GeneChip was created by Affymetrix using a photolithographic process. By doing microarray experiments, GeneChip is performed to analyze the expression levels of genes in samples. GeneChips are constructed from the sequence repetition of the multiple probes in which a hundred repetitions is usually sufficient. Photolithography and in situ solid-phase oligonucleotide DNA synthesis were used to create the GeneChip array. In contrast, silica microbeads are used in Illumina microarrays. Several copies of an oligonucleotide probe are coated on the silica beads that are placed in microwells targeted at specific genes in the genome. The matrix component can be presynthesized oligonucleotides or PCR-amplified cDNA inserts obtained from expressed sequence tags (ESTs) using high-speed robotics. The density characteristics, tiny features, and the ability to analyze multiple samples simultaneously are some of Illumina's strengths and also is one of the cheapest available techniques [113, 114].

Today, different advancements have been achieved using these platforms in the area of pharmacogenomics, toxicogenomics, gene discovery, and discriminating between responders and nonresponders to prescribed drugs. Several commercial arrays containing pharmacogenetic content are available such as Agena's iPLEX PGx Pro panel, Infinium Global Screening Array (GSA), VeraCode ADME core panel, Affymetrix's Drug Metabolizing Enzymes and Transporters panel, and ADMET arrays, which could be used for detection of PGx variants which are related to drug response based on PharmGKB or PGx guidelines [44, 47, 57, 87, 115].

A large number of variations are included in some commercial arrays and most of them are interested in research in personalized medicine and clinical activities. For example, the AmpliChipTM CYP450 test from Affymetrix microarray technology was authorized by FDA to analyze 27 CYP2D6 alleles (including seven duplications) and three CYP2C19 alleles linked to distinct metabolizing phenotypes [116].

For novel gene and pharmacological target identification, increasing the number of targets on the array to include anonymous ESTs, ESTs with functional orthologues, and homologies to known genes of other animal models can be useful. When assessing microarray approaches, some major considerations should be evaluated: (1) array content, (2) ability to change the content, (3) array expenditure, (4) cycle times for DNA sample, (5) sample size of each array, and (6) technician associated with creating the data. The expression data of each sample should be compared to a publicly available database, and expression profiles will be used to study the biological effect of cytotoxic agents, therapeutic drugs, environmental toxins, and adverse effects of different drugs used for genetic disorders, especially cancers. Using gene expression profiles, it is now possible to identify patients at risk and leukemia subtypes with poor prognoses that may end up failing therapy [57, 113, 115, 117].

Chine et al. analyzed the expression pattern of breast cancer MCF-7 cells chosen for antidoxorubicin resistance or treated with doxorubicin using DNA microarray. They observed transient alterations in the expression of a significant number of genes in MCF-7 cells treated with doxorubicin. Some of these genes such as XRCC1 and microsomal epoxide hydrolase 1 have a critical role in drug resistance, which may lead to accelerated doxorubicin metabolism and reduce medication availability. According to this data, they were able to define the treatment plan and anticipate clinical results for that patient [114]. Stanford University and the University of Florida collaborated in 2012 to establish an SNP microarray panel from 120 genes, including 25 genes involved in drug metabolism and 12 drug transporter genes based on PharmGKB. In total, 256 SNPs are screened including 252 "PGXs SNPs," two quality-control duplicates, and two sex markers. This panel is used for disease risk prediction, improving patient care based on genetic information, and providing pharmacogenetic data in clinical activities [117].

Clinical PGx implementation studies frequently employ these types of arrays. The VeraCode ADME core panel was used in Vanderbilt Electronic Systems for Pharmacogenetic Assessment study and the PREDICT project to generate extensive information for PGx and precision medicine. The St. Jude's Children's Research Hospital used the DMET array through the PG4KDS protocol (step-by-step approach to deploying gene/drug pairings, collecting data, and getting patient and family permission) in 1559 patients that the role of four genes (*CYP2D6*, *CYP2C1*, *TPMT*, and *SLCO1B1*) has been highlighted in PGx implementation [57, 118]. The genome-wide and pharmacogene coverage of the commercial microarray panel has been investigated in the comprehensive study consisting of 15 Affymetrix genome-wide and 18 Illumina arrays. They analyzed more than 20,000 variants in 3146 genes and demonstrated these panels provide low coverage for genome-wide study, but they could be implemented as complementary assays in pharmacogenomics investigations [115].

The developments in DNA microarray allow comparative measurement of all genes and their products and also evaluate the alteration of gene expression levels in response to drug treatment. In combination with PGx approaches in the preclinical phase of drug discovery, this high throughput provides insight into the cytotoxic effects of drugs before they are clinically tested. Besides this, linking the in vivo PK/PD experiments and modeling/stimulation data with an expression profile database would shed light on knowledge of pharmacological mechanisms and therapeutic effects and accelerate the speed of drug discovery. Therefore, comprehensive information was obtained across any chemical compounds and drug targets chosen from monitoring in the combination of expression profile and chemical genomics.

## 8.10   Conclusion

NGS in combination with innovative technologies such as DNA microarray and transcriptome sequencing created a new window for the investigation of genetic disorders. The discovery of the human genome, alongside the development of high-throughput technologies, provides a strong potential for the detection of complex genetic variants, especially with the advent of the era of personalized medicine. In this era, genetic profiling as a useful diagnostic tool enables to offer pharmacological therapy with greater efficacy and fewer unwanted side effects. In recent years, the variety of genotyping technologies for PGx has grown dramatically and keeps rising.

NGS technologies are becoming more common in clinics and PGx research studies and, as costs fall, make it a routine part of medical care and treatment. These technologies provide a lot more information and also are easier, faster, and more targeted than other types of genetic testing, which may be extremely useful when attempting to figure out what's wrong with a patient. The widespread adoption of DNA technologies in various clinical settings will be due to the quick development of component and bioinformatics tools, as well as the lower cost and technical innovation that will allow for testing of a larger number of drug-related genes and biomarkers.

Despite some limitations of DNA technologies such as time-consuming for data analysis, huge storage capacity, the requirement for complicated bioinformatics processes, identifying the high number of VUS and their management, poor coverage of some sequences by various platforms, and the limitation of functional analysis through bioinformatics tools for variants, the use of actionable pharmacogenetic variations and PGx testing in clinical practice and researches is growing. These approaches provide opportunities for PGx variant discovery, more accurate prediction of specific drug phenotypes in individuals, and, as a result, more appropriate genotype-based treatment modifications and a promising future for pharmacogenomics-guided medicine.

## References

1. Madian AG, et al. Relating human genetic variation to variation in drug responses. Trends Genet. 2012;28(10):487–95.
2. Guchelaar H-J. Pharmacogenomics, a novel section in the European Journal of Human Genetics. Berlin: Nature Publishing Group; 2018. p. 1399–400.
3. Suarez-Kurtz G, Parra EJ. Population diversity in pharmacogenetics: a Latin American perspective. Adv Pharmacol. 2018;83:133–54.
4. Evans WE, Johnson JA. Pharmacogenomics: the inherited basis for interindividual differences in drug response. Annu Rev Genomics Hum Genet. 2001;2(1):9–39.
5. Schärfe CPI, et al. Genetic variation in human drug-related genes. Genome Med. 2017;9(1):1–15.
6. Rabbani B, Tekin M, Mahdieh N. The promise of whole-exome sequencing in medical genetics. J Hum Genet. 2014;59(1):5–15.
7. Rabbani B, et al. Next-generation sequencing: impact of exome sequencing in characterizing Mendelian disorders. J Hum Genet. 2012;57(10):621–32.
8. Diaz-Horta O, et al. Whole-exome sequencing efficiently detects rare mutations in autosomal recessive nonsyndromic hearing loss. PLoS One. 2012;7(11):e50628.
9. Kingsmore S, Saunders C. Deep sequencing of patient genomes for disease diagnosis: when will it become routine? Sci Transl Med. 2011;3:87ps23.
10. Bamshad MJ, et al. Exome sequencing as a tool for Mendelian disease gene discovery. Nat Rev Genet. 2011;12(11):745–55.
11. Metzker M. Sequencing technologies-the next generation. Nat Rev Genet. 2010;11(1):31–46.
12. Bielinski SJ, et al. Preemptive genotyping for personalized medicine: design of the right drug, right dose, right time—using genomic data to individualize treatment protocol. In: Mayo clinic proceedings. Elsevier; 2014.
13. Gabrovska P, et al. Gene expression profiling in human breast cancer-toward personalised therapeutics? Open Breast Cancer J. 2010;2:46–59.
14. Rabbani B, et al. Next generation sequencing: implications in personalized medicine and pharmacogenomics. Mol Biosyst. 2016;12(6):1818–30.
15. Hintzsche JD, Robinson WA, Tan AC. A survey of computational tools to analyze and interpret whole exome sequencing data. Int J Genomics. 2016;2016:7983236.
16. Slatko BE, Gardner AF, Ausubel FM. Overview of next-generation sequencing technologies. Curr Protoc Mol Biol. 2018;122(1):e59.
17. Dhawan D. Chapter 2 – Clinical next-generation sequencing: enabling precision medicine. In: Verma M, Barh D, editors. Progress and challenges in precision medicine. Academic Press: New York; 2017. p. 35–54.
18. Health, N.I.o., *NCI dictionary of cancer terms*. National Institutes of Health, Bethesda, MD, USA; 2018.
19. Gilissen C, et al. Genome sequencing identifies major causes of severe intellectual disability. Nature. 2014;511(7509):344–7.
20. Nones K, et al. Genomic catastrophes frequently arise in esophageal adenocarcinoma and drive tumorigenesis. Nat Commun. 2014;5(1):1–9.
21. van El CG, et al. Whole-genome sequencing in health care: recommendations of the European Society of Human Genetics. Eur J Hum Genet. 2013;21(6):580–4.
22. Choi J, Tantisira KG, Duan QL. Whole genome sequencing identifies high-impact variants in well-known pharmacogenomic genes. Pharmacogenomics J. 2019;19(2):127–35.
23. Yang W, et al. Comparison of genome sequencing and clinical genotyping for pharmacogenes. Clin Pharmacol Ther. 2016;100(4):380–8.
24. He Y, Hoskins JM, McLeod HL. Copy number variants in pharmacogenetic genes. Trends Mol Med. 2011;17(5):244–51.
25. Martis S, et al. Multi-ethnic cytochrome-P450 copy number profiling: novel pharmacogenetic alleles and

mechanism of copy number variation formation. Pharmacogenomics J. 2013;13(6):558–66.

26. Santos M, et al. Novel copy-number variations in pharmacogenes contribute to interindividual differences in drug pharmacokinetics. Genet Med. 2018;20(6):622–9.

27. Ashley EA, et al. Clinical assessment incorporating a personal genome. Lancet. 2010;375(9725):1525–35.

28. Daly AK. Genome-wide association studies in pharmacogenomics. Nat Rev Genet. 2010;11(4):241–6.

29. Welch JS, et al. Use of whole-genome sequencing to diagnose a cryptic fusion oncogene. JAMA. 2011;305(15):1577–84.

30. Jorgenson E, Witte JS. A gene-centric approach to genome-wide association studies. Nat Rev Genet. 2006;7(11):885–91.

31. Badiee A, et al. Evaluation of five different cDNA labeling methods for microarrays using spike controls. BMC Biotechnol. 2003;3(1):1–5.

32. Lam HY, et al. Performance comparison of whole-genome sequencing platforms. Nat Biotechnol. 2012;30(1):78–82.

33. Zook JM, et al. Integrating human sequence data sets provides a resource of benchmark SNP and indel genotype calls. Nat Biotechnol. 2014;32(3):246–51.

34. Klein H-G, Bauer P, Hambuch T. Whole genome sequencing (WGS), whole exome sequencing (WES) and clinical exome sequencing (CES) in patient care. LaboratoriumsMedizin. 2014;38 https://doi.org/10.1515/labmed-2014-0025.

35. Suwinski P, et al. Advancing personalized medicine through the application of whole exome sequencing and big data analytics. Front Genet. 2019;10:49.

36. Yang Y, et al. Clinical whole-exome sequencing for the diagnosis of mendelian disorders. N Engl J Med. 2013;369(16):1502–11.

37. McVean GA, et al. An integrated map of genetic variation from 1,092 human genomes. Nature. 2012;491(7422):56–65.

38. Tennessen JA, et al. Evolution and functional impact of rare coding variation from deep sequencing of human exomes. Science. 2012;337(6090):64–9.

39. Lek M, et al. Analysis of protein-coding genetic variation in 60,706 humans. Nature. 2016;536(7616):285–91.

40. Schwarz UI, Gulilat M, Kim RB. The role of next-generation sequencing in pharmacogenetics and pharmacogenomics. Cold Spring Harb Perspect Med. 2019;9(2):a033027.

41. Dewey FE, et al. Distribution and clinical impact of functional variants in 50,726 whole-exome sequences from the DiscovEHR study. Science. 2016;354(6319):aaf6814.

42. Kozyra M, Ingelman-Sundberg M, Lauschke VM. Rare genetic variants in cellular transporters, metabolic enzymes, and nuclear receptors can be important determinants of interindividual differences in drug response. Genet Med. 2017;19(1):20–9.

43. Katsila T, Patrinos GP. Whole genome sequencing in pharmacogenomics. Front Pharmacol. 2015;6:61.

44. Tafazoli A, et al. Applying next-generation sequencing platforms for Pharmacogenomic testing in clinical practice. Front Pharmacol. 2021;12(2025):693453.

45. Chua EW, et al. Cross-comparison of exome analysis, next-generation sequencing of amplicons, and the iPLEX(®) ADME PGx panel for Pharmacogenomic profiling. Front Pharmacol. 2016;7:1.

46. Mueller JJ, et al. Massively parallel sequencing analysis of mucinous ovarian carcinomas: genomic profiling and differential diagnoses. Gynecol Oncol. 2018;150(1):127–35.

47. Pereira M, et al., Application of next-generation sequencing in the era of precision medicine; 2017.

48. Costain G, et al. Clinical application of targeted next-generation sequencing panels and whole exome sequencing in childhood epilepsy. Neuroscience. 2019;418:291–310.

49. Raffan E, Semple RK. Next generation sequencing—implications for clinical practice. Br Med Bull. 2011;99(1):53–71.

50. Khoja L, et al. Pembrolizumab. J Immunother Cancer. 2015;3:36.

51. Prasad V, Kaestner V, Mailankody SJJO. Cancer drugs approved based on biomarkers and not tumor type—FDA approval of pembrolizumab for mismatch repair-deficient solid cancers. JAMA Oncol. 2018;4(2):157–8.

52. Jørgensen JT. Twenty years with personalized medicine: past, present, and future of individualized pharmacotherapy. Oncologist. 2019;24(7):e432–40.

53. Demkow U. Chapter 11 - next generation sequencing in pharmacogenomics. In: Demkow U, Płoski R, editors. Clinical applications for next-generation sequencing. Boston: Academic Press; 2016. p. 217–40.

54. Cohn I, et al. Genome sequencing as a platform for pharmacogenetic genotyping: a pediatric cohort study. npj Genomic Med. 2017;2(1):19.

55. Zhang Q, et al. Clinical application of whole-exome sequencing: a retrospective, single-center study. Exp Ther Med. 2021;22(1):753.

56. Carter TC, He MM. Challenges of identifying clinically actionable genetic variants for precision medicine. J Healthc Eng. 2016;2016:3617572.

57. van der Lee M, et al. Technologies for pharmacogenomics: a review. Genes (Basel). 2020;11:12.

58. Cousin MA, et al. Pharmacogenomic findings from clinical whole exome sequencing of diagnostic odyssey patients. Mol Genet Genomic Med. 2017;5(3):269–79.

59. Rennert H, et al. Development and validation of a whole-exome sequencing test for simultaneous detection of point mutations, indels and copy-number alterations for precision cancer care. Npj. Genomic Med. 2016;1(1):16019.

60. Dong L, et al. Clinical next generation sequencing for precision medicine in cancer. Curr Genomics. 2015;16(4):253–63.

61. Tilleman L, et al. Pan-cancer pharmacogenetics: targeted sequencing panels or exome sequencing? Pharmacogenomics. 2020;21(15):1073–84.

62. Guidugli L, et al. Clinical utility of gene panel-based testing for hereditary myelodysplastic syndrome/acute leukemia predisposition syndromes. Leukemia. 2017;31(5):1226–9.

63. Conesa A, et al. A survey of best practices for RNA-seq data analysis. Genome Biol. 2016;17(1):1–19.

64. Hoelzer D, et al. Acute lymphoblastic leukaemia in adult patients: ESMO clinical practice guidelines for diagnosis, treatment and follow-up. Ann Oncol. 2016;27:v69–82.

65. Wang ET, et al. Alternative isoform regulation in human tissue transcriptomes. Nature. 2008;456(7221):470–6.

66. Denoeud F, et al. Annotating genomes with massive-scale RNA sequencing. Genome Biol. 2008;9(12):1–12.

67. Maher CA, et al. Transcriptome sequencing to detect gene fusions in cancer. Nature. 2009;458(7234):97–101.

68. Carey MF, Peterson CL, Smale ST. Chromatin immunoprecipitation (chip). Cold Spring Harb Protoc. 2009;2009(9):pdb.prot5279.

69. Soldner F, et al. Parkinson's disease patient-derived induced pluripotent stem cells free of viral reprogramming factors. Cell. 2009;136(5):964–77.

70. Matigian N, et al. Disease-specific, neurosphere-derived cells as models for brain disorders. Dis Model Mech. 2010;3(11–12):785–98.

71. Atz M, et al. Methodological considerations for gene expression profiling of human brain. J Neurosci Methods. 2007;163(2):295–309.

72. Monoranu CM, et al. pH measurement as quality control on human post mortem brain tissue: a study of the BrainNet Europe consortium. Neuropathol Appl Neurobiol. 2009;35(3):329–37.

73. Courtney E, et al. Transcriptome profiling in neurodegenerative disease. J Neurosci Methods. 2010;193(2):189–202.

74. Janitz M. Next-generation genome sequencing: towards personalized medicine. Hoboken, NJ: John Wiley & Sons; 2011.

75. Cai J, et al. ATRX mRNA expression combined with IDH1/2 mutational status and Ki-67 expression refines the molecular classification of astrocytic tumors: evidence from the whole transcriptome sequencing of 169 samples. Oncotarget. 2014;5(9):2551.

76. Kridel R, et al. Whole transcriptome sequencing reveals recurrent NOTCH1 mutations in mantle cell lymphoma. Blood J Am Soc Hematol. 2012;119(9):1963–71.

77. Greif P, et al. Identification of recurring tumor-specific somatic mutations in acute myeloid leukemia by transcriptome sequencing. Leukemia. 2011;25(5):821–7.

78. Creighton CJ, Reid JG, Gunaratne PH. Expression profiling of microRNAs by deep sequencing. Brief Bioinform. 2009;10(5):490–7.

79. Wang L, McLeod HL, R.M.J.N.E.J.o.M. Weinshilboum. Genomics and drug response. N Engl J Med. 2011;364(12):1144–53.

80. Evans WE, McLeod HL. Pharmacogenomics—drug disposition, drug targets, and side effects. N Engl J Med. 2003;348(6):538–49.

81. Iyer L, et al. UGT1A1* 28 polymorphism as a determinant of irinotecan disposition and toxicity. Pharmacogenomics J. 2002;2(1):43–7.

82. Tukey RH, Strassburg CP, Mackenzie PI. Pharmacogenomics of human UDP-glucuronosyltransferases and irinotecan toxicity. Mol Pharmacol. 2002;62(3):446–50.

83. Wang D, et al. Common CYP2D6 polymorphisms affecting alternative splicing and transcription: long-range haplotypes with two regulatory variants modulate CYP2D6 activity. Hum Mol Genet. 2014;23(1):268–78.

84. Chhibber A, et al. Transcriptomic variation of pharmacogenes in multiple human tissues and lymphoblastoid cell lines. Pharmacogenomics J. 2017;17(2):137–45.

85. Zanger UM, Schwab M. Cytochrome P450 enzymes in drug metabolism: regulation of gene expression, enzyme activities, and impact of genetic variation. Pharmacol Ther. 2013;138(1):103–41.

86. Gordon AS, et al. PGRNseq: a targeted capture sequencing panel for pharmacogenetic research and implementation. Pharmacogenet Genomics. 2016;26(4):161–8.

87. Gulilat M, et al. Targeted next generation sequencing as a tool for precision medicine. BMC Med Genet. 2019;12(1):81.

88. Han SM, et al. Targeted next-generation sequencing for comprehensive genetic profiling of pharmacogenes. Clin Pharmacol Ther. 2017;101(3):396–405.

89. Mamanova L, et al. Target-enrichment strategies for next-generation sequencing. Nat Methods. 2010;7(2):111–8.

90. Klein K, et al. A new panel-based next-generation sequencing method for ADME genes reveals novel associations of common and rare variants with expression in a human liver cohort. Front Genet. 2019;10:7.

91. Mardis ERJN. A decade's perspective on DNA sequencing technology. Nature. 2011;470(7333):198–203.

92. Navin N, Hicks J. Future medical applications of single-cell sequencing in cancer. Genome Med. 2011;3(5):1–12.

93. Tang F, et al. mRNA-Seq whole-transcriptome analysis of a single cell. Nat Methods. 2009;6(5):377–82.

94. Dean FB, et al. Comprehensive human genome amplification using multiple displacement amplification. Proc Natl Acad Sci. 2002;99(8):5261–6.

95. Lasken RS. Single-cell genomic sequencing using multiple displacement amplification. Curr Opin Microbiol. 2007;10(5):510–6.

96. Baslan T, et al. Genome-wide copy number analysis of single cells. Nat Protoc. 2012;7(6):1024–41.

97. Van Gelder RN, et al. Amplified RNA synthesized from limited quantities of heterogeneous cDNA. Proc Natl Acad Sci. 1990;87(5):1663–7.

98. Nagano T, et al. Single-cell hi-C reveals cell-to-cell variability in chromosome structure. Nature. 2013;502(7469):59–64.

99. Guo H, et al. Single-cell methylome landscapes of mouse embryonic stem cells and early embryos analyzed using reduced representation bisulfite sequencing. Genome Res. 2013;23(12):2126–35.

100. Tang X, et al. The single-cell sequencing: new developments and medical applications. Cell Biosci. 2019;9(1):1–9.

101. Qiu S, et al. Single-neuron RNA-Seq: technical feasibility and reproducibility. Front Genet. 2012;3:124.

102. Usoskin D, et al. Unbiased classification of sensory neuron types by large-scale single-cell RNA sequencing. Nat Neurosci. 2015;18(1):145–53.

103. Wang J, et al. Genome-wide single-cell analysis of recombination activity and de novo mutation rates in human sperm. Cell. 2012;150(2):402–12.

104. Tang F, et al. Tracing the derivation of embryonic stem cells from the inner cell mass by single-cell RNA-Seq analysis. Cell Stem Cell. 2010;6(5):468–78.

105. Xue Z, et al. Genetic programs in human and mouse early embryos revealed by single-cell RNA sequencing. Nature. 2013;500(7464):593–7.

106. Treutlein B, et al. Reconstructing lineage hierarchies of the distal lung epithelium using single-cell RNA-seq. Nature. 2014;509(7500):371–5.

107. Shalek AK, et al. Single-cell RNA-seq reveals dynamic paracrine control of cellular variation. Nature. 2014;510(7505):363–9.

108. Navin NE. Cancer genomics: one cell at a time. Genome Biol. 2014;15(8):1–13.

109. Van Loo P, Voet T. Single cell analysis of cancer genomes. Curr Opin Genet Dev. 2014;24:82–91.

110. Navin NE. The first five years of single-cell cancer genomics and beyond. Genome Res. 2015;25(10):1499–507.

111. Lee M-CW, et al. Single-cell analyses of transcriptional heterogeneity during drug tolerance transition in cancer cells by RNA sequencing. Proc Natl Acad Sci. 2014;111(44):E4726–35.

112. Xu X, et al. Single-cell exome sequencing reveals single-nucleotide mutation characteristics of a kidney tumor. Cell. 2012;148(5):886–95.

113. Fajriyah R. Paper review: an overview on microarray technologies. Bull Appl Math Math Educ. 2021;1(1):21–30.

114. Chin K-V, Kong ANT. Application of DNA microarrays in pharmacogenomics and Toxicogenomics. Pharm Res. 2002;19(12):1773–8.

115. Lemieux Perreault LP, et al. Pharmacogenetic content of commercial genome-wide genotyping arrays. Pharmacogenomics. 2018;19(15):1159–67.

116. de Leon J, Susce MT, Murray-Carmichael E. The AmpliChip CYP450 genotyping test: integrating a new clinical tool. Mol Diagn Ther. 2006;10(3):135–51.

117. Johnson JA, et al. Implementing personalized medicine: development of a cost-effective customized pharmacogenetics genotyping array. Clin Pharmacol Ther. 2012;92(4):437–9.

118. Hoffman JM, et al. PG4KDS: a model for the clinical implementation of pre-emptive pharmacogenetics. Am J Med Genet C Semin Med Genet. 2014;166C(1):45–55.

# Precision Medicine Initiatives

9

Forough Taheri, Monika Frenzel, Pirooz Ebrahimi,
Negar Sarhangi, Mandana Hasanzad,
and Mahsa M. Amoli

**What Will You Learn in This Chapter?**
This chapter gives you an update on various regional, national, and international initiatives in the field of precision medicine conducted through public or private sectors or as inter-sectorial cooperation efforts.

**Rationale and Importance**
A vast range of projects, initiatives, and collaborative programs are currently underway or scheduled in the field of precision medicine which includes many countries and sectors all over the world. Although we were not able to fully cover all running activities in this short chapter, we made the most effort to at least give a snapshot of ongoing projects. Comprehensive coverage of those initiatives definitely can prevent many cases of double-work among the professional community while helping the organizers and designers of research programs both in private and public sectors to assign academic, financial, and workforce resources in a smarter way to complement, enhance, and advance the previous achievements through a big-picture perspective provided by such a survey.

F. Taheri · M. M. Amoli (✉)
Metabolic Disorders Research Center, Endocrinology and Metabolism Molecular-Cellular Sciences Institute, Tehran University of Medical Sciences, Tehran, Iran
e-mail: amolimm@tums.ac.ir

M. Frenzel
French National Research Agency, Paris, France

P. Ebrahimi
Department of Pharmacy, Health and Nutritional Sciences, University of Calabria, Arcavacata, Italy

N. Sarhangi
Endocrinology and Metabolism Research Center, Endocrinology and Metabolism Clinical Sciences Institute, Tehran University of Medical Sciences, Tehran, Iran

M. Hasanzad
Personalized Medicine Research Center, Endocrinology and Metabolism Clinical Sciences Institute, Tehran University of Medical Sciences, Tehran, Iran

Medical Genomics Research Center, Tehran Medical Sciences, Islamic Azad University, Tehran, Iran

## 9.1 Introduction

A large number of public- and private-funded initiatives have been launched or are currently underway regarding various aspects of precision medicine including data sharing, legal and information privacy, incorporating and integrating various large databases, research and medical aspects, gene sequencing, and all other required preconditions, platforms, and infrastructures.

Countries on many continents in the world have been aware of the tremendous potential of precision medicine for a long time, and many countries went ahead preparing legal and legislative laws, data-sharing facilities, and technical structures in the form of funding national and regional programs.

However, many challenges exist in this field. For example, patients and public engagement in programs, pricing and reimbursement pathways, and translation into clinical practice are some of those challenges to be faced.

The field of precision medicine is growing and changing at a very fast pace; the challenge in such a field is mainly that a "gold standard" for clinical trials at a large scale cannot be feasible. However, several choices are being available by various agencies that include creative approaches like some new models of evidence development and sharing the risks among technology developers, payers, and healthcare systems [1]. For example, the Centers for Medicare and Medicaid Services (CMS) which is a US government agency had published 22 policies by December 2016 regarding the "coverage with evidence development" [2], but as mentioned before, no one-size-fits-all approach can exist, because the threshold of documenting and managing evidence in each case and situation would be due to risks of the tests and financial impact on the stakeholders [3].

## 9.2    Precision Medicine Initiatives

As mentioned, many countries have been launching initiatives and programs in precision medicine and will be mentioned here, reflecting a subset of currently running activities. For example, Australian Genomics Health Alliance[1] has developed the national framework for translating -omics discoveries into clinical research and practice. This framework is intended to provide incorporating advice based on the results from genomics research and clinical testing/trials. The program called "Australian genomics' work" is focused on 24 prioritized activities and projects addressing the strategic necessities throughout the entire value chain of research in genomics and the health system. Some of the

main categories are information management in genomics, training, and workforce, public knowledge regarding genomics, as well as local genomic priorities.

In Belgium, the Belgian Medical Genomics Initiative[2] is designed mainly to (1) improve individual efforts toward research in understanding the biology of diseases, (2) improve approaches for prediction of clinical consequences from genomic data to achieve a leading role for the country to facilitate the incorporation/integration of genomic information systems into clinical healthcare in Belgium, and (3) play a booting and bolstering role in preparing the next generations of researchers in the genomics field and improving the knowledge of medical practitioners regarding the fast-evolving trends in medical genomics, and promoting public engagement.

GenomeCanada[3] is a large-scale initiative comprising several platforms which support a precision medicine structure. One of the main purposes of GenomeCanada is to arrange the funding for the operations of personalized health in clinics augmenting its benefit. This strategy is presently concentrated on the clinical diagnosis of rare diseases. Development of the necessary organization, facilities, guidelines, and partnerships is the main focus point of the work which provides a wider national initiative specially designed for precision medicine.

Estonian Program for Personal Medicine (ePerMed)[4] intended to sequence the genome of 5000 people; the current cohort (as of November 2021) is 200,000 individuals. The program aimed to develop an Estonian genotyping array and offered to all the population between ages 35 and 65.

"Plan France Médecine Génomique 2025"[5] is launched in response to the request of France's prime minister in April 2015 from the Aviesan Alliance that asked for examinations and implementation of methods to provide large-scale access to the genetic diagnosis in France.

---

[1] https://www.australiangenomics.org.au

[2] http://www.belspo.be

[3] https://www.genomecanada.ca/

[4] https://www.geenivaramu.ee/en

[5] https://www.sensgene.com

This plan addresses three main objectives:

1. Preparation of the integration of genomic medicine in the current healthcare stream and disease management procedures. The goal is to ensure fair access to genomic medicine for all patients. Currently, four pilot projects are running focusing on cancer (MULTISARC trial), rare diseases (DEFIDIAG – cognitive impairment), common diseases (GLUCO-GEN – diabetes), and general population (POPGEN).
2. To set up the structure of a national sector for genomic medicine to serve patients. This sector should be capable of enhancing scientific and technological innovation, as well as encouraging industrial efforts and economic growth.
3. Create a place for France as one of the world's leading countries committed to personalized medicine while creating the capability to export the practical genomic medicine knowledge within the medical and industrial sectors.

In Israel, the Bench-to-Bedside Project[6] has been started by the "Weizmann Institute of Science" in cooperation with "Clalit Health Systems" which is Israel's largest Health Maintenance Organization (HMO). These institutes are conducting a joint research program that will facilitate the integration and sharing of patients' data as a unified information database for scientific research and clinical care activities. The program, which is the central feature of the new Bench-to-Bedside Project, is aimed to be a major step in linking the scientific research findings with a large data repository that stores and provides medical records for more than half of the country's population.

Many public and private institutions in South Korea, including Macrogen, Samsung Genome Institute, and the Korea University, have been cooperating to create "the Cancer Precision Medicine Diagnosis and Treatment Enterprise"

(K-MASTER) with the goal of the advancement of cancer solutions available in the country [4].

The K-MASTER project group has been aiming to create an optimal domestic integrated platform that is convergent even at a large scale to develop a global leading precision medical diagnosis and treatment scheme of cancer precision medicine so that rapid clinical application and multicenter utilization would be possible. Development and advancement of cancer precision diagnostic NGS panel, cancer genome profiling of 10,000 people, network construction of participating hospitals nationwide, and a clinical genome analysis process management system are among the main objectives of this project [5].

Luxembourg Centre for Systems Biomedicine (LCSB)[7] is in close collaboration with the focus on stratification and early diagnosis of various forms of Parkinson's disease. LCSB cooperates with the "Luxembourg Institute of Health" (LIH) and its integrated database called the "Integrated BioBank of Luxembourg" (IBBL) through the framework of the Personalized Medicine Consortium (PMC) of Luxembourg.

The LCSB was founded in 2009 as the first biomedical research center of the University of Luxembourg. This facility is located at the new university campus in Esch-sur-Alzette since September 2011.

The main goal of the LCSB is to close the gap between systems biology and medical research and accelerate biomedical research in this way. The main focus of LCSB is on the study of neurodegenerative diseases while special emphasis is placed on Parkinson's disease.

Currently, around 230 employees and 16 research groups are working for the LCSB which is organized into three departments: (1) computation department (responsible for Bioinformatics Core, Molecular Systems Physiology, Computational Biology, Biomedical Data Science, Systems Biochemistry, and Systems Control), (2) experimental department (working on Chemical Biology, Integrative Cell Signaling, Experimental Neurobiology, Developmental and Cellular Biology, Eco-systems Biology,

---

[6]www.weizmann.ac.il

[7]wwwfr.uni.lu

Enzymology and Metabolism, Molecular and Functional Neurobiology), and (3) translation department with the clinical orientation (conducting Medical Translational Research, as well as Clinical and Experimental Neuroscience activities).

In Singapore, the "Personalized OMIC Lattice for Advanced Research and Improving Stratification" (POLARIS) program[8] was established by A*STAR in 2013 as a pilot project for the application of clinical genomics knowledge in the diagnosis and treatment of medical diseases in that country and the region [6].

The main goals of the project are to act as a pilot phase of TGFBI gene testing for the diagnosis of the disease and assess the family risk regarding the stromal corneal dystrophies and then to implement a panel including 90 genes related to gastrointestinal cancers. "Pharmacogenomics Network" in Thailand[9] was initiated to implement pharmacogenomics as a tool for identifying the risks of using the top ten drugs which have been associated with SJS/TEN as an integrated part of the nationwide pharmacovigilance program [3].

In South Africa, the South African Medical Research Council (SAMRC) launched the South African Precision Medicine Program aiming to build innovative tools and capabilities to create a precision medicine environment. It supports genomic research to contribute to a more proactive and preventive approach to health. The SAMRC launched in 2017, in collaboration with Innovate UK, a joint call to enable precision medicine projects that develop affordable gene-based diagnostics for noncommunicable diseases and child health, targeted specifically at the South African population. Research is fostered to cover genome-wide investigations addressing the burden of disease in Southern Africa and assessing the impact of genomic variants on the health of the indigenous populations of Africa.

Furthermore, in 2019, the SAMRC Genomics Centre[10] opened. This national research and innovation platform is mandated to further develop the SAMRC Precision Medicine program for South Africa by operationalizing the first high-throughput sequencing center focusing on the analysis of the population's genetic diversity. The program also involves initiating innovative research projects and managing SAMRC funding streams, external grants, and other key SAMRC initiatives to develop precision medicine approaches. Also in 2019, the SAMRC launched an open call for researchers based at South African universities, science councils, and other public research organizations, focused on pharmacogenomics in precision medicine.[11] Many diverse aims and goals are defined for this type of program all over the world. The focal points of any project are determined by the regional, national, and population needs and necessities. However, future works need to be focused on emphasizing the practical importance of the ability of precision medicine to affect the populations and dedicated target groups, hence the term "precision public health." The initial stages and drive toward precision public health are already in progress, but more efforts are needed to realize the development of a robust evidentiary foundation for practical use. Also increasing the level of understanding about how precision medicine might cause the historical gaps of access to fair and widespread public healthcare to narrow or widen would be another point of emphasis. Do the strata of societies that did not have access to precision medicine services in the past benefit from the new signs of progress? And what policies could be implemented to ensure appropriate access? These types of questions have to be addressed within the larger and ever-changing context of healthcare reforms and the pertinent revisions proposed in form of medical aid program updates.

The funds dedicated to "lifestyle medicine" have been at relatively lower levels in the past, but in fact, some key initiatives exist. "The Common Fund initiative on the Molecular

[8] https://www.a-star.edu.sg

[9] http://www.thailandpg.org/

[10] https://www.samrc.ac.za/innovation/genomics-centre

[11] https://www.samrc.ac.za/media-release/first-high-through-put-genomics-sequencing-centre-africa-decode-genes

Transducers of Physical Activity in Humans" introduced by the "National Institutes of Health" is one of the largest ones which is launched to determine the optimal recommendations for the state and quality of physical activities for people at various stages of life. The program aims to provide "precisely targeted [exercise] regimens" for any individual according to their particular health needs [7]. "The Molecular Transducers of Physical Activity Consortium" (MoTrPAC) aims to discover the molecular mechanisms regarding how exercise improves and maintains the health of the body's tissues and organs.

As another example, the Food4Me research project, a program funded by the European Union, is aimed to explore the effects and consequences of genetic traits on the personalization of diet. Findings suggest that when the interventions in the form of personalized diets are supported with genetic data, they would lead to improvements in the consumption of healthy foods [8] and educate people about how to sustain their choices regarding healthy food consumption [9].

Interestingly, there is little evidence of clinical relevance for the most loci selected for intervention through the Food4Me project (variants at APOE, TCF7L2, FTO, MTHFR, and FADS1) [10] suggesting that the success of the Food4Me intervention did not depend on the quality of the genetic data given to the participants.

The findings of precision medicine initiatives almost always are assumed to help with improvements of people's health; however, the emphasis in the majority of cases is to always generate new scientific knowledge. For example, the "Million Veteran Program" under the "Precision Medicine Initiative (PMI)" in the United States is designed "to improve our understanding regarding the process by which the genetic characteristics, environmental factors, and human behaviors could affect our health" [11]. Although this objective will certainly be realized, the practical use of this information to optimize the efforts regarding the prevention and treatment of any particular malady is not clear yet. There are only a small number of precision medicine initiatives that are focused on complex diseases and could lead to a practical and clear action plan for clinical practices [12].

## 9.2.1 The Precision Medicine Initiative (PMI) (United States)

The PMI program launched in 2015 is a nationwide initiative designed by the US government with the vision to gradually abandon the "one-size-fits-all" approach to healthcare delivery and to provide disease prevention and treatment strategies customized based on certain characteristics of any person, including their biological traits, environment, as well as lifestyle [13].

The Office of the National Coordinator for Health IT (ONC) supports the PMI initiative in several ways including improving the conditions for innovative collaboration through pilot projects and testing of standards that support health IT interoperability for research, designing and implementing standards and policies to support the security and privacy of information for cohort participants, and revising/improving the standards that facilitate participant-driven data contribution.

As of November 2021, there are various activities underway by ONC in the framework of the PMI initiative:

"Sync for Science" is a private/public collaboration program that aims to develop a secure, simplified, and scalable access route for individual citizens to use their electronic health record data and share them with the researchers.

"Sync for Genes" is a project to create standard ways to share genomic information among laboratories, researchers, patients, and providers as well as enhance the development and application of standards for genomic data which are supported by the industry.

"Advancing Standards for Precision Medicine" is a project which aims to facilitate aggregation, curate, sharing, and synthesizing the healthcare data through the focus on standards of mobile healthcare, wearable devices, and sensor data as well as social factors having determined effects on the health data.

There are several completed activities that were done by the ONC in the past, including the "API Privacy and Security Considerations," which is a report about the key considerations for the security and privacy of "healthcare application programming interfaces" (APIs) required for "Sync for Science" project. And the "Data Security Principles" is a guideline providing some best practices in data management within the PM realm as well as the security issues. This guide was generated for the benefit of the organizations that work within the PMI initiative or use its resulting data [14].

In 2015, when the PMI with an initial budget of $215 million was announced by the US federal government, the plans to implement PM accelerated to a new level [15, 16]. The PMI was focused on cancer in the short term, whereas the initiative was planned to cover all areas of health and healthcare. The major aspect of the PMI is that a cohort including one million American residents was registered in a program called "All of Us." One of the goals was to manage the information so that accredited researchers can access the data stored in the biobanks and samples that would help with the achievement of several major goals, including pharmacogenomics objectives, innovative clinical trials, interactions between genes and the environment, mobile health technologies, stratification of risk categories, health empowerment, etc. [12]

There are a few federal programs and projects that are related to the PMI: the "ONC Patient-Centered Outcomes Research" is conducted through collaboration with several federal agencies. A series of significant projects have been conducted by the ONC which lead to progress regarding the creation of standards, policies, and services necessary for enhancing the data infrastructures for further research. Those measures ensure that the decisions made by patients are considered as a part of research and healthcare recommendations.

"All of Us Research Program" is a national effort by the National Institutes of Health (NIH) to collect genetic and medical data from at least one million people in the United States to help researchers to develop personalized treatments and methods of disease prevention through precision medicine.

"Precision FDA" is a cloud-based portal that allows science experts from various fields of industry, government, academia, and other partners to form close relationships to facilitate innovations.

The "Million Veteran Program" is another effort the ONC has participated in, which is one of the world's largest databases about medical information with adequate security. Health information and blood samples are being safely collected from one million veteran volunteers; the gathered and genetically generated information enables the experts to know how genes affect certain medical conditions mainly related to military illnesses.

The "Big Data Science Initiative" is launched by the collaboration of ONC, the US Department of Energy, and the Department of Veteran Affairs. The main goal of the project is to assist the stakeholders to advance the healthcare and next-generation supercomputing designs [14].

### 9.2.2 Electronic Medical Records and Genomics (eMERGE)

The National Human Genome Research Institute (NHGRI) has provided various funding programs to identify and eliminate the issues and problems in turning discoveries in genomic science into actual clinical practice. Those initiatives are focused on evidence generation, documentation, implementation, and health policies in genomics and precision medicine.

The eMERGE network [17] was the first major NHGRI consortium intended to explore the applications of DNA repositories in connection with electronic medical record systems. Its objective was to advance the discoveries while addressing the policy issues regarding national strategies for precision medicine research. An electronic library of phenotyping called "Phenotype KnowledgeBase" was created by the eMERGE program and integrated with practically usable variants of electronic medical record (EMR) systems for clinical care procedures.

Recently, many pilot projects have been initiated to examine the usage of various pharmacogenetic achievements to provide know-how for treatments using EMR-integrated clinical decision support tools [3].

### 9.2.3 The Clinical Sequencing Evidence-Generating Research (CSER)

The Clinical Sequencing Evidence-Generating Research (CSER) consortium [18] was formed in 2011 to create methods for the integration of genome sequencing science into routines within the clinical medicine field. The main purposes included advancing the process of discovery and interpretation of new genomic variants and investigating the practical impact of genome sequencing knowledge on outcomes of healthcare procedures.

A body of evidence about the clinical applications of genome sequencing advantages has been defined, generated, and analyzed by the consortium; interactions among healthcare practitioners, clinical laboratories, patients, and their family members and those which influence the application of genome sequencing methods within the clinical schemes are investigated and documented; and real-world issues and barriers for integrating the clinical, genomic, and healthcare data utilization throughout a national healthcare apparatus with the purpose of better decision-making are identified and addressed [3].

### 9.2.4 Implementing GeNomics in PracTicE (IGNITE)

The IGNITE [19] initiative was formed in 2013 as a "pragmatic clinical trials network" funded by the NIH in the United States. This initiative is specially designed to support the implementation of genomics in healthcare.

The aim is to examine and mitigate the challenges regarding the wide-range clinical implementations of genomic medicine. As such, some evidence of its real-world utility can be developed. IGNITE has investigated and expanded various models for genomic medicine in practical usage that integrate genomic data perfectly into the electronic medical records, and several tools for helping the decision-making process at the point of care have been deployed [3].

### 9.2.5 ALL of Us

"All of Us" Research Program is a large initiative conducted by the NIH. This program is an effort to enhance the PM procedures through collecting and analyzing a wide range of data from at least one million Americans. The main point of the "All of Us" initiative is to accelerate the pace of health research and medical breakthroughs which leads to the possibilities of individualized prevention, treatment, and healthcare for all Americans.[12]

The program is designed to collect various types of genetic, environmental, and clinical as well as lifestyle information. The participants consent to the data being collected on a regular and ongoing basis throughout their lifetime. The data are being collected through a wide variety of methods including surveys, electronic health records, wearable technologies, and bio-specimens.

Diversity is one of the core values of the *All of Us* Research Program. The participants are from very different age groups, ethnicities, races, and regions of the country. Also, a great amount of diversity exists in terms of health status, disability, education, socioeconomic status, sexual orientation, and gender identity. Particularly, the populations who are traditionally not presented adequately in biomedical research are considered here. All the participants can access their information at will and the program goals are defined such that the participation is based on engagement and full transparency. Also, a diverse group of researchers ranging from prominent university labs to citizen scientists is involved [20].

---

[12] https:llallofus.nili.gov

Besides the biobank, the All of Us Research Program has established a "Data and Research Support Center"[13] that aims to securely store and organize the data while providing authenticated access to one of the world's largest datasets for precision medicine practitioners and researchers [21].

### 9.2.6    100,000 Genomes

UK's 100,000 Genomes Project was initiated with the vision to sequence 100,000 genomes from 70,000 patients suffering from cancer or rare diseases. Acquired information involves phenotypic, genomic, and other clinical data. Another purpose of the program is to transform the healthcare system of the United Kingdom into a precision medicine scheme by creating the capability, legacy, and capacity of personalized medicine.

The 100,000 Genomes Project is conducted by 13 national health system (NHS) centers around the country which is dedicated to genomic medicine. A specially commissioned data center that is secure behind an NHS firewall and approved by the government is responsible for sequencing the genomes and collecting electronic health record data from the patients.

Academic researchers need to become members of a "Genomics England Clinical Interpretation Partnership" to access the data.

A clinical data capture system including predetermined "Human Phenotype Ontology" (HPO) terms for each category of diseases and a standardized reporting system for results of imaging and lab tests has been created for the rare disease diagnostics. Clinicians can decide about the relevant gene panels based on each patient's phenotypes. Animal models are included within the 100,000 Genomes Project as a reference to validate the variant forms and also to test potential treatment procedures [20].

### 9.2.7    France: The Plan for Genomic Medicine (PFMG)[14] and Health Data Hub

The French prime minister issued a request in 2015 that would eventually lead to a large plan with the purpose of integrating genomics into the French healthcare system as well as the industrial sector. The Aviesan (French National Alliance for Life Sciences and Health) responded to that request.

The national plan: the 2025 France Genomic Medicine Initiative (PFMG2025) is currently implemented aiming to introduce precision medicine into the care pathway and develop a national framework.

A 10-year plan for the next decade of genomic medicine was generated as the report of the project. The first phase included the integration of genomic sequencing into routine clinical practice. The second goal was to develop a national genomic sector that connects healthcare and science within the healthcare system to facilitate access to innovation for everyone. To achieve this goal, within the PFMG, the further three essential aspects will be linked from the very beginning: research, economic models (e.g., expenditure by the healthcare system or development of the private sector), and translation of scientific approaches into healthcare.

The initiative placed France among the world-leading countries in the field of genomic medicine while establishing a distinctive industrial and medical sector for genomic medicine in the country. A national database covering the genetic and clinical data was created. To be operational, the required structures for information exchange between clinical decision-makers, diagnostics laboratories, and the patients had to be implemented.

The main focus of the PFMG lies in the patient and the physician. The structure of the PFMG

---

[13] https://allofus.nih.gov

[14] https://pfmg2025.aviesan.fr/

aims to allow the introduction of genome sequencing into the care pathway, starting with the prescription of genome analysis until the medical report. The operational implementation focuses on (1) development of a network of high-throughput sequencing platforms, (2) development of standards for accessing genomic diagnosis, (3) central analyzer of data (CAD, process and exploit the large volume of genomic data, for clinical and research use), and (4) training and education activities. Four pilot research projects are implemented to demonstrate the first application and to overcome potential technological, clinical, and regulatory hurdles. The pilot projects focus on the further medical needs: cancer (MULTISARC trial), rare diseases (DEFIDIAG – cognitive impairment), common diseases (GLUCOGEN – diabetes), and aspects related to the general population (POPGEN). The latter intends to list the common genetic variants present in the French population. Furthermore, a Center of Reference, Innovation, and Expertise (CRefIX) was established that develops and harmonizes best practices and standards. This center also supports technological developments and innovations through academic and industrial collaborations, before ensuring their deployment in clinical practice.[15]

In 2019, France started the Health Data Hub (HDH)[16] aiming to act as a platform allowing the efficient use of existing health databases particularly for medical research purposes, and for improvement in the long term of the quality of care. The HDH follows the idea that research can improve and develop efficient new approaches for the diagnosis, treatment, and prevention through the processing and comparison of large datasets.

## 9.2.8 Japan: Initiative on Rare and Undiagnosed Diseases

Diagnosis for some patients does not happen, since their disease is so rare that physicians are not familiar with it; also some diseases are completely new with unknown conditions. Japan's "Initiative on Rare and Undiagnosed Diseases" (IRUD) is designated to help those patients. Undiagnosed patients need to have at least one of the admission criteria to be accepted into IRUD: the disease must occur as a familial condition, must affect more than one organ, and has to be started through a congenital onset.

A program is used by the IRUD to analyze and evaluate the patients' genetic information which includes a wide range of data from human diseases and their associated murine phenotypes.

IRUD started in July 2015 and identified about ten new diseases only in the first 2 years of work. Up to now, the diagnosis success rate has been around 30%. International IRUD collaborations have been started and the samples of patients in other countries have been analyzed as well.

## 9.2.9 GenomeCanada[17]

The "GenomeCanada" precision health strategy is structured in four phases:

Phase 1: the Foundation – Applied research projects about human health on large scale based on precision health are initiated. The Genome Canada investments include a Human Health Competition for applied genomics and proteomics research in 2004, "the 2012 Genomics and Personalized Health Competition" and the "2017 Genomics and Precision Health Competition." Almost all the range of precision health fields including disease prevention and

---

[15] https://www.icpermed.eu/en/2025-France-Genomic-Medicine-Initiative-A-comprehensive-approach.php

[16] https://www.health-data-hub.fr/

[17] https://www.genomecanada.ca

health maintenance, prognosis, treatment, and early detection are covered within the two last competitions.

Phase 1 of this Canadian strategy aims to identify the areas in which genomics could directly impact a patient's health condition and build the improvement capacity within the research community. Various research hospitals around the country are beginning to benefit from the results of this phase, but a comprehensive national system available to all Canadians is yet to be realized.

Phase 2: Into the Clinic – Results of the previous phase are implemented at the clinical level for rare diseases. This phase is called and funded by the "All for One" program within the GenomeCanada program. The main objective is to improve the well-being and health of Canadians who are entangled with serious genetic conditions by facilitating their access to a timely and accurate diagnosis based on the best genomic knowledge.

The program is focused on six domains to advance the precision health schemes:

1. Provide access to genome-wide sequencing techniques
2. Data governance to address policies and economic and social issues
3. A national data ecosystem for data sharing
4. Achieving the required scale for application and insight through creating a rare disease cohort
5. Patient and community engagement
6. Engagement and education with healthcare professionals

Phase 3: Beyond Rare Disease – Other features of the disease would be considered in clinical applications. The program is designed to address rare illnesses, but it would be a starting point for creating healthcare services that build a learning system and lead to an advanced new generation of healthcare delivery. When a suitable foundation is laid, the other areas such as cancer treatment and pharmacogenomics would benefit as well. The need to establish new systems and mechanisms to collect and integrate

data from all over the country would be needed as initiatives in oncology, pharmacogenomics, and various other disciplines advance further. The further organizing of data, accessing and managing the information, as well as patients' consent can benefit from the experiences gained through the program dedicated to rare diseases.

Phase 4: A cohort including the population on large scale would be created as a national database. The future well-being and health (and also wealth) of the nation will be increasingly dependent on the management of data assets on a large scale. Essential importance has been attached to the delivery of personalized medicine since future innovations and economical development would be propelled by the powerful engines of these national datasets. Genomic sequencing of a large percentage of the country's population (e.g., more than a hundred thousand citizens) in this phase leads to the adoption of precision health and medicine in Canada. Such a large cohort would be a reliable sample for ancestry data, demographic distribution, diseases, and regional coverage [22].

### 9.2.10 World Economic Forum (WEF): Center for the Fourth Industrial Revolution

In March 2017, the WEF launched the "Center for the Fourth Industrial Revolution" with a focus on emerging technological areas. The program has nine program areas including precision medicine. This precision medicine project is designed to shape the path of precision medicine and lead it to promote societal benefits with the lowest possible risks. This will be realized through collaborations in the design and test of management approaches using various precision medicine pilot projects.

The program is conducted through three phases. First, the current situation with precision medicine is assessed by a survey, key stakeholders are identified, and the obstacles in the implementation of precision medicine are described. Next, cooperation with the government is initiated to determine some of the areas of high priority and find ways to transform them into pilot projects. Finally, the pilot projects are designed and conducted [20].

The usual challenges which have been identified through various precision medicine projects are:

- Generating the evidence of efficacy
- IT infrastructure and data sharing
- Legislation and regulatory considerations
- Integration into clinical practice
- Design and creation of reimbursement pathways as well as pricing
- Education and social engagement at the patient and public levels [20]

## 9.2.11 European Initiatives Dedicated to Personalized Medicine

Several joint actions in the field of personalized medicine were established and are running in Europe with support from the European Commission. From 2010 to 2011, several preparatory workshops were organized by the European Commission, followed by a concluding conference in 2011. The outcome of these reflections fed into the Horizon 2020 framework program.

A first coordination and support action (CSA) "PerMed" was funded (2011–2015) and published the first Strategic Research and Innovation Agenda (SRIA) for personalized medicine: "Shaping Europe's Vision for Personalised Medicine."[18] After several strategic workshops in 2016, the "International Consortium for Personalised Medicine – ICPerMed" was established in the very same year.[19] Currently, ICPerMed brings together 46 public and private "not-for-profit" health research funding and policy organizations from 30 countries and seven regions of four continents. ICPerMed acts as a platform aiming to support the personalized medicine science base and foster the alignment of strategies in personalized medicine policy development and funding.

ICPerMed is in the very center of the so-called ICPerMed family.[20] The ICPerMed family includes several coordinations and support actions funded or co-funded by the European Commission:

- The CSA ICPerMed Secretariat provides administrative support to the ICPerMed.
- The European Research Area Network (ERA-Net Cofund) for Personalised Medicine,[21] ERA PerMed, aligns research and funding activities and is implementing the strategic recommendation of ICPerMed into concrete funding activities. ERA PerMed is the largest ERA-Net in the health sector with around 42 participating funding organizations, including regional and national funders, in 32 countries and five continents.[22] In 2023, after launching in a total of five annual calls, it is expected that over 100 M€ will be invested in personalized medicine research through the ERA PerMed program, e.g., in the field of cancer, neurosciences, infectious diseases, and other medical fields.
- CSA Regions4PerMed[23] ("interregional coordination for a fast and deep uptake of personalized health") and CSA SAPHIRe[24] ("Securing the Adoption of Personalised Health in Regions") focus on the value European regions can bring to personalized medicine developments and on the challenge for considering the needs of remote and sparsely populated regions and those with reduced capacities for innovation and implementation.
- The CSA HEcOPerMed,[25] "Healthcare- and pharma-economics in support of the

---

[18] https://www.icpermed.eu/media/content/PerMed_SRIA.pdf

[19] https://www.icpermed.eu/index.php

[20] https://www.icpermed.eu/en/related-initiatives.php

[21] https://erapermed.isciii.es/

[22] https://ec.europa.eu/info/sites/default/files/research_and_innovation/funding/documents/ec_rtd_he-personalised-medicine.pdf

[23] https://www.regions4permed.eu/

[24] https://www.saphire-eu.eu/

[25] https://hecopermed.eu/

International Consortium for Personalised Medicine – ICPerMed," is focusing on health economics reflections and models in personalized medicine.

- The CSA PERMIT,[26] "PERsonalised MedicIne Trials," addresses questions related to personalized medicine clinical studies and trial design.

- Four CSAs focus on international collaboration with (1) the Caribbean and Latin American countries (EULAC PerMed,[27] "Widening EU-LAC policy and research cooperation in Personalised Medicine"), (2) China (SINO-EU PerMed,[28] "Fostering the cooperation between Europe and China in Personalised Medicine" and IC2PerMed,[29] "Integrating China on the International Consortium for Personalised Medicine"), and (3) Africa (EU-Africa PerMed,[30] "Building Links Between Europe and Africa in Personalised Medicine").

The preparations for a European Partnership for Personalised Medicine – EP PerMed, in the frame of the new framework program of the European Commission "Horizon Europe," already started. The partnership will most probably start by the end of 2023 and will foster translational research and innovation and provide a platform to share evidence, demonstrate solutions, and support activities in policy, regulatory science, and health economics. The vision of the European Partnership for Personalised Medicine is "to improve health outcomes within sustainable healthcare systems through research, and the development and implementation of personalized medicine approach for the benefit of patients, citizens, and society."[31]

Many other joint activities are taking place in Europe, for example, the Beyond 1 Million Genomes (B1MG) project aims to support the development of a network of genetic and clinical data across Europe. It provides furthermore coordination and support to the 1+ Million Genomes Initiative (1+MG) in that 23 European countries committed to giving cross-border access to one million sequenced genomes by 2022.[32] B1MG aims to create a long-term impact on sharing data and enable access to Beyond 1 Million Genomes by the development of a data-sharing infrastructure, legal guidance, and sharing best practices. They aim to allow scientists and clinicians to study the genotypic and phenotypic data from over one million people.[33]

## 9.3 Initiatives Related to Pancreatic Pathology

### 9.3.1 PanCAN (United States)

The precision medicine initiative called the Pancreatic Cancer Action Network (PanCAN) is conducted by a nongovernment organization dedicated to improving the condition of everyone involved with pancreatic cancer through many various ways.

There have been substantial efforts to define a concept of PM for pancreatic cancer patients. Many well-known professional and academic groups have tried to discover genetic and molecular alterations related to this type of cancer.

Meanwhile, many countries from all over the world have dedicated funds to investigate the biological features of tumors and create the best options to treat pancreatic cancer patients. A precision medicine program called "Know Your Tumor®" is initiated by the Pancreatic Cancer Action Network (PanCAN) in the United States.[34] This program has suggested some personalized treatment strategies based on a procedure of molecular profiling of tumors [23]. It was revealed that genetic/molecular alterations

---

[26] https://ecrin.org/activities/permit-project

[27] https://www.eulac-permed.eu/

[28] https://www.sino-eu-permed.eu/

[29] https://www.ic2permed.eu/

[30] https://www.euafrica-permed.eu/

[31] https://ec.europa.eu/info/sites/default/files/research_and_innovation/funding/documents/ec_rtd_he-personalised-medicine.pdf

[32] https://digital-strategy.ec.europa.eu/en/policies/1-million-genomes

[33] https://b1mg-project.eu/

[34] https://www.pancan.org/

observed in 50% of pancreatic patients are "actionable" and some targeted therapy could be devised for them. A small cohort of the patients was selected, and methods of personalized and matched molecular targeted therapies were applied instead of the standard treatments and care procedures. Results showed that the progression-free survival rate of those patients was significantly improved for this cohort after the new therapy strategy applied [23].

Accordingly, the Precision PromiseSM[35] project was started as a PanCAN initiative. PromiseSM is a response-adaptive clinical trial platform that includes several treatment disciplines that simultaneously test different therapeutic possibilities. The defined goals for the initiative to be achieved by 2030 are shifting the 5-year survival rate from 10% to 20%, raising $80 M annually since the organization does not receive any government funds and all the progress they make is funded by donations, and growing awareness of PanCAN in the United States from 14% to 30%.

The Precision-Panc[36] initiative for pancreatic cancer in the United Kingdom pursues a similar approach.

The PRIMUS (Pancreatic canceR Individualized Multi-arm Umbrella Studies) program is a research umbrella initiative that covers many clinical studies investigating the treatment strategies as well as efforts to develop biomarkers used in prognosis. This program has included three phases in the past [24].

The PRIMUS-004, a significant pancreatic cancer trial, is designed to provide more target-specific and effective customized treatment of tumors for each patient. The project is run by Precision-Panc, a therapeutic development program dedicated to pancreatic cancer which is led by the University of Glasgow. The "Cancer Research UK" provides the major source of funds for this trial, and a PM approach toward pancreatic cancer in the United Kingdom has been promised through the program [24].

### 9.3.2 Enhanced Pancreatic-Cancer Profiling for Individualized Care (EPPIC) (Canada)

The Canadian EPPIC (Enhanced Pancreatic-Cancer Profiling for Individualized Care),[37] an initiative funded by Terry Fox Research Institute, has started the COMPASS trial project which was the first prospective study for advanced "pancreatic ductal adenocarcinoma" (PDAC) disease. This trial aimed to integrate the PDAC subtypes with chemotherapy responses.

The EPPIC team has defined a goal to sequence the genetic material related to metastatic pancreatic tumors in 400 patients in four states of Canada (as of March 2018). They hope to provide a deeper understanding of pancreatic cancer bio-mechanisms to design individualized treatment strategies and to help with creating new treatment options.

This project is currently underway in the form of two clinical trials called PanGen and COMPASS in Toronto and Vancouver and is scheduled to be expanded so that it includes eligible patients in four other regions (Kingston, Ottawa, Calgary, and Edmonton). Genomic sequencing process and bioinformatics analyses of tumor samples are conducted at the Ontario Institute for Cancer Research (OICR) and the BC Cancer Genome Sciences Centre [25].

### 9.3.3 The PancREatic Cancer OrganoiDs rEsearch (PRECODE) (Germany)

The central mission of PRECODE (PancREatic Cancer OrganoiDs rEsearch) Network by definition is to train the next generation of innovative and creative researchers in the field of pancreatic cancer and establish the significance of pancreatic organoid research within the European Union.

This will be done through a series of connected doctorate programs with three main characteristics of being international, interdisciplinary, and

---

[35] https://www.pancan.org
[36] https://www.precisionpanc.org/

[37] https://www.tfri.ca

inter-sectorial. Organoids can be considered as small micro-organs with the ability to recapitulate the organization and the associated functions. The organoids as micro-organs can be used to reduce the need to conduct experiments on animal models and help to reach the project goals faster and easier. Organoids from pancreatic cancer will help us to discover effective therapeutic drugs. Organoids can be isolated from a variety of clinical sources relatively easily. This feature makes organoids almost a perfect tool for PM [26].

### 9.3.4   The Australian Pancreatic Cancer Genome Initiative (APGI) (Australia)

The Australian Pancreatic Cancer Genome Initiative (APGI)[38] is mainly focused on epigenomic and genomic variations in tumor tissues. The team of the project published the proceedings of the IMPaCT (Individualized Molecular Pancreatic Cancer Therapy) trial in 2015 which demonstrated that biomaterials could be collected and screened to find actionable molecular targets [27].

All different initiatives mentioned above are associated with or comprise comprehensive databases (biobanks) that include clinical and genomic annotation data. Researchers in all those programs and initiatives hope to somehow contribute to the understanding of biological features of diseases in each individual to find new biomarkers and create personalized treatment procedures and/or strategies for every patient [28].

### 9.4   Initiatives with the Most Focus on Diabetes

The European Federation of Pharmaceutical Industries and Associations, European academic institutions, and the European Commission launched a joint initiative in 2008. This program

with a budget of €5.6 billion called the "Innovative Medicines Initiative" (IMI) covers around 100 different projects which mainly concentrate on major illnesses.

A common objective for those programs is to discover and validate useful biomarkers which help us to stratify populations of patients into subgroups so that the treatment can be more effective than without biomarker classification. In some of the projects, diagnostic reclassification is also a point of focus. Those PM initiatives within the IMI, unlike most other initiatives, focus on the integration of multiple biomarkers, such as metabolites, proteins, genomic transcripts, genotypes, and meta-genomics sequences.

The UK Biobank cohort which is established in 2005 currently includes around half a million adults, and the vast majority of them have provided self-reported data on their well-being, health status, and lifestyle as well as non-fasting blood samples. Those samples are being genotyped using techniques such as targeted metabolomics analyses and genome-wide arrays.

For example, research conducted using UK Biobank data has suggested the possibility of genetic alterations associated with obesity to influence individuals' susceptibility to some types of modifiable lifestyle characteristics [29–32] that might be relevant to diabetes disease [12].

In China which has been significantly advancing in decoding genomes [33], the government announced a bold initiative in 2016 with the vision to become a global superpower in precision medicine. It is a 15-year precision medicine initiative with a 9 billion US dollar funding [34]. Three core objectives are defined for the Chinese initiative focused on (1) creating a large cohort from all seven main regions of China including millions of participants such that it can represent the country's population, (2) eight disease-specific cohorts (immune system disorders, psychosomatic, neurological, metabolic, respiratory, cerebrovascular, cardiovascular, and seven common malignant tumors) that will cover 0.7 million

---

[38]https://www.pancreaticcancer.net.au/

participants in total, and (3) a clinical cohort (N = 50,000 patients) suffering from 50 rare diseases [35].

### 9.4.1 Nordic Precision Medicine Initiative (NPMI) (North Europe)

The "Nordic Precision Medicine Initiative" (NPMI) was formed in 2015 with the perspective to integrate the volumes of genetic and other biomedical data from one million citizens of Norway. Several national precision and genomic medicine projects are being pursued in Northern Europe. The project FinnGen in Finland was launched back in 2017 to link previously recorded digital healthcare data existing for half a million Finnish citizens (10% of the total population of that country) with the new genomic sequencing data. The project is conducted by a public-private joint venture.

In Sweden, the GAPS (Genomic Aggregation Project in Sweden) initiative was launched to integrate the genomic data from 160,000 genotyped samples which comprise a wide range of diseases including diabetes [30 REF 05] [36], and GMS (Genomic Medicine Sweden) is making a big effort to promote clinical genomics into a standard at the national level. A company named "deCODE genetics" in Iceland has created a phenotype and genetic information database which covers the entire population of the country (n = 334,000). The database will provide the required information for genetics research and validation of drugs targets to treat a wide range of diseases, including diabetes [12].

### 9.4.2 EPMPP, FinnGen, GAPS, deCODE

The Estonian government recently started the Estonian Personalized Medicine Pilot Project (EPMPP) (2015–2018), which is aimed to implement personalized medicine on a national scale. As such, the government has provided 5.9 billion US dollars to facilitate genotyping of 100,000 Estonian citizens.

The "Estonian Genome Center" at the University of Tartu in Estonia has launched a program to curate a biobank based on the country's population which is suitable for PM research regarding complex diseases, especially diabetes. The program aims to advance and promote the development of genetic research activities and implement the genomic data within the fields of clinical practice which ultimately leads to improvements in public health [12].

### 9.4.3 Saudi Human Genome Project (SHGP)

The very high prevalence of type 2 diabetes (about 20%) in Saudi Arabia lead the Saudi government and King Abdulaziz City for Science and Technology to initiate a public project called the "Saudi Human Genome Project" (SHGP), with the aim of sequencing the genomes of 100,000 citizens to identify the genetic basis, possible biomarkers, and potential treatment of prevention methods of some complex and monogenic diseases among that population [12].

## 9.5 Initiatives Related to Alzheimer's Disease (AD)

Alzheimer's disease as a neurodegenerative illness is among the most prominent pathological issues of the current century. This disease, like the other main health issues, has been particularly addressed by many national and international initiatives within the precision medicine schemes since its genetic roots of it demand the prevention, diagnosis, and treatment methods to be personalized and customized for certain groups of population or (at an ideal level) the individuals.

### 9.5.1 The Genetic Frontotemporal dementia Initiative (GENFI)

GENFI[39] is a consortium of research sites across Europe (Sweden, Germany, Italy, Portugal, France, Spain, Netherlands, Belgium, United Kingdom) and Canada. The main goal of this initiative is to investigate and interpret the genetics of frontotemporal dementia (FTD), mainly in individuals carrying mutations in progranulin (GRN), the chromosome 9 open reading frame 72 (C9ORF72), and microtubule-associated protein tau (MAPT) genes [21].

### 9.5.2 The Dominantly Inherited Alzheimer Network (DIAN)

The "Dominantly Inherited Alzheimer Network" (DIAN)[40] is an international research organization established in 2008. DIAN involves a series of institutions all over the United States, South America, Australia, Asia, and Europe. The data from individuals with early-onset familial Alzheimer's disease and noncarrier family members (recruited as control subjects) are recorded and analyzed within the scheme of big single research [37].

### 9.5.3 Alzheimer Precision Medicine Initiative (APMI)

An initiative called "Alzheimer Precision Medicine Initiative" (APMI) and the cohort program associated with it are launched in 2016. The projects are organized and facilitated by a coordinating center within the academic sector which is run by the "Sorbonne University Clinical Research Group in Alzheimer Precision Medicine (Sorbonne University GRC APM)." The main goal is to transform research into conventional clinical diagnostics, healthcare, and drug development regarding Alzheimer's disease.

The APMI and the cohort program associated with it (APMI-CP)[41] have been conceptually linked to the PMI and the "All of Us" Research Program in the United States [38]. The APMI is an international network of leading scientists, researchers, and clinicians from various disciplines devoted to the transformation of neuroscience, psychiatry, and neurology. Its objective is to adopt PM practices based on complex systems theory [39] (using systems neurophysiology [38], systems biology [38, 40], and systems pharmacology), integrative disease modeling (IDM) guided by the biomarkers [41, 42], and "big data science." The initiative is intended to facilitate healthcare solutions for protein pathogenesis of the brain, protein misfolding disorders, and neurodegenerative diseases like Alzheimer's [38, 41, 43].

The APMI International Working Group (APMI-IWG) was created to attract international class experts in the fields of bioinformatics, neuroimaging and biophysics, neurogenetics, biomarkers, blood-based and cerebrospinal fluid (CSF), laws and regulatory affairs, pharmaceutical industry management, clinical trial development, and preclinical studies.

Many pioneering research programs regarding the integration of precision medicine into neuroscience fields are conducted through the APMI framework. Some examples of funded PM-oriented research programs are the following:

- The "MIDAS" research program as a roadmap to PM for Alzheimer's disease is designed to achieve several goals: (1) to define alterations in structure and function of the brain connectivity related to or caused by the APOE gene profile and regional brain atrophy; (2) to determine a series of levels that represent the aging population strata based on the genetic patterns caused by single nucleotide polymorphisms (SNPs) associated with neuroimaging phenotypes, that is, regional brain atrophy and functional/structural connectivity; and (3) as an objective in subsequent phases, to generate quantitative models of disease mechanisms

[39] http://genfi.org.uk/
[40] https://dian.wustl.edu/about/
[41] https://www.apmiscience.com/

and progression based on the obtained data from preclinical stages of AD and to identify the most useful neuroimaging and biological biomarkers, to be integrated into the staging model of AD which was defined.

- The "PHOENIX" research program is dedicated to exploring the systems biology and neurophysiology of Alzheimer's disease. This project is designed to address and solve the complexities of AD through "deconstructing the disease" into some biological subsets which are guided by a comprehensive matrix of systems neurophysiology and systems biology.
- The "POSEIDON" research program is aimed to provide a preclinical understanding of AD. This project uses a combined MEG-fMRI approach to assess the changes in the neuronal network at an early stage.
- The "VISION" research program is basically designed to evaluate the process and results of retinal amyloid imaging which helps to improve amyloid screening, as well as prediction of physiological disease progression pathology, cognitive decline, and conversion to prodromal Alzheimer's disease and tracking amyloid progression [21].

### 9.5.4  Dementias Platform United Kingdom (DPUK)

Dementias Platform United Kingdom (DPUK)[42] is a private-public partnership established in 2014 by the Medical Research Council (MRC) in the United Kingdom. The major objective is to advance the detection of dementia disorders at early stages, support various research efforts for creating innovative treatments, and ultimately find some ways to prevent the disease. The DPUK is coordinating and assembling one of the world's largest populations of study participants in the dementia research fields with more than two million participants aged 50 and above (including people who are at risk of developing dementia, subjects that early-stage dementia has been diag-

nosed for them, and the general population). Twenty-two study groups are collaborating within the United Kingdom creating a coordinated research environment and maximizing the potential of cohort studies in this field in the United Kingdom.

An interesting feature of DPUK is that the electronic data are shared through a portal (the "DPUK Data Portal"[43]) which allows for rapid testing of new research designs.

"Integration and Analysis of heterogeneous Big Data for Precision Medicine"[44] is one of the projects initiated within this environment using the EU funds which have proposed a number of treatments for different groups of patients. This project is aimed to design a comprehensive conceptual plan to represent various sources of data. The data are combined from large pools of genomics data, medical records, and imaging databases so better individualized diagnosis and treatment strategies would be possible for Alzheimer's disease [21].

### 9.5.5  EU Joint Program on Neurodegenerative Disease Research (EU JPND)

The EU JPND initiative[45] was established in 2009 through the collaboration of 24 countries to address the increasing burden of ND on European society. The key idea around which the JPND is created is that dementia is a global challenge and any country alone cannot solve it. Therefore, some data-driven, collaborative efforts at large scales are required to create reliable knowledge and breakthroughs regarding the ND. JPND Scientific Advisory Board has identified precision medicine as the key approach to realizing this goal. Furthermore, it is acknowledged that "omic" technologies need to be utilized to enhance dementia syndrome stratification on a robust biological level within the precision medicine scheme [21].

---

[42] https://www.dementiasplatform.uk/

[43] https://portal.dementiasplatform.uk/
[44] iASiS is available at http://project-iasis.eu/
[45] http://www.neurodegenerationresearch.eu/

### 9.5.6 European Prevention of Alzheimer's Dementia (EPAD)

The European Prevention of Alzheimer's Dementia (EPAD) program[46] [44] is a European collaboration initiative in which the academic and private sectors are involved. The main purpose of the program is to create a platform for designing and conducting phase II proof-of-concept clinical trials in the secondary prevention of AD [21].

### 9.5.7 The AETIONOMY Project

An innovative project called AETIONOMY[47] has been formed with participants from at least 13 European countries. The project aims at developing "taxonomy of mechanisms" based on the biological pathways that play various roles in the pathophysiology of diseases. The highest goals of the project are defined as (1) to provide guidelines for the classification of disease in main and secondary classes and (2) to generate data that could be used for developing knowledge-based models and ontological information regarding the disease.

In summary, the project consortium will be a powerful body that facilitates the exploration of the major mechanisms of neurodegeneration processes. Moreover, the project activities will promote the development of new preventive approaches [21].

### 9.5.8 The Women's Brain Project (WBP)

The Women's Brain Project (WBP)[48] was initiated in 2016 with the perspective to inspire a global discussion about gender determinants of the vulnerability of females to the brain and mental diseases. WBP involves various regulators, drug developers, scientists, and policymakers who work together to propose solutions. The project provides support for artificial intelligence, as well as social and basic clinical research to identify methods and tools for better diagnosis, care, and treatment of conditions of the brain and mental health of women. The WBP proposes a personalized approach to developing and performing medical treatments, prevention strategies, and caregiving services using technological improvements based on sex and gender differences [21].

The findings and substantial outcomes of the research projects thus far developed under the umbrella of the APMI ("MIDAS," "PHOENIX," "VISION," and "POSEIDON") will help and inform the experts to design, prioritize, and perform controlled clinical intervention trials whether in pharmacological or nonpharmacological based on identified intermediate endophenotypes as well as planning and conducting candidate surrogate biomarker studies and systems-based diagnostic. In general, the APMI ultimately leads to the development of prospective longitudinal studies designed to analyze a comprehensive multimodal biomarker array (in vivo) that serves to facilitate better risk assessment and prediction of cognitive decline conditions in different subsets of people.

### References

1. Ramsey SD, Sullivan SD. A new model for reimbursing genomebased cancer care. Oncologist. 2014;19(1):1–4.
2. Carlson JJ, Chen S, Garrison LP Jr. Performance-based risk-sharing arrangements: an updated international review. PharmacoEconomics. 2017;35(10):1063–72.
3. Ginsburg GS, Phillips KA. Precision medicine: from science to value. Health Aff. 2018;37(5):694–701.
4. https://www.lek.com/insights/ei/promise-precision-medicine-asia-pacific
5. http://k-master.org/intro_1_e.php
6. https://www.a-star.edu.sg/polaris/
7. National Institutes of Health. Molecular transducers of physical activity in humans, frequently asked questions (FAQs) [Internet], 2018. Available from https://commonfund.nih.gov/MolecularTransducers/FAQs. (Accessed 20 Mar 2018).
8. Livingstone KM, Celis-Morales C, Navas-Carretero S, et al. Food4Me study. Effect of an internet-based, personalized nutrition randomized trial on dietary

---

[46] http://ep-ad.org/

[47] https://www.aetionomy.eu/en/background.html

[48] http://womensbrainproject.com/

changes associated with the Mediterranean diet: the Food4Me study. Am J ClinNutr. 2016;104:288–97.

9. Celis-Morales C, Livingstone KM, Marsaux CF, et al. Food4Me study. Effect of personalized nutrition on health-related behaviour change: evidence from the Food4Me European randomized controlled trial. Int J Epidemiol. 2017;46:578–88.

10. Celis-Morales C, Livingstone KM, Marsaux CF, et al. Design and baseline characteristics of the Food4Me study: a web-based randomised controlled trial of personalised nutrition in seven European countries. Genes Nutr. 2015;10:450.

11. Gaziano JM, Concato J, Brophy M, et al. Million veteran program: a megabiobank to study genetic influences on health and disease. J Clin Epidemiol. 2016;70:214–23.

12. Fitipaldi H, McCarthy MI, Florez JC, Franks PW. A global overview of precision medicine in type 2 diabetes. Diabetes. 2018;67(10):1911–22.

13. https://www.healthit.gov/topic/onc-hitech-programs

14. https://Iwww.healthit.gov/topiclscientific-initiativeslprecision-medicine

15. Reardon S. Giant study poses DNA data-sharing dilemma. Nature. 2015;525:16–7.

16. Hudson K, Lifton R, Patrick-Lake B. The precision medicine Initiative cohort program – building a research foundation for 21st century medicine. Bethesda, MD: National Institutes of Health; 2015.

17. National Human Genome Research Institute. Electronic Medical Records and Genomics (eMERGE) Network [Internet]. Bethesda (MD): National Institutes of Health; [last updated 2017 Nov 3; cited 2018 Mar 21]. Available from: https://genome.gov/27540473/electronicmedical-records-and-genomicsemerge-network/

18. Clinical Sequencing Evidence- Generating Research Consortium. Clinical Sequencing Evidence-Generating Research [Internet]. Bethesda (MD): CSER; [cited 2018 Mar 21]. Available from: https://cser-consortium.org/

19. Implementing Genomics in Practice. Get started [Internet]. Durham (NC): IGNITE, Duke University; [cited 2018 Mar 21]. Available from: https://www.ignite-genomics.org

20. https://www.ncbi.nlm.nih.govlbocksINBK5072151.

21. Hampel H, Vergallo A, Perry G, Lista S, Alzheimer Precision Medicine Initiative. The Alzheimer precision medicine initiative. J Alzheimers Dis. 2019;68(1):1–24.

22. https://www.genomecanada.ca/en/programs/precision-health-strategy-canada-think-big-start-small-learn-fast

23. Pishvaian MJ, Bender RJ, Halverson D, Rahib L, Hendifar AE, Mikhail S, Chung V, Picozzi VJ, Sohal D, Blais EM, et al. Molecular profiling of patients with pancreatic cancer: initial results from the know your tumor Initiative. Clin Cancer Res. 2018;24:5018–27.

24. Dreyer S, Jamieson N, Morton JP, Sansom O, Biankin A, Chang D. Pancreatic cancer: from genome discovery to PRECISION-Panc. Clin Oncol. 2019;32:5–8.

25. https://www.tfri.ca/our-research/research-project/enhanced-pancreatic-cancer-profiling-for-individualized-care-(eppic)

26. https://precode-project.eu/

27. Regel I, Mayerle J, Ujjwal Mukund M. Current strategies and future perspectives for precision medicine in pancreatic cancer. Cancers. 2020;12(4):1024.

28. Armitage E, Barbas C. Metabolomics in cancer biomarker discovery: current trends and future perspectives. J Pharm Biomed Anal. 2014;87:1–11.

29. Rask-Andersen M, Karlsson T, Ek WE, Johansson Å. Gene-environment interaction study for BMI reveals interactions between genetic factors and physical activity, alcohol consumption and socioeconomic status. PLoS Genet. 2017;13:e1006977.

30. Celis-Morales CA, Lyall DM, Gray SR, et al. Dietary fat and total energy intake modifies the association of genetic profile risk score on obesity: evidence from 48 170 UK biobank participants. Int J Obes. 2017;41:1761–8.

31. Tyrrell J, Wood AR, Ames RM, et al. Gene-obesogenic environment interactions in the UK biobank study. Int J Epidemiol. 2017;46:559–75.

32. Young AI, Wauthier F, Donnelly P. Multiple novel gene-by-environment interactions modify the effect of FTO variants on body mass index. Nat Commun. 2016;7:12724.

33. Cyranoski D. China's bid to be a DNA superpower. Nature. 2016;534:462–3.

34. Cyranoski D. China embraces precision medicine on a massive scale. Nature. 2016;529:9–10.

35. Bu D, Peng S, Luo H, et al. Precision medicine and cancer immunology in China: from big data to knowledge in precision medicine. Science. 2018;Suppl:S35–8.

36. Bergen SE, Sullivan PF. National-scale precision medicine for psychiatric disorders in Sweden. Am J Med Genet B Neuropsychiatr Genet. 7 July 2017 [Epub ahead of print]. https://doi.org/10.1002/ajmg.b.32562.

37. Morris JC, Aisen PS, Bateman RJ, Benzinger TL, Cairns NJ, Fagan AM, Ghetti B, Goate AM, Holtzman DM, Klunk WE, McDade E, Marcus DS, Martins RN, Masters CL, Mayeux R, Oliver A, Quaid K, Ringman JM, Rossor MN, Salloway S, Schofield PR, Selsor NJ, Sperling RA, Weiner MW, Xiong C, Moulder KL, Buckles VD. Developing an international network for Alzheimer research: The Dominantly Inherited Alzheimer Network. Clin Investig (Lond). 2012;2:975–84.

38. Hampel H, Toschi N, Babiloni C, Baldacci F, Black KL, Bokde ALW, Bun RS, Cacciola F, Cavedo E, Chiesa PA, Colliot O, Coman CM, Dubois B, Duggento A, Durrleman S, Ferretti MT, George N, Genthon R, Habert MO, Herholz K, Koronyo Y, Koronyo-Hamaoui M, Lamari F, Langevin T, Leh'ericy S, Lorenceau J, Neri C, Nistic'o R, Nyasse-Messene F, Ritchie C, Rossi S, Santarnecchi E, Sporns O, Verdooner SR, Vergallo A, Villain N, Younesi E, Garaci F, Lista S, Alzheimer Precision Medicine Initiative (APMI). Revolution of Alzheimer precision

neurology. Passageway of systems biology and neurophysiology. J Alzheimers Dis. 2018;64:S47–S105.

39. Lista S, Khachaturian ZS, Rujescu D, Garaci F, Dubois B, Hampel H. Application of systems theory in longitudinal studies on the origin and progression of Alzheimer's disease. Methods Mol Biol. 2016;1303:49–67.

40. Castrillo JI, Lista S, Hampel H, Ritchie CW. Systems biology methods for Alzheimer's disease research toward molecular signatures, subtypes, and stages and precision medicine: application in cohort studies and trials. Methods Mol Biol. 2018;1750:31–66.

41. Hampel H, O'Bryant SE, Durrleman S, Younesi E, Rojkova K, Escott-Price V, Corvol JC, Broich K, Dubois B, Lista S, Alzheimer Precision Medicine Initiative. A precision medicine Initiative for Alzheimer's disease: the road ahead to biomarker-guided integrative disease modeling. Climacteric. 2017;20:107–18.

42. Younesi E, Hofmann-Apitius M. From integrative disease modeling to predictive, preventive, personalized and participatory (P4) medicine. EPMA J. 2013;4:23.

43. Hampel H, O'Bryant SE, Castrillo JI, Ritchie C, Rojkova K, Broich K, Benda N, Nistic'o R, Frank RA, Dubois B, Escott-Price V, Lista S. PRECISION MEDICINE - the Golden Gate for detection, treatment and prevention of Alzheimer's disease. J Prev Alzheimers Dis. 2016;3:243–59.

44. Ritchie CW, Molinuevo JL, Truyen L, Satlin A, Van der Geyten S, Lovestone S. Development of interventions for the secondary prevention of Alzheimer's dementia: the European prevention of Alzheimer's dementia (EPAD) project. Lancet Psychiatry. 2016;3:179–86.

# Economic Aspects in Precision Medicine and Pharmacogenomics

**10**

Marziyeh Nosrati, Shekoufeh Nikfar, and Mandana Hasanzad

**What Will You Learn in This Chapter?**

Precision medicine intervention is a novel health technology that created a new paradigm in the health systems. In comparison with conventional interventions, precision medicine causes improved treatment outcomes and reduced side effects. However, it has remarkably high prices, which impact the economy of the health systems. Therefore, it is necessary for health-system decision-makers to apply evidence answering whether it is worth funding such costly interventions. One of the tools, providing such evidence, is economic evaluation studies. However, regarding the special characteristics of precision medicine, applying economic evaluation studies would be associated with challenges. This chapter generally explains the importance of applying economic evaluations in the field of precision medicine, the associated challenges, and the available solutions.

**Rationale and Importance**

The remarkable high price of precision medicine causes limited access to them. Considering the limited budget in health systems, it is necessary to decide about the resource allocation to a new health intervention based on evidence that examines the opportunity costs. Economic evaluation is a crucial tool to determine and compare the value of various interventions, especially the expensive ones, and plays a remarkable role in the decision-making process in healthcare systems. However, because of the features of precision medicine, conducting an economic evaluation of precision medicine intervention is associated with challenges. Therefore, economic evaluation of precision medicine intervention and challenges and solutions would be important issues addressed in this chapter.

M. Nosrati
Department of Pharmacoeconomics and
Pharmaceutical Administration, Faculty of Pharmacy,
Tehran University of Medical Sciences, Tehran, Iran

S. Nikfar (✉)
Personalized Medicine Research Center,
Endocrinology and Metabolism Clinical Sciences
Institute, Tehran University of Medical Sciences,
Tehran, Iran

M. Hasanzad
Personalized Medicine Research Center,
Endocrinology and Metabolism Clinical Sciences
Institute, Tehran University of Medical Sciences,
Tehran, Iran

Medical Genomics Research Center, Tehran Medical
Sciences, Islamic Azad University, Tehran, Iran

## 10.1 The Importance of Economic Evaluation in the Field of Precision Medicine

In recent decades, personalized medicine intervention (PMI) has been introduced as a new paradigm in the diagnosis and treatment of diseases, having caused significant development in the health system, to tailor protocols and treatments of patients [1].

PMI has become a reality in practice through targeted therapies. Using genetic tests and evaluating biomarkers, patients who could potentially benefit from the medicine would be identified and delivered the treatment. The most important consequence of such a method is to increase the safety and effectiveness of treatment. Thus, the occurrence of prescribing unsafe and ineffective interventions would be minimized and healthcare outcomes are highly likely to improve. Consequently, the resource of the healthcare system would be more efficiently consumed [2].

Although PMIs have considerable benefits, they are often remarkably costly [3]. Because of the growth of their application in health systems [4], one of the main concerns about them is to provide sufficient evidence for their assumed economic impact [5]. In other words,

regarding their significant costs, the rationale behind conducting an economic evaluation of PMIs is critically important which has been addressed in the previous literature [2]. Another reason why sufficient evidence is required is to justify insurance companies to allocate a portion of their limited resources to these costly interventions [4].

The exact economic value of PMIs would affect the extent of their application and implementation [6]. Economic evaluation (EE) studies answer the main question of whether it is cost-effective to increase the clinical effectiveness of an intervention for its additional cost [3]. Achieving this purpose, in such studies, a ratio is calculated which has been named the "incremental cost-effectiveness ratio" (ICER). This ratio is obtained through the following formula:

$$ ICER = \frac{\Delta cost \left(cost\ of\ the\ intervention - cost\ of\ the\ comparator\right)}{\Delta effectiveness \left(effectiveness\ of\ the\ intervention - effectiveness\ of\ the\ comparator\right)} $$

According to the above formula, ICER indicates how much it will cost to achieve greater outcomes from later (and possibly more costly) intervention in comparison with the previous one [7]. Conducting these studies for PMIs provides sufficient evidence applicable to making appropriate decisions about using them in healthcare systems [4].

## 10.2 Challenges of EEs in the Field of Precision Medicine

The results of systematic review studies on EE of PMIs have demonstrated that although these interventions are cost-effective, they are not cost-saving. Evidence suggests that the results of the economic evaluation would be dependent on factors such as the prevalence of a gene (or allele) in the population, the accuracy of genetic tests (false positive and negative), and the costs associated with tests [3, 8]. Also, these studies have shown that the reliability of the EE of PMIs needs

to be improved. Evidence has revealed that there are some considerable challenges in conducting EE in the field of precision medicine [6]. Some of the issues related to the EE of PMIs are listed below:

1. Lack of consistent methods for conducting economic evaluations
2. Measuring the real value of personalized medicine
3. Inadequate available data [3]
4. The increasing complexity of EE of PMIs [3, 9]

### 10.2.1 Lack of Consistent Methods for Conducting Economic Evaluations

One of the considerations about EE of PMIs is the lack of reliable methods for conducting these studies regarding the specific characteristics of precision medicine clinical trials. Economic evaluations are mostly designed and carried out based on clinical trials [10]. Therefore, changing the specifications of the clinical studies would affect the required methods for the implementation of the EEs.

Clinical trials of PMIs are complex. Since treatment strategies are a test-and-treat strategy instead of a treat-all therapy, the approach, in which the study population has been examined for biomarkers before randomization of patients, has been adopted in conducting trials to ensure that these patients benefit from the intervention [9].

In the field of precision medicine, clinical trials have been mainly conducted based on biomarkers. In these studies, it is necessary to evaluate many separate subgroups based on the existence or nonexistence of a specific biomarker [11]. These trials have different types, some of which are mentioned below.

### 10.2.1.1 Enrichment Design

The common type of novel clinical trial is one in which the included patients should have the biomarker and they are randomly assigned to one of the two intervention or control groups. This type of study is useful in cases where biomarker-positive patients benefit from therapeutic intervention and is usually performed in the third phase of clinical trials. The positive point of this type of study lies in its power to find the benefits of treatment because, in these clinical trials, the only response to treatment is assessed. Thus, the required sample size is minimized [11].

### 10.2.1.2 Randomized All Designs

In these clinical trials, all patients included, regardless of the type of biomarker, are randomly assigned to one of the two intervention or control groups. These trials are especially applicable when the results of previous studies cannot prove the (in)effectiveness of the treatment in biomarker-negative patients. These studies examine whether the existence of biomarkers in the intervention and control groups is related to the treatment effect.

In the two types of clinical trials described above, the types of medicine and disease in the studied population are the same, and the relationship between the type of biomarker and response to treatment is assessed. Patients are evaluated only based on the type of biomarker and response to treatment [12].

### 10.2.1.3 Basket Clinical Trial

In this study, the effectiveness of medicine would be measured in the study population with a specific biomarker and different types of diseases [12]. The application of these studies is in the area of cancer disease [10]. In most cases, the patient inclusion criterion is the existence of one biomarker that is common to several types of cancer.

An important advantage of these studies is the efficiency of their protocol because the effect of a drug in several different diseases is simultaneously studied. Also in these trials, the protocol is always open to adding a new type of disease (tumor), which in turn keeps the study up-to-date and does not require conducting another study to evaluate new cases. In addition, such studies are more effective in finding tumor efficacy signals in all types of tumors [13].

### 10.2.1.4 Umbrella Trials

In this method, the number of medicines in a specific type of disease with a variety of biomarkers would be assessed [14, 15]. All participants in the umbrella study have the same type of cancer histology or organ involvement [13]. Such studies are conducted with at least one of the following purposes: (1) finding appropriate and safe doses of drugs and (2) evaluating the effectiveness of a known new compound in comparison with the standard treatment [12].

Conducting clinical studies with a biomarker evaluation approach in the study population can also lead to changes in economic evaluations because these studies are designed and implemented based on clinical trials. Thus, in EE of PMI, genetic characteristics or (non)existence of biomarkers in the hypothetical cohort have to be considered. As previously described, in clinical trials of precision medicine, participants' selection would be based on the type of biomarker rather than that of disease. In particular, in basket trials, patients have a specific type of biomarker and various diseases. Therefore, in economic evaluations based on these studies, the hypothetical cohort has different diseases and remarkable heterogeneity. Consequently, the researchers have been challenged by how the costs and outcomes would be aggregated in this extremely heterogeneous population. According to researchers in the field of pharmacoeconomics, no economic evaluation study has been conducted on a population with one type of biomarker with different types of diseases and one cost-effectiveness ratio calculated. This necessitates further studies and finding a method specifically designed to calculate the incremental cost-effectiveness ratio in a population with the above characteristics [10].

## 10.2.2 Measuring the Real Value of Personalized Medicine

One of the most important outcomes measured after the administration of treatment is the quality of life or utility. This outcome is converted to the QALY (quality-adjusted life year) in EEs. QALY incorporates two aspects of the effects of an intervention. One aspect is the effect on the quality of life and the other one is the effect on length of life [7]. However, in EE of PMI, the question is whether improving the health utility following the PMI is the best way to assess the value thereof.

Not all of the benefits of PMI can be assessed through health utility. Although there are different methods for measuring outcomes in economic evaluation studies, QALY is the one most commonly used. Drummond et al. have concluded that health economists have mostly considered QALY as a usual method for assessing the value of interventions. However, QALY cannot assess all the benefits of PMI [15] because it would focus mainly on the consequences of quality of life in terms of health. The World Health Organization defines health as "the perfect social, mental, and physical condition, not just the absence of disease or disability." Most preference-driven health status measurement tools focus mainly on the health aspects mentioned in the recent definition. Therefore, focusing on QALY

in deciding on resource allocation leads to inefficiency and non-optimization of the decision because QALY cannot assess all the beneficial aspects of an intervention. Consequently, it is not possible to solely assess the value of more complex interventions through QALY because their benefits are not limited to health. Regarding the limitations of QALY discussed above, some of the alternative methods considered for assessing the real value of the PMI are explained below.

### 10.2.2.1 Willingness to Pay

One of the alternative methods suggested for the assessment of the value of PMI is willingness to pay (WTP). In such a method, eligible patients to receive PMI would be asked how much they are willing to pay. WTP would be preferable because, unlike QALY, it takes into account various benefits affecting the respondents' choice.

On the other hand, the limitations of WTP are the following:

1. Patients' response depends on their ability to pay.
2. Patients will easily not be able to quantify the value of health and quality of life [16].
3. In the field of precision medicine, patients' lack of knowledge makes it difficult for them to imagine the consequent condition [17].

### 10.2.2.2 Benefits Beyond the Health-Related Quality of Life

Applying genetic tests to select the approach of prevention and/or treatment in the field of precision medicine can lead to outcomes beyond the clinical ones. Many researchers and reviewers have believed that measuring the costs and outcomes of precision medicine has many challenges [8].

Reviewing evidence has demonstrated the composite structures used to measure the positive outcomes of health interventions. These composite structures are based on four types of frameworks:

1. Benefits affecting health and well-being (personal health, mental health, potential health, and strength)

2. Benefits arising from the process of providing healthcare (process utility)
3. Benefits beyond the affected people (overflow effects, externalities, selection value, and distributive benefits)
4. Benefits outside the healthcare sector [16]

In an article that specifically focuses on the challenges of conducting EE of PMI, it has been pointed out that process utility and capability theory would be the methods that must be considered for the assessment of the real value of PMI [15]. According to Foster et al., the personal utility method could be applied to identify a group of individuals who are most likely to benefit from the use of the precision medicine approach [17].

### Capability Theory

Capability is the individual's potential to do a specific or a combination of tasks that a person can do or a situation that can exist. Capability theory, which examines the competence of individuals, is categorized in the first type of framework which is benefits that affect well-being (individual well-being, psychological well-being, overall well-being, empowerment, and power). In this theory, people focus on choice and control, and it measures what people are capable to do beyond performance [16].

### Process Utility

One of the considerations raised in the assessment of the health intervention value is to examine both the outcome and the process of achieving it. It would be important to take into consideration that health service consumer has preferences not only for health outcomes but also for the accompanying conditions of those outcomes. This (dis)utility achieved from the actual process of receiving health service is called process utility. Failure to consider this type of utility in therapeutic interventions that have the same outcomes but have different therapeutic processes can lead to ignoring several benefits and suboptimal outcomes of the interventions. Process utility considers the patient's experience and satisfaction with the treatment process, which can be related to the quality of the provided service [16].

### Personal Utility

One of the key features of the health service delivery process is the value of awareness and information in a way that is distinct from the value of information for decision-makers. In various studies, the benefits of information generated by healthcare interventions, often in the field of genetic testing, have been specifically examined. Some of the consequences of this awareness include anxiety (or eliminating anxiety) and requiring testing for other family members.

In a study by Grosse, the utility of genetic information can be considered from three perspectives: the public health perspective, the clinical perspective (on the effectiveness of this genetic information in diagnosis and treatment selection), and the individual perspective, which means examining the value of these tests for each person and showing the advantages and disadvantages of performing them outside of the medical field [16, 17]. Examples of personal utility are improved confidence from awareness, people's sense of control, autonomy, and self-identity [16].

## 10.2.3 Inadequate Available Data

Generally, required data for economic evaluations will be obtained from clinical studies [7, 8]. For some types of PMI such as pharmacogenomic tests, clinical studies play a remarkable role in the uptake of information about the clinical utility of the tests, changes in health status, and resource utilization and costs. The major challenges related to the data needed for conducting economic evaluations of PMI are their availability and reliability [7].

One of the main barriers to practically applying PMI is the data gap about its effectiveness. In other words, before routinely applying a biomarker-driven approach to choose an appropriate treatment, clinical studies must prove the clinical utility of the test [18]. Clinical utility means "the probability that a test can change and improve patient outcomes." This case refers to the correlation between genetic tests and the treatment [15]. Designing and implementation of adaptive clinical studies (which are carried out more quickly than conventional trials) is one of the solutions to overcome this limitation. A distinctive feature of these studies is that if the intervention proves to be effective (or ineffective), the study will be stopped as soon as possible. Thus, using adaptive clinical studies can provide the required data about the effectiveness of PMI, but this method is currently being applied only to a limited extent [6].

Another considerable challenge in the issue of data for EE of PMI is the quality of evidence. As mentioned before, economic evaluation is carried out based on the final results of clinical trials. However, in the field of precision medicine, it is necessary to obtain more evidence by conducting prospective observational studies to support the results of these trials because clinical studies of PMI have a limited population (usually subsets of patients from trials with large numbers of participants). Thus, complementary data in practice (real-world data) can increase the validity of the results of the clinical trials. In other words, performing EE of PMI using real-world data can lead to higher quality and reliability of the results of these studies, although the use of this type of data in EEs is very limited [1, 6].

### 10.2.4  Increasing Complexity of EEs of PMI

Many studies have addressed the challenges of EEs of PMI [6, 8, 15, 19]. Evidence concludes that EEs of PMI are associated with considerable complexity, because such studies evaluate more than one technology (at least two, one intervention, and one test). To better understand and minimize the complexities of EE of PMI, it is necessary to consider the following aspects.

#### 10.2.4.1  Study Question

The first step in conducting an EE is defining the research question. In other words, in the beginning, the characteristics of the intervention arm, comparator arm, study population, time horizon, and perspective of the study should be completely and clearly stated. However, it would not be eas-

ily possible in the field of precision medicine regarding the complexity of the EEs of PMI. Sometimes for one intervention, more than one type of test may be needed to be performed. Moreover, it is also considered that applying a genetic test or assessing a biomarker would be performed, with different types of kits available in the market with different prices. Such cases cause complications in explicitly stating a research question in EE of PMI [4, 8, 15]. Therefore, it is recommended that the following items should be identified before conducting cost-effectiveness studies of a PMI:

- What intervention needs to be evaluated?
- What is the biomarker(s) of the intervention?
- What are the types of laboratory methods used to evaluate biomarker(s)?
- Which laboratory methods are available in the market?

And then, following the answers to these questions, the comparator arm should be determined correctly [15].

#### 10.2.4.2  Sensitivity/Specificity and Predictive Value of the Test

The sensitivity and specificity of the tests, as well as their predictive value, should be considered in the modeling of EE of PMI. Sensitivity and specificity are important issues and should be considered in cost evaluation because test errors lead to false-positive and false-negative results which affect health status and costs. False-positive results cause unnecessary costs and side effects in patients who received the medicine incorrectly and unnecessarily. False-negative results can on the one hand cause loss of health and on the other hand lead to either increased or decreased costs [5, 6].

The concept of the predictive value of tests depends on the prevalence of biomarkers or mutation in a population [5], and in these studies, biomarkers with real prevalence associated with the increased therapeutic response should be examined [8].

### 10.2.4.3  Cascading Decision in EE of PMI

PMI mainly includes genetic sequencing and assessing of biomarkers, followed by decisions about the choice of treatment based on the results of these tests. The multiplicity of types of tests required in a treatment method, the sequenced or simultaneous performance of the test, and making decisions based on the results lead to complex care pathways in PMI. In other words, in these interventions, we encounter sequential or cascade testing, which affects the EE of PMI modeling, because, in these studies, a cascade modeling must be designed. Considering the risk of obtaining random results (as described in the previous sections) and the possibility of false positives or negatives and the possible costs of performing each test, PMI modeling involves many complexities [3, 15, 19, 20].

In general, complexities in economic evaluations lead to uncertainty in the results. Since health systems make decisions about resource allocation to technologies (a method of treatment or a medicine) based on the results of economic evaluations, they should be aware of the level of uncertainty in the study results. EE of PMI has considerable complexity, and performing sensitivity analysis in such situations shows the degree of uncertainty in the results and can demonstrate how likely the decision about resource allocation to PMI would be cost-effective [15].

## References

1. Terkola R, Antoñanzas F, Postma M. Economic evaluation of personalized medicine: a call for real-world data. New York: Springer; 2017. p. 1065–7.
2. Fugel H, Nuijten M, Postma M, Redekop K. Economic evaluation in stratified medicine: methodological issues and challenges. Front Pharmacol. 2016;7:113.
3. Kasztura M, Richard A, Bempong N-E, Loncar D, Flahault A. Cost-effectiveness of precision medicine: a scoping review. Int J Public Health. 2019;64(9):1261–71.
4. Husereau D, Marshall DA, Levy AR, Peacock S, Hoch JS. Health technology assessment and personalized medicine: are economic evaluation guidelines sufficient to support decision making? Int J Technol Assess Health Care. 2014;30(2):179–87.
5. Doble B. Budget impact and cost-effectiveness: can we afford precision medicine in oncology? Scand J Clin Lab Invest. 2016;76(sup245):S6–S11.
6. Shabaruddin FH. Economic evaluations of personalized medicine: existing challenges and current developments. Pharmacogenomics Pers Med. 2015;8:115–26.
7. Rascati K. Essentials of pharmacoeconomics. Philadelphia, PA: Lippincott Williams & Wilkins; 2013.
8. Phillips KA, Deverka PA, Marshall DA, Wordsworth S, Regier DA, Christensen KD, et al. Methodological issues in assessing the economic value of next-generation sequencing tests: many challenges and not enough solutions. Value Health. 2018;21(9):1033–42.
9. Garattini L, Curto A, Freemantle N. Personalized medicine and economic evaluation in oncology: all theory and no practice? Expert Rev Pharmacoecon Outcomes Res. 2015;15(5):733–8.
10. Nosrati M, Nikfar S. Conducting economic evaluation based on basket clinical trial in the area of precision medicine. Taylor & Francis: New York; 2021. p. 169–71.
11. Janiaud P, Serghiou S, Ioannidis JP. New clinical trial designs in the era of precision medicine: an overview of definitions, strengths, weaknesses, and current use in oncology. Cancer Treat Rev. 2019;73:20–30.
12. Food U, Maryland DAJF. Master Protocols: efficient clinical trial design strategies to expedite development of oncology drugs and biologics Guidance for industry-draft guidance; 2018.
13. Cecchini M, Rubin EH, Blumenthal GM, Ayalew K, Burris HA, Russell-Einhorn M, et al. Challenges with novel clinical trial designs: master protocols. Clin Cancer Res. 2019;25(7):2049–57.
14. Faulkner E, Annemans L, Garrison L, Helfand M, Holtorf A-P, Hornberger J, et al. Challenges in the development and reimbursement of personalized medicine-payer and manufacturer perspectives and implications for health economics and outcomes research: a report of the ISPOR personalized medicine special interest group. Value Health. 2012;15(8):1162–71.
15. Annemans L, Redekop K, Payne K. Current methodological issues in the economic assessment of personalized medicine. Value Health. 2013;16(6, suppl):S20–S6.
16. Engel L. Going beyond health-related quality of life for outcome measurement in economic evaluation: Health Sciences: Faculty of Health Sciences; 2017.
17. Grosse SD, McBride CM, Evans JP, Khoury MJ. Personal utility and genomic information: look before you leap. Genet Med. 2009;11(8):575–6.
18. Antoniou M, Jorgensen A, Kolamunnage-Dona R. Biomarker-guided adaptive trial designs in phase

II and phase III: a methodological review. PLoS One. 2016;11(2):e0149803.

19. Marshall DA, Grazziotin LR, Regier DA, Wordsworth S, Buchanan J, Phillips K, et al. Addressing challenges of economic evaluation in precision medicine using dynamic simulation modeling. Value Health. 2020;23(5):566–73.

20. Rogowski W, Payne K, Schnell-Inderst P, Manca A, Rochau U, Jahn B, et al. Concepts of 'Personalization' in personalized medicine: implications for economic evaluation. PharmacoEconomics. 2015;33(1):49–59.

# Ethical, Legal and Social Aspects of Precision Medicine

# 11

Maria Josefina Ruiz Alvarez ⓘ, Erich Griessler ⓘ, and Johannes Starkbaum ⓘ

**What Will You Learn in This Chapter?**
In this chapter, the importance of the ethical, legal, and social aspects (ELSA) under the view of PM research and implementation is illustrated, which is focused on understanding ELSA basic concepts related to the opportunity of sharing genomic and health-related data, ensuring continued progress in our knowledge of human health and well-being. Overall, the citizens' data and risk to privacy, informed consent, and data regulation are being discussed.

**Rationale and Importance**
The purpose of this chapter is to provide a framework for the ethical, social, and legal aspects and to focus on specific issues that are key to the development and interventions of PM. At present, the healthcare system faces significant challenges in adopting a safe and effective, personalized medicine approach on prevention, diagnostic, and therapeutic. Health data must be collected, processed, stored, and distributed under the legal requirements in place within that country or region. In this line, Ethics on Health is the set of principles or values to guide every con-

duct in the health field, specifically oriented to the use of health data and health technologies. The PM ethical aspects illustrated in this chapter reflect the application of new technologies with the allocation of needed resources.

The most important regulations briefly described in this chapter are included in Box 11.1.

> **Box 11.1 The Selected Ethical Lines Followed in This Chapter**
> **Universal Declaration of Human Rights**[1] – 1948. Selected articles dealing with the key topics illustrated in this chapter
>
> Article 27 guarantees the rights of every individual «to share in scientific advancements and its benefits» (including to freely engage in responsible scientific inquiry as well as «to the protection of the moral and material interests resulting from any scientific production of which [a person] is an author»
>
> Article 4: Human genome no give rise to financial gains

M. J. R. Alvarez (✉)
Research Coordination and Support Service, Istituto Superiore di Sanità, Rome, Italy
e-mail: mariajose.ruizalvarez@iss.it

E. Griessler · J. Starkbaum
Institut für Höhere Studien – Institute for Advanced Studies (IHS), , Vienna, Austria

[1]This section on https://www.ohchr.org/Documents/Publications/ABCannexesen.pdf

Article 5e: Outlines the idea of consent

Article 7: Outlines the idea of confidentiality

Article 12: States benefit from advances in genetics and medicine available at all about dignity and human rights

**International Declaration on Human Genetics. Data – UNESCO (2003)[2]** Human genetic data have a special status because:

1. They can be predictive of genetic predispositions concerning individuals
2. They may have a significant impact on the family, including offspring, extending over generations, and in some instances, on the whole, group to which the person concerned belongs
3. They may contain information on the significance of which is not necessarily known at the time of the collection of the biological samples
4. They may have cultural significance for persons or groups

**Declaration of Helsinki (1° v: 1964; last: 2013)[3]** The Declaration of Helsinki lays down ethical principles for medical research involving human subjects, including the importance of protecting the dignity, autonomy, privacy, and confidentiality of research subjects, and obtaining informed consent for using identifiable human biological material and data. Convention on Human Rights and Biomedicine (Oviedo Convention)[4] is the only international legally binding instrument exclusively concerned with human rights in biomedicine, and its Additional Protocol concerning Genetic Testing for Health Purposes. Oviedo Convention contains specific provisions relating to genetics (Articles 11 to 14), particularly predictive genetic tests and interventions on the human genome

## 11.1 Ethical, Social, and Legal Aspects of PM: Research and Healthcare Context

This chapter starts with this sentence of Archbishop Desmond Tutu, the Nobel Peace Prize laureate who helped end apartheid in South Africa, who has died this year (quoted by Jance)[5]:

> …….. My dream is that by including all peoples in understanding and reading the genetic code we will realize that all of use belong in one global family – that we are all brothers and sisters…..

Nowadays, the concept of bioethics is moving fast and combines the values of science, medicine, law, and philosophy under healthcare.

Usually, ethical and legal aspects are followed in parallel not only because they are related but also because they answer diverse complementary issues. Legal is linked to law or policy, including professional codes of practice. Not always an ethically right situation concurs with what is legally permissible, and this opens the floor for a discussion to revise the law under the ethics of the issue.

There are universal and unintentional biases such as the herd mentality that tend to reinforce the "values" of the society we live in. These are not necessarily the ones that are the most defensible or even the ones that are in line with what we think our ethical principles are, and they are very often difficult to identify and correct. These aspects are not discussed in this chapter. The fundamental concepts and principles that typically influence bioethical reasoning such as respect for autonomy, prevention of evil, equity, etc. are briefly illustrated.

---

[2]This section on https://en.unesco.org/about-us/legal-affairs/international-declaration-human-genetic-data

[3]This section on https://www.wma.net/policies-post/wma-declaration-of-helsinki-ethical-principles-for-medical-research-involving-human-subjects/

[4]This section on https://rm.coe.int/168007cf98

[5]https://www.quotemaster.org/author/Desmond+Tutu

The respect for autonomy allows human beings to be able to make decisions for themselves, and it is one of the pillars of ethical and legal requirements to accept medical treatment or diagnostic tests. Citizens have to understand all the relevant information about the risks and benefits, not being pressured to have to accept this test or procedure and with sufficiently autonomous choices.

However, persons need to reflect on how far their individual decision should be respected when weighed against other individual considerations. It is the harm principle, so long as individuals' choices do not pose a threat of serious harm to others, they should be respected [1].

## 11.2 Data Sharing: Clinical Care and Research

The World Economic Forum during a Global Precision Medicine Council[6] in 2019 elaborates a synthesis of the key policy and governance gaps for the PM implementation and their possible solutions to overcome them:

1. Data sharing and interoperability
2. Ethical use of technology
3. Societal trust and engagement
4. Access and fair pricing
5. Responsive regulatory systems

One of the main obstacles described in the application of genomics is data-sharing complexity. It can be also applied to all new technologies on PM [2]. With the PM advances, data sharing is crucial and several guidances are being developed. However, there is still a fragmentation in the policy landscape across specific organizations and data types. The difficulty in sharing the data are grouped on:

- Lack of policy harmonization
- Lack of structural support

- Legal and ethical hurdles
- Cultural barriers

However, the bias in the selection of data is really crucial as it can make results not generalizable, facilitate the dissemination of existing prejudices, and also exacerbate disparities in healthcare. All these consequences reduce the reliability of the related technologies.

The most adequate way to move the PM toward a greater competent data-sharing ecosystem is to involve policymakers. Central policy themes had to incorporate, as priorities, the privacy, consent, and data quality and the crucial interoperability, attribution, and public engagement. From the side of public engagement, a simplification of informed consent procedures, privacy-preserving data processing, and encouraging data quality are needed. From the scientific side, reevaluating interoperability, attribution, and facilitating participatory governance are also requested [3].

Considering privacy and confidentiality, clear rules are needed as data sharing has important implications for the population (individual and group). To encourage transparency and public trust in the use of data, citizens should have a role as accountable for their data and have the possibility to decide on some aspects of management as the control over data access and distribution practices. This citizen participation should be a rule, above all, for genetically isolated populations, disease groups, ethnic groups, minority groups, or specific communities. Examples are hereditary cancers or genetic diseases, where the information to the family is crucial as there is the risk of developing the related disease.

Data sharing, in the context of new diagnostic and therapies, requires clear rules from protecting patient data and other legal risks to technical difficulties and institutions' unwillingness to collaborate. Indeed, data infrastructures for biomedical research are working to combine multiple data formats for a complete exchange of data without losing information under a harmonized ethical and legal framework [4].

Lastly, by increasing the data protection rules and surveillance, the risk of data leakage due to secondary use of data without consent will be

---

[6] WEF_Global_Precision_Medicine_Council_Vision_Statement_2020.pdf

avoided and with this, the risk of data misuse that induces stigmatization and discrimination.

Bioethical research is being further applied following the innovation tools on PM research.

## 11.3 Patient Recruitment

Subject enrollment starts with the individual selection within the pool of eligible subjects, followed by the engagement and the retention. Recruitment must follow the fair distribution of burdens and benefits, ensure the social value of research, enhance scientific validity, minimize risk to subjects, protect the vulnerable, and enhance benefits to participants.

## 11.4 Informed Consent (IC)

The requirements of informed consent in the context of PM will be revised under the ethical view in this section and under the legal framework under the GPDR section.

The IC follows the ethical requirements deriving from the Declaration of Helsinki[7]: "Right to human dignity and right to the integrity of individuals." Briefly, the three most important objectives of the informed consent under the Clinical Trials Regulation are first, to provide full details about the study for the appropriate participant's information; second, the future use of data and the disclosure of the research results to patients; and lastly, the possibility of unexpected new information and considerations regarding genetics. Based on this information, patients decide whether they want to participate in the study or not.

Details of the study include the length of their participation, the number of participants, participants' duties and rights (number of visits, filling of forms, treatment compliance, etc.), procedures that encompass their participation and risks of such procedures, and the possibility to be randomly assigned to any of two groups: control and experimental.

Participants in research projects must give their own informed consent under clear and precise information, without technicalities, and ideally in a limited but complete way, with a space for asking questions and enough time for a mediated decision to participate.

They need to understand that if they can conclude, at any moment, their participation in the study without impact on the quality of healthcare provided to them, there will be no loss of benefits; thus, they are otherwise entitled.

The information must include [5]:

- Alternative treatments related to their medical condition.
- Any foreseeable risks to the participating subject or to others, such as the fetus in case a woman gets pregnant (if there are any because of the gender unbalance).
- Adverse events that can occur and what to do in case of occurrence, whom to contact, where to go, and who will cover medical costs. They have to receive specific information on what costs will not be covered and which are covered as participants because if there is a liability policy for lesions due to the trial, that information should also be provided to participants.
- What will happen after the end of the trial (control visit, valuation of new symptoms).

Another important aspect of this "consent process" is the exact information on the individual and public benefits of sharing their data. Whether there are not individual benefits, this must be stated clearly too.

Essential for a right informed consent is an ideal ethical framework[8]: providing information about how the data privacy is being implemented and on how data sharing promotes public health and addressing the supervision of risk of patients.

*Dynamic informed consent*: It is a tool that is based on a personalized communication platform that aims to facilitate the consent process. This

---

[7]https://www.who.int/bulletin/archives/79(4)373.pdf

[8]Framework for Responsible Sharing of Genomic and Health-Related Data http://genomicsandhealth.org

tool facilitates participants to make autonomous and informed choices on whether or not to participate in research projects. It is very useful to support continuous two-way communication between researchers and participants.

However, improving consent models and their application will require a mix of traditional and innovative educational approaches to engaging the public more generally.

## 11.5  Pediatric Informed Consent

Children and teenagers face up to different levels of participation [6]:

1. Unable to contribute own views after unappropriated information process
2. Can form views and express opinions, but cannot make independent decisions after an appropriate information process
3. Has intellectual capacity and maturity to make own decisions, but is still considered minor in her/his domestic legal systems after an appropriate information process

The role of parents in making decisions on their behalf must consider the child's best interest, without forgetting their individuality. Parents should consider children's both current and long-term welfare being conscious of their evolving developmental capacities.

Although clinical studies are needed to find personalized diagnostic and treatment, there are some core recommendations to include children or younger adults in research:

1. Only if pertinent and safe.
2. Use of information that is easy to understand for participants.
3. Encourage participation through the assent process.
4. Research ethics committees (REC) should include experts in childhood (psychology, health, etc.) when reviewing projects that involve children and promote their participation through young persons' advisory groups.

An example of an essential adequately conceived and conducted clinical research in children is the field of rare diseases. Genomic testing is able to improve the diagnosis and classification of a personalized treatment for children. Thereby, clinicians and researchers should evaluate the benefits and risks of new applied technologies. Last but not least, the inclusion of the children and young patients is important on the making process and the design of projects and new medical devices.

To sum up:

- Information provided to children and adolescents should be given in a language appropriate to their age.
- Level and assent should be obtained from minors who are in the range of age that allows them to assent.
- Accept the voluntariness of their participation: it will not occur if they do not want to, even if their parents have authorized their participation in the study.

## 11.6  Unsolicited or "Incidental" Findings

Clinicians and researchers have to face up with unexpected findings. The situation is more and more frequent with the new technologies such as genetic information. Questioning about how to deal with incidental findings is of huge importance not only for diagnostic testing [7].

These can be related to the different levels of adult-onset disorders and can have family implications, even reproductive implications on carrier status for recessive disorders.

The whole exome sequencing (WES) and the most recent PCR-free whole genome sequencing (WGS) [8] bring also to deal with other legal aspects such as the identification of nonpaternity or adoption or the identification of areas of the excessive absence of heterozygosity (SNP array), as an indicator of consanguinity and raise suspicions of abuse/incest.

Following the recommendation of the European Society of Human Genetics[9], the following options for reporting these data can be extracted: to give the opportunity to opt-out; to prefer targeted/hypothesis-based approaches (sequencing and/or analysis) in the clinical setting; and, the most transparent, to develop protocols for the return guide of unsolicited results.

These reports should be made by health professionals, especially if the data suggests serious or preventable health problems. With the advancement of new technologies and genetic treatments, the development of specific guidelines on what, how, and to whom (patient, family, social services) unsolicited information must be disclosed is essential. Information that should be given without interference in the autonomy of the rights of the family and the needs of health interests.

*FAIR Data Principles* are a set of guiding principles in order to make data findable, accessible, interoperable, and reusable [9]. Scientific data management and stewardship follow these principles as guidance in all the sectors of the digital ecosystem. They are rules that describe how research outputs should be organized so they can be more easily accessed, understood, exchanged, and reused. Major funding bodies, including the European Commission, promote FAIR data to maximize the integrity and impact of their research investment.[10]

## 11.7 Genetic Service and Biobanks

Subsequent to the needs of researchers to go back to participants whenever their data or specimens are used, the biobanks have been structured.

The ISO 20387 published to help organizations get the most out of the standard requires biobanks to implement quality management [10]. These requirements deal with improving the quality of biological material and data collections that are stored and shared, enhancing the outcomes of collaboration, strengthening trust between partners, and advancing research and development. Briefly, the biobanks describe challenges to follow it is:

- To obtain consent for the storage and use in unspecified future research
- Preservation of donors' confidentiality
- Interpretation and management of incidental findings
- Ownership and control of tissues
- Acknowledgment and management of cultural sensitivities
- Reporting of results
- Community participation
- Benefit-sharing
- Return of materials to communities and disposal of unused material

The OECD Guidelines on Human Biobanks and Genetic Research Databases, 2009, provide "guidance for the establishment, governance, management, operation, access, use and discontinuation of human biobanks and genetic research databases ('HBGRD'), which are structured resources that can be used for the purpose of genetic research and which include: (1) human biological materials and/or information generated from the analysis of the same; and (2) extensive associated information" [11].

HBGRD foster that data and materials be rapidly and widely available to researchers with respect to human rights and freedoms, and the protection of participants' privacy and the confidentiality of data.

In addition, HBGD also minimize risks to participants, their families, and potentially identifiable populations, develop and maintain clearly documented operating procedures for all the processes (procurement, collection, labeling, registration, processing, storage, tracking, retrieval, transfer, use and destruction, data and/or information), and, finally, ensure that the results of research conducted using its resources, regardless of the outcome, are made publicly available.

[9] https://www.eshg.org/index.php?id=home
[10] https://ec.europa.eu/info/sites/default/files/turning_fair_into_reality_1.pdf

## 11.8    Institutional Review Boards (IRB)

The IRB has the aim to review and monitor biomedical research involving human subjects. In accordance with FDA regulations, an IRB " has the authority to approve, require modifications in (to secure approval), or disapprove research. This board is responsible for the protection of the rights and welfare of human research subjects."

Independent review by an IRB or equivalent is an important part of a system of protections aiming to ensure that ethical principles are followed and has an important role in protecting research participants from possible harm and exploitation.

## 11.9    The General Data Protection Regulation (GDPR) and Related Definitions

GDPR is the acronym commonly used for General Data Protection Regulation (2018)[11], which unites all data protection legislation across the member states of the European Union. It also includes Switzerland, Norway, Liechtenstein, and Ireland.

It considers the basic definition of personal data "is any information relating to an identified or identifiable natural person (data subject). In other words, any information that obviously relates to a particular person and can be used to identify them."

Biological materials, alone or with the associated data, can allow the identification of persons (directly or through a code controlled by a third party, e.g., clinical care).

## 11.10    Reidentification and Pseudonymization

Pseudonymization is "when data is masked by replacing any identified or identifiable information with artificial identifiers." Even if patient's names are reidentified and/or replaced with an

identification code, they can be identified and are not anonymous. Reidentification of single participants in genome-wide association studies is possible [12].

## 11.11    Blockchain

Blockchain is a decentralized list of digital archives linked by cryptography. Each record or block contains a cryptographic hash of the previous block ad example mathematical algorithm, timestamp, and data of that transaction. Blockchain technology offers a secure open ledger to record digital transactions, managed by a peer-to-peer network [13].

Blockchain databases are designed to be only-ever-created, and not edited or deleted. For this reason, it is used in healthcare to increase the security of various transactional activities in the healthcare space. It can decrease bureaucracy and manual inefficiencies, improve the quality of care and privacy of patient data, and ensure up-to-date fields with a high level of security and privacy of data, and data is encrypted in blockchains and can only be decrypted with the patient's private key [14]. Their use in the healthcare sector needs a comprehensive guide through a functional and technical understanding.

It has been described that blockchains permit healthcare stakeholders to collaborate without the control of central management and support immutable audit trails useful to record critical information. It can use and identify the different sources of data ensuring the robustness and availability of data.

## 11.12    Personal Data

According to the GDPR regulation, if personal data is also sensitive data, it requires "special protection." It distinguishes between the concept of "personal data" and "sensitive data."

Sensitive data is a set of special categories that should be handled with extra security (Article 6 of the GDPR), including some exceptions described (Table 11.1).

---

[11] https://www.gdpreu.org/the-regulation/key-concepts/personal-data/

**Table 11.1** GDPR identifiers' considered personal data

| Personal data | No personal data |
|---|---|
| Name and surname | Information deceased person |
| Email address | Properly anonymized data |
| Phone number | Information about public authorities and companies |
| Home address | |
| Date of birth | |
| Racial or ethnic origin (S) | |
| Gender | |
| Political, religious, or philosophical opinions (S) | |
| Credit card numbers | |
| Data held by a hospital/ doctor and health data (S) | |
| Photograph identifiable | |
| Identification card number | |
| A cookie ID | |
| Internet Protocol (IP) address | |
| Location data | |
| Advertising identifier of phone | |
| Code assigned and any set of information related to the code | |
| Genetic data (S) | |
| Biometric data (S) | |
| Sexual life or sexual orientation data (S) | |

Sensitive data are indicated with (S)

Health data are personal data on the physical or mental health of a natural person, including the provision of healthcare services, which reveal information about his or her health status (Art. 4 GDPR).

Genetic data are the acquired genetic characteristics of a natural person which give unique information about the physiology or the health of that natural person and which result, in particular, from an analysis of a biological sample from the natural person in question (e.g., DNA, RNA analysis, and other data that may be inferred from samples' analysis).

Health and genetic data need to be processed with additional conditions and security measures.

Some countries established additional security measures [15], such as storing of the database, protecting data by enforcing a double authentication factor (e.g., token, double password), allowing to send these data via email just as an attachment, restricting access to facilities, and if possible implementing biometric control access system.

The company/sponsor or site that collects personal data from data subjects is called a "data controller." It is the natural or legal person who determines the processing of personal data (decides what data to collect, how long to store the data, how to analyze the data, etc.) and ensures compliance with principles of GDPR (security measures, data subjects rights, etc.).

The company (CRO, laboratories, service providers, principal investigator, or monitors) that is employed to process that data is called a "data processor," natural or legal person who processes personal data and is responsible in case of infringement of the instructions or infringement of the specific obligations under GDPR (bound by a contract, Article 28).

GDPR Regulation applies in the context of the activities of an establishment of a controller or a processor in the Union, regardless of whether the processing takes place in the Union or not. GDPR Regulation applies always to EU individuals.

## 11.13 International Transfers of Personal Data: Regulation (EU) 2018/1725

EU data protection rules apply to the European Economic Area (EEA: EU Member States and Iceland, Liechtenstein, and Norway). As the controller for the processing of personal data, EU institutions (EUIs), bodies, offices, and agencies are accountable for the transfers within and outside the European Economic Area.

EU works to facilitate the use of the range of alternative transfer tools to protect data protection rights when data are transferred to countries whose domestic law does not ensure an adequate level of data protection.

A brief description of the main related articles of this Regulation (EU) 2018/1725 has been illustrated in the following text [16].

…" Art. 45: Adequacy decision European: Commission has decided that the third country in question ensures an adequate level of protection. Such a transfer shall not require any specific authorization. The EUI or its processor can provide appropriate safeguards, by using transfer tools according to Article 48, such as standard contractual clauses for transfers, or for transfers from a processor of an EUI to sub-processors by also using transfer tools within the meaning of Article 46 of the GDPR (Binding corporate Rules; Code of conducts; Certifications; Standard Contractual clauses). Article 49 describes certain specific situations or derogations.

Concerning the transparency of dates, Article 13 illustrates the information that has to be included: briefly, the identity and contact details of the controller and the data protection officer (Article 37), the purposes of the processing and the legal basis, the recipients of the personal data, the fact that the controller intends to transfer personal data to a third country and reference to the appropriate or suitable safeguards, and time of storing and rights of the data subject, including the right to withdraw consent at any time and right to complain. The right not to be subject to a decision based solely on automated processing is also regulated by Article 22.

Article 7 Rec.33/34 illustrated consent as a legal basis for data processing. Shortly, it must be a freely given, specific, intelligible, and easily accessible form, clearly distinguishable from the other written declarations, and very explicit regarding special categories of data and about how data will be processed. It is important to underline that this consent can be withdrawn at any time, and data subjects should be allowed to give their consent not only to the full project but also to certain areas of scientific research, or to parts of research projects.

It is also described as a "secondary use" of data in accordance with Article 89 "further processing for […] scientific research purposes […] shall, not be considered to be incompatible with the initial purposes." It refers solely to situations where the sponsor may want to process the data of the clinical trial subject "outside the scope of the protocol," but "exclusively" for scientific purposes. However, patients must be informed of this possibility. Article 28 of clinical trial regulation indicates at the time of the request for informed consent for participation in the clinical trial. The GDPR writes that "it would require another specific legal ground than the one used for the primary purpose."

Concerning the data collected, "it must be limited to what is necessary in relation to the purpose and stored in a form which permits identification of data subjects for no longer than is necessary for the purposes for which the personal data are processed" (Article 5).

Lastly, regarding the personal data breach, GDPR indicates the notification to data protection authorities and affected individuals following their discovery. It is considered as a "breach of security, leading to the accidental or unlawful destruction, loss, alteration, unauthorized disclosure of, or access to, personal data transmitted, stored or otherwise processed."

As PM becomes further incorporated into clinical practice, the regulation of these important ethical, social, and legal aspects should be harmonized among different countries and adapted to the continually evolving science and technology.

## 11.14 Equal Access to Personalized Care

Lack of diversity in research contributes to health disparities in PM. It can be affected by a bias in the collection of data during the involvement of minority groups weakly represented in the healthcare system. It has been recognized that including minorities and avoiding structural racism is needed on the integration of biased data, and on the analysis of the results.

Some investigators theorize that the differences in health outcomes' race-associated are really due to the effects of "structural racism" and recommend that research studies need be available not only for patients living in countries where targeted therapies are subsidized. In some countries, it is the only possibility to access the clinical services through which these therapies might be offered.

It is evident the influence that racism still has overall in healthcare, as described in a complex disease hospital algorithm applied [17], where black people were selected with less frequency to improve care than white people. In this line, three levels of structural racism on health have been defined [18] with the aim of understanding the mechanism and fight against it: institutionalized, personally mediated, and internalized.

Institutionalized racism refers to material conditions and access to power, personally mediated to prejudice and discrimination and internalized to stigmatized races on own abilities and intrinsic worth.

Scientific racism, conscious or unconscious, is a fact to be avoided. Otherwise, it gives the risk of the perpetuation of inaccurate notions of human populations such as the real implication of phenotypes, the already well-documented fact that 99.9% of humans are identical, the difference among gene frequency and gene expression, and the biological roots of behavior and physiology.

The implementation of PM initiatives requires ethnic data diversity and appropriate ethnic racial representation in their cohorts. In this way, a trusting relationship with these minority groups has to be consolidated to succeed in their research objectives, such as the collection and integration of health data [19]. Inequalities can be eradicated if the research efforts will be addressed to avoid this inaccurate idea.

## 11.15  National Strategy or National Plan on Personalized Medicine

PM as a global concern in research and implementation on healthcare is linked to the healthcare systems. These are heterogeneous worldwide and depend on the region and country. However, healthcare needs to be comparable, and the best solutions are often found on a transnational level with the development of common strategies, standards, and frameworks with cross-border collaborations and interactions.

The WHO defines a national health strategy[12] as "a document or set of documents that lay out the context, vision, priorities, objectives and key interventions of the health sector, multisectoral or disease programmed, as well as providing guidance to inform more detailed planning documents." A strategy provides the "big picture" and the road map for how goals and objectives are to be achieved. A national health plan is a document or set of documents that provide details on how objectives are to be achieved, the time frame for work, who is responsible, and how much it will cost. This may come in the form of a multi-year plan, supported by annual operational plans that allow for adjustment as a program.

In this direction, national genomic or PM strategy and national plan should be developed, mainly to allow detailed knowledge of the genetic background and the distribution of rare and common variants varies across populations and for admixed populations (underrepresented).

## 11.16  Examples of International Networks Working on Sharing Data

The Global Alliance for Genomics and Health (GA4GH) is a policy-framing and technical standards-setting organization, seeking to enable responsible genomic data sharing within a human rights framework. Their strategic plan (GA4GH Connect) aims to drive the uptake of standards and frameworks for genomic data sharing within the research and healthcare communities in order to enable responsible sharing of clinical-grade genomic data by 2022. [20].

The plan follows four lines of work: from the development of standards, tools, and frameworks to overcome technical and regulatory hurdles, the identification of world genomic data initiatives sourced that provide guidance on standards development, providing mechanisms and recommendations to create internal consistency and technical

---

[12] https://www.who.int/ehealth/publications/overview.pdf

alignment across GA4GH Work Streams and product deliverables, and finally, facilitating two-way dialogue with the international community, including national initiatives, major healthcare centers, and patient advocacy groups.

Medical information commons (MIC) are networked environments to shared resources in diverse health, medical, and genomic data on large populations [21].

The American College of Medical Genetics and Genomics[13] is an interdisciplinary professional membership organization of the entire medical genetics team including clinical geneticists, clinical laboratory geneticists, and genetic counselors. They elaborate on guidelines, technical standard, and position statements on laboratory and clinical genomic data sharing to improve genetic healthcare. Recently, they have published recommendations for reporting incidental findings in clinical exome and genome sequencing.

The National Institute of Allergy and Infectious Diseases (NIAID)[14] supports and complies with the data-sharing policies, including the NIH Genomic Data Sharing (GDS) Policy. Genomic summary results (GSR) generated with NIH funding should be made freely available on the Internet with no access restriction.

The current development of European Countries' recommendations for dealing against in-equalities in health care (EU, study policies, 2018)[15] reflects the growing interest in national and European authorities in personalized medicine and other personalized approaches to health. It is also evidenced by the development of national plans by some countries, as well by the foundation in 2016 of the International Consortium for PM (ICPerMed)[16] and the umbrella Coordination and Research supporting initiatives. All of them are supported by the European Commission funds.

## 11.17 Intellectual Property of PM

Following the WIPO[17] definition of intellectual property (IP), it refers to creations of the mind, such as inventions, literary and artistic works, designs, symbols, names, and images used in commerce. WIPO is the global forum for intellectual property (IP) services, policy, information, and cooperation.

Gene sequences and their expression patterns due to the capacity to better identify and personalize detriment of tumor types become of considerable economic value to them discovered through protection as intellectual property rights.

Not all the inventions can be patented, for example, diagnostic tests based on purely natural principles or phenomena cannot be patented, as in the case of Myriad Genetics [22]. This company has discovered and commercialized several genetic tests for the risk of developing the disease, assessed the risk of disease progression, and guided treatment decisions across medical specialties After several legal procedures, the Myriad patents have been revoked (USPTO) or strongly limited in their scope (EPO). The Supreme Court of the United States concluded that genomic DNA is not admissible for patents, while synthetic DNA remains patentable.

The loss of intellectual property protection and the consequent loss of economic returns can make access to the market difficult. In terms of PM, the proven clinical utility seems to facilitate better protection of intellectual property [23].

The European Commission Directorate General Research and Innovation, the European Innovation Council and SMEs Executive Agency (EISMEA), and the European Union Intellectual Property Office (EUIPO)[18] are working together on developing intellectual property (IP) management. Among diverse activities, they are elabo-

---

[13] https://www.acmg.net/ACMG/Education-and-Events/

[14] https://www.niaid.nih.gov/research/genomic-data-sharing

[15] https://www.europarl.europa.eu/RegData/etudes/IDAN/2020/646182/EPRS_IDA(2020)646182_EN.pdf

[16] International Consortium for Personalised Medicine (ICPerMed). Available: http://www.ICPerMed.EU [Accessed 06 Sep 2020]

[17] https://www.wipo.int/about-wipo/en/

[18] https://ec.europa.eu/info/news/commission-and-european-union-intellectual-property-office-commit-closer-collaboration-intellectual-property-support-market-uptake-research-results-2021-nov-10_en

rating the Code of Practice for smart use of IP (expected by the end of 2022). The code of practice for the smart use of IP (ERA policy action 7) is a bottom-up initiative with the aim of providing support to R&I stakeholders via recommendations and practical examples on how to handle challenges related to intellectual assets in the current R&I context such as " increasing awareness, harmonizing rules and procedures, fostering cooperation of industry with research organizations/universities, and providing support and guidance on intellectual assets management".

## 11.18 Health Technology Assessment (HTA)

HTA is a multidisciplinary process that uses stated clearly and in detail methods to determine the value of health technology at different points in its life cycle. HTA drives the decision-making to promote an equitable, efficient, and high-quality health system [24].

From the ethical view, the objective of this assessment is to decrease or eliminate the risk of factors that can contribute to health disparities in PM. Factors usually relate to the high cost of new technologies and applied treatment that limit access to these new services. Although the benefit of PM technologies has been demonstrated, not all healthcare systems are able to support this high reimbursement. This economic limitation causes a disparity in the provision of care to those who can afford it. In addition to the cost of the new technologies, the health system also needs to invest in literacy and continuous medical training in these technologies. To solve this risk, new protocols and national guidelines on HTA are being developed and trying to support it. The major points of this assessment of health technology are:

- follow the specific rules and regulations of the health care system in which decisions are made.
- be accountable to the health care system within which they operate.

- the Coverage/reimbursement of a product within a determined health care system follows the basis of effectiveness, costs, and system affordability (value for money, priorities, and values within the system)
- demonstrate the evidence on safety, relative effectiveness, economics, and budgetary impact; social, ethical, legal, and organizational impact.

HTA requires the participation of all appropriate interest groups and must follow the condition to be equitable and efficient.

The ethics of HTA is represented by transparency, timeliness, and accountability. Moreover, a good assessment process seeks to benefit more patients and society regardless of the outcome of the assessment. The outcomes (such as recommendations) extend the ethics of professional practice and consider ethical principles of justice, benefit, and harm.

Potential ethical issues during the HTA process [25] can be related to the next considerations:

the scope of the HTA and the choice of research methods; the existence of driving forces behind the plan to perform the assessment (relevant reasons for performing/not performing an HTA on the topic, interests of the technology producers); the chance of related technologies to be morally contentious; the interests of the content expert group should be discussed openly in order for the work to be conducted in an objective and independent way; the choice of endpoints in the assessment; and morally relevant issues related to the selection of meta-analyses and studies has to be carefully considered [26].

## 11.19 Patient Empowerment/ Involvement

As explained previously in this book, public engagement is needed to evaluate the points of view of citizens toward novel technologies and programs and to ascertain their acceptability, potential ethical issues, and challenges in implementation and scalability, helping the health system in the decision-making process.

The population's characteristics and credence limited their engagement. It has been demonstrated in several studies, such as the results obtained from the survey used for the assessment of public attitudes toward donating and sharing own genomic information and data (project "Your DNA, Your Say," part of the Regulatory and Ethics Work Stream of the Global Alliance for Genomics and Health). The final report showed that the profile of people unwilling to donate their genomic information that more likely to be older, of lower education background, childless, and identifying themselves as part of an ethnic minority [27].

Citizens should also participate in the research process; therefore, their educational training and socioeconomic hurdles should be properly addressed. Nowadays, literacy and engagement of citizens is an emerging policy priority in the national governmental strategies and plans [28]. Citizens' experience should directly be included in the implementation of PM in the health system. Indeed, they should be already involved in the initial step of identifying policy priorities, as well as in the policy planning and implementation phases.

## 11.20  Incentives: Consideration

Incentives can be the payments or gifts offered to subjects as reimbursement for their participation. There is clear evidence that people are more likely to contribute in research projects when they receive an economic incentive.

Incentivizing patients has been a practice in the United States for almost 200 years. In the last century, compensation became more frequent and the ethics of payment and the potential effects on research is already opened. The financial incentive is described as the primary factor encouraging healthy participants to enroll in phase I trials and becomes less so in phase II and phase III trials, principally in developing countries. All the socioeconomic-related aspects create a potential bias on research [5].

The participants should be aware of the conditions under which they will receive partial or no payment. Usually, incentives should not be high enough to exert a coercive or undue influence in their decision to participate in the study and include transportation costs, food, general checkup, and compensation for work hours lost during visits.

## 11.21  Direct-to-Consumer Genetic Testing

Nowadays, with the era of the direct-to-test easily acquired on the web, direct-to-consumer genetic testing is also increasing. These tests, as products of PM, collect both risks and benefits.

The practical benefits are evident. However, also the psychological effects of the home tests result, without the right interpretation and right oversight. It is needed a link with clinically right advised care. Other risks are, not with minor importance, the incorrectly reported data and the misinterpretation of positive and negative results.

One example described is an app to predict sexuality based on findings of a massive study on the genetics of same-sex sexual behavior, without the right interpretation of the results of this study [29].

## 11.22  Work and Genetic Discrimination

Article 9 of GDPR, regarding the process of sensitive data on the work, indicated that processing is prohibited unless one exception applies. One of these exceptions is the "Purposes of preventive or occupational medicine, for the assessment of the working capacity of the employee, medical diagnosis, the provision of health or social care or treatment or the management of health or social care systems and services on the basis of Union or Member State law or pursuant to contract with a health professional and subject to the professional secret."

Following the concept of genetic information as sensible data, the US Equal Employment

Opportunity Commission (EEOC)[19] defines "genetic information" as all information about the following:

- Individual's genetic tests.
- Genetic tests of an individual's family members' information about the manifestation of a disease or disorder in an individual's family members.
- Family medical history is included in the definition of genetic information because it is often used to determine whether someone has an increased risk of getting a disease, disorder, or condition in the future.
- Individual's request for, or receipt of, genetic services.
- Participation in clinical research that includes genetic services by the individual or a family member.
- The genetic information of a fetus carried by a pregnant woman or by a woman family member.
- The genetic information of any embryo legally held by the individual or family member using an assisted reproductive technology.

Title II of the Genetic Information Nondiscrimination Act of 2008 (GINA)[20], which prohibits genetic information discrimination in employment, took effect on November 21, 2009.

In regard to genetic testing, the potential for using this to deny a job due to a person's predisposition to a present or future medical problem has led many countries to adopt legal measures. Several EU Member States have introduced legislation prohibiting genetic discrimination such as France, Sweden, Finland, and Denmark. Others have prohibited or restricted the collection of genetic data from employees without their explicit consent as seen in Austria, the Netherlands, Luxembourg, Greece, and Italy.

## 11.23  Notes About Equity

Since the adoption of the Oviedo Convention,[21] developments in biomedicine and society have participated in increasing disparities in access to healthcare. Indeed, new innovative treatments and diagnostic may not be accessible to everyone due to their high price. The Council of Europe is addressing these developments through its Committee on Bioethics concerning the Oviedo Convention and has published a Strategic Agenda[22] to guide the answers to new ethical challenges in human rights and shared European values.

Nowadays, several grades of unequal society with historically marginalized minorities and communities are observed worldwide. Thereby, speaking about higher quality, safe, and equitable healthcare means speaking about removing involved obstacles like poverty, discrimination, lack of access to goods, fair pay, education, and safe environments.

Implementation of PM tools in the health system attempts to explore health equity as a fair and just opportunity to be as healthy as possible. Considering equality means each citizen or community receives the same resources or opportunities. However, equity recognizes that each citizen has different requirements and, in this way, needs to receive the essential resources and opportunities to reach an equal outcome.

This process includes preventative care and also personal care treatments. And the health system needs to monitor how to design systems or public health activities. In this way, it would be feasible to provide what each population needs to maximize quality care and outcomes for populations. In this direction, the health system has to increase actions supported by resources and infrastructure, focusing on system redesign.

PM studies will be completed following all applicable laws and regulations including the International Conference on Harmonisation (ICH) Guideline for Good Clinical Practice

---

[19] https://www.eeoc.gov/genetic-information-discrimination

[20] https://www.oregon.gov/gov/policies/diversity/Documents/docs/Genetic%20Information%20Nondiscrimination%20Act%20of%202008.pdf

[21] This section on https://rm.coe.int/168007cf98

[22] https://rm.coe.int/strategic-action-plan-final-e/1680a2c5d2

(GCP) the ethical principles that have their origins in the Declaration of Helsinki,[23] the updated version of the General Data Protection Regulation (EU) 2016/679 (GDPR), and other applicable privacy laws.

## 11.24 ELSA Research Versus ELSA on PM Research

Ethical Legal and Social Aspects (ELSA) research must not be confused with the compliance of research projects with the aforementioned ethical and legal requirements such as international treaties, GDPR, and ethics guidelines. Instead, ELSA research is an inter- and transdisciplinary research area in which researchers from the social sciences and humanities, law, and theology address and critically reflect broad questions about the ethical, legal, and social aspects of science, technology, and innovation, often in the area of biomedicine.[24] Box 11.2 provides some exemplary questions raised by ELSA research. Being in an interdisciplinary research area, ELSA researchers often cooperate with researchers from other disciplines, either from the social science and humanities or the natural sciences, and – being transdisciplinary – also with a broad set of stakeholders in research and innovation including patients and citizens.

> **Box 11.2 Examples for ELSA Questions**
> ELSA research raises issues such as the following:
>
> - What is the impact of certain research and innovation on fair and equal access?
> - Are there groups particularly affected or excluded by new technology (e.g., because of economic and educational status, gender, being part of a minority or from a disadvantaged geographical area, etc.)?
> - To what extent does healthcare innovation implicitly or explicitly continue or even strengthen existing inequalities? What can be done to avoid this?
> - To what extent does a particular area of research and innovation impact sensitive areas, e.g., human dignity, equality, autonomy, privacy, and animal rights?
> - How do different stakeholders such as health professionals, policymakers, patients, and the public perceive and evaluate such innovations and their potential impact?
> - How can they be discussed and deliberated?
> - How does a certain innovation affect certain professions and cooperation between different stakeholders (researchers from different disciplines and areas, different healthcare practitioners, industry, policymakers, patient organizations, communities)?
> - Does a new technology necessitate new regulation? In what way?

ELSA research answers the critique from inside and outside the research system as well as from society of the dominant mode of doing research and innovation which might not sufficiently address the abovementioned questions. It originates from a critical bottom-up movement from various disciplines such as philosophy, bioethics, technology assessment, and science and technology studies (STS). However, ELSA research also has important roots in science policymaking and research funding. In the US American context, research on Ethical Legal and Social Implication (ELSI) of biomedicine has its

---

[23] current official version: Fortaleza, 2013; https://www.wma.net/policies-post/wma-declaration-of-helsinki-ethical-principles-for-medical-researchinvolving-human-subjects/

[24] This section on ELSA research builds heavily on work of Hub Zwart, Laurens Ladeweerd and Arian von Roij who describe the origins of ELSA research (Zwaart et al. 2014).

origins in the Human Genome Project. As such, ELSA has always been related to new types of data and hopes for more personalized medicine. It has political roots in the movement for more public involvement in research and innovation policy. It has been adopted and adapted in EU Framework Programmes since 1994 as looking into the ethical, legal, and social aspects of emerging technologies [30]. The concept of ELSA is still widely used in the medical field but has been replaced in EU research and innovation policy by cognate concepts like Responsible Research and Innovation [31] or Open Science. These frequent changes demarcate slight but important semantic shifts in emphasis of the thematic areas targeted by research funding programs and also impact on the research topics which are funded and subsequently addressed.

Being an inter- and transdisciplinary endeavor, ELSA research has challenges for both researchers from natural science and medicine on the one side and social science and humanities, law, and theology on the other side. These challenges are not uncommon for inter- and transdisciplinary research and, apart from the difficulty of getting to know one another, to understand what ELSA is about,[25] and finding a shared language also includes the right degree of proximity and autonomy in the relationship between natural scientists and ELSA researchers.

ELSA research is also highly relevant for personalized medicine. While individual characteristics of patients have always been considered in medicine, new developments in genomics and data sciences have fuelled the emergence of practices subsumed under the term precision medicine, with high hopes for more individualized treatment of patients. From an ELSA perspective, this trend raises several questions on issues such as anonymization, genetic discrimination, and data governance [32]. Thus, several authors emphasize the importance to consider patient and citizen's perspectives in (precision) medicine [33, 34].

Precision medicine is in many ways interrelated with trends toward digitalization and big data. Scholars [35] comment critically on this and claim the rise of data-driven rather than a knowledge-driven science. Other authors [36] raise awareness of new forms of biases that may appear with the widespread use of health data. Thus, with the rising importance and quantities of data in medicine, new ethical [37], legal, and policy debates [38] emerge. Machine learning- and artificial intelligence (AI)-based support systems add another layer to these debates on bias and challenges to privacy [39].

Precision medicine and related infrastructures like biobanks are repeatedly framed as public goods in scientific discourse, which is to sway some socio-ethical concerns [40]. Proponents of such communitarian ethics emphasize societal benefits and community management over individual concerns, which also translates in respective practices for open forms of informed consent, which allow wide applications with collected health data [41]. Critics claim that these models undermine established notions of informed consent and thus individual control over data [42]. In line with this development and the broader trend for self-quantification through digital technologies, such as apps, health management becomes, to some degree, an individual endeavor. Some authors argue that personalized medicine does not necessarily fuel the ongoing trend toward individualization of responsibilities in health, but that solidarity-driven approaches to health do not contradict this trend [34].

We conclude that ELSA provides a critical reflection on developments in medicine and research and innovation more broadly. While other policy documents gained more prominence in the EU policy landscape in the last decade, ELSA is still a timely and useful concept, especially in medicine. With its origins in the Human Genome Project in the United States and the public engagement discourse in Europe, it has always been tied to precision medicine to some extent. The recent advances in data-driven medicine and AI intensify the need for reflection as paradigmatic foundations of medical research and innovation begin to change.

---

[25]The Societal Readiness Thinking Tool is a practical guide for researchers to identify ELSA questions in their research (Bernstein et al. 2022). It can be downloaded from the Internet at https://newhorrizon.eu/thinking-tool

# References

1. Sinnott-Armstrong W. Consequentialism. The Stanford Encyclopaedia of Philosophy (Summer 2019 Edition), Edward N. Zalta (ed.) https://plato.stanford.edu/archives/sum2019/entries/consequentialism/
2. Knoppers BM, Harris JR, Budin-Ljøsne I, Dove ES. A human rights approach to an international code of conduct for genomic and clinical data sharing. Hum Genet. 2014;133(7):895–903. https://doi.org/10.1007/s00439-014-1432-6.
3. Blasimme A, Fadda M, Schneider M, Vayena E. Data sharing for precision medicine: policy lessons and future directions. Health Aff (Millwood). 2018;37(5):702–9. https://doi.org/10.1377/hlthaff.2017.1558.
4. Canham S, Ohmann C, Boiten JW, Panagiotopoulou M, Hughes N, David R, Sanchez Pla A, Maxwell L, Aerts J, Facile R, Griffon N, Saunders G, van Bochove K, Ewbank J. EOSC-life report on data standards for observational and interventional studies, and interoperability between healthcare and research data. Zenodo. 2021; https://doi.org/10.5281/zenodo.5810612.
5. Ahmad W, Al-Sayed M. Human subjects in clinical trials: Ethical considerations and concerns. J Transl Sci. 2018;4 https://doi.org/10.15761/JTS.1000239.
6. Botkin JR, Belmont JW, Berg JS, Berkman BE, Bombard Y, Holm IA, Levy HP, Ormond KE, Saal HM, Spinner NB, Wilfond BS, McInerney JD. Points to consider: ethical, legal, and psychosocial implications of genetic testing in children and adolescents. Am J Hum Genet. 2015;97(1):6–21. https://doi.org/10.1016/j.ajhg.2015.05.022.
7. van El CG, Cornel MC, Borry P, Hastings RJ, Fellmann F, Hodgson SV, Howard HC, Cambon-Thomsen A, Knoppers BM, Meijers-Heijboer H, Scheffer H, Tranebjaerg L, Dondorp W, de Wert GM, ESHG Public and Professional Policy Committee. Whole-genome sequencing in health care. Recommendations of the European Society of Human Genetics. Eur J Hum Genet. 2013;21 Suppl 1(Suppl 1)):S1–5. PMID: 23819146; PMCID: PMC3660957
8. Meienberg J, Bruggmann R, Oexle K, Matyas G. Clinical sequencing: is WGS the better WES? Hum Genet. 2016;135(3):359–62. https://doi.org/10.1007/s00439-015-1631-9.
9. Wilkinson MD, Dumontier M, Aalbersberg IJ, et al. The FAIR Guiding Principles for scientific data management and stewardship [published correction appears in Sci Data. 2019 Mar 19;6(1):6]. Sci Data. 2016;3:160018. Published 2016 Mar 15. https://doi.org/10.1038/sdata.2016.1810.1038/sdata.2016.18. Erratum in: Sci Data. 2019 Mar 19;6(1):6.
10. Borisova AL, Pokrovskaya MS, Meshkov AN, Metelskaya VA, Shatalova AM, Drapkina OM. ISO 20387 biobanking standard. Analysis of requirements and experience of implementation. ISO 20387. Klin Lab Diagn. 2020;65(9):587–92. https://doi.org/10.18821/0869-2084-2020-65-9-587-592.
11. OECD Guidelines on human biobanks and genetic research databases OECD 2009. https://www.oecd.org/sti/emerging-tech/44054609.pdf
12. Jacobs K, Yeager M, Wacholder S, et al. A new statistic and its power to infer membership in a genome-wide association study using genotype frequencies. Nat Genet. 2009;41:1253–7. https://doi.org/10.1038/ng.455.
13. Abu-Elezz I, Hassan A, Nazeemudeen A, Househ M, Abd-Alrazaq A. The benefits and threats of blockchain technology in healthcare: a scoping review. Int J Med Inform. 2020;142:104246. https://doi.org/10.1016/j.ijmedinf.2020.104246.
14. Dimitrov DV. Blockchain applications for healthcare data management. Health Inform Res. 2019;25(1):51–6. https://doi.org/10.4258/hir.2019.25.1.51.
15. Boccia S, Federici A, Siliquini R, Calabrò G, Ricciardi W, Health, Expert. Implementation of genomic policies in Italy: the new national plan for innovation of the health sys-tem based on omics sciences. Epidemiol Biostat Public Health. 2017;14:e12782–1.
16. Regulation (EU) 2016/679 of the European Parliament and of the Council of 27 April 2016 on protecting natural persons concerning the processing of personal data and on the free movement of such data, and repealing Directive 95/46/EC (General Data Protection). http://data.europa.eu/eli/reg/2016/679/oj
17. Ledford H. Millions affected by racial bias in healthcare algorithm. Nature. 2019;574:608. https://media.nature.com/original/magazine-assets/d41586-019-03228-6/d41586-019-03228-6.pdf
18. Jones CP. Levels of racism: a theoretic framework and a gardener's tale. Am J Public Health. 2000;90(8):1212–5. https://doi.org/10.2105/ajph.90.8.1212.
19. Geneviève LD, Martani A, Shaw D, et al. Structural racism in precision medicine: leaving no one behind. BMC Med Ethics. 2020;21:17. https://doi.org/10.1186/s12910-020-0457-8.
20. Sharon FT. The global alliance for genomics & health. Genet Test Mol Biomarkers. 2014;18:375–6. https://doi.org/10.1089/gtmb.2014.1555.
21. Bollinger JM, Zuk PD, Majumder MA, et al. What is a medical information commons? J Law Med Ethics. 2019;47(1):41–50. https://doi.org/10.1177/1073110519840483.
22. Supreme Court of the United States, Association for Molecular Pathology et al. v. Myriad Genetics, Inc., et al., June 13, 2013 https://www.unipv-lawtech.eu/files/SupremeCourtMyriad-.pdf
23. Love J, Blair E. Intellectual property in personalised medicine. Intellect Prop Rights. 2016;4. https://doi.org/10.4172/2375-4516.1000164. Open Access

24. O'Rourke B, Oortwijn W, Schuller T, International Joint Task Group. The new definition of health technology assessment: a milestone in international collaboration. Int J Technol Assess Health Care. 2020;36(3):187–90. https://doi.org/10.1017/S0266462320000215.

25. Saarni SI, Braunack-Mayer A, Hofmann B, van der Wilt GJ. Different methods for ethical analysis in health technology assessment: an empirical study. Int J Technol Assess Health Care. 2011;27:305–12. https://doi.org/10.1017/S0266462311000444.

26. EUnetHTA Joint Action 2, Work Package 8. HTA Core Model ® version 3.0 (Pdf); 2016. Available from www.htacoremodel.info/BrowseModel.aspx.

27. Middleton A, Milne R, Thorogood A, Kleiderman E, Niemiec E, Prainsack B, Farley L, Bevan P, Steed C, Smith J, Vears D, Atutornu J, Howard HC, Morley KI. Attitudes of publics who are unwilling to donate DNA data for research. Eur J Med Genet. 2019;62(5):316–23. https://doi.org/10.1016/j.ejmg.2018.11.014.

28. Boccia S, Federici A, Siliquini R, Calabrò G, Ricciardi W. Implementation of genomic policies in Italy: the new national plan for innovation of the health sys-tem based on omics sciences. Epidemiol Biostat Public Health. 2017;14:e12782–1.

29. Maxmen A. Gay gene' app provokes fears of a genetic wild west. Nature. 2019;574:609. https://media.nature.com/original/magazine-assets/d41586-019-03228-6/d41586-019-03228-6.pdf

30. Zwart H, Landeweerd L, van Rooij A. Adapt or perish? Assessing the recent shift in the European research funding arena from 'ELSA' to 'RRI'. Life Sci Soc Policy. 2014;10:11. https://doi.org/10.1186/s40504-014-0011-x.

31. Von Schomberg R. Towards responsible research and innovation in the information and communication technologies and security technologies fields. Brussels: European Commission, 2011. https://doi.org/10.2139/ssrn.2436399.

32. Owen SG, Shyong Tai E, Sun S. Precision medicine and big data. The application of an ethics framework for big data in health and research. Asian Bioethics Rev. 2019;11:275–88. https://doi.org/10.1007/s41649-019-00094-2.

33. Kerr A, Hill R, Till C. The limits of responsible innovation: exploring care, vulnerability and precision medicine. Technol Soc. 2018;52:24–31. https://doi.org/10.1016/j.techsoc.2017.03.004.

34. Prainsack B. Personalized medicine: empowered patients in the 21st Century? NYU Press Scholarship Online. 2017; https://doi.org/10.18574/nyu/9781479814879.001.0001.

35. Kitchin R. Big data, new epistemologies and paradigm shifts. Big Data Soc. 2014:1–12. https://doi.org/10.1177/2053951714528481.

36. Ferryman K, Pitcan M. Fairness in precision medicine; 2018. https://kennisopenbaarbestuur.nl/media/257243/datasociety_fairness_in_precision_medicine_feb2018.pdf

37. Hummel P, Braun M. Just data? Solidarity and justice in data-driven medicine. Life Sci Soc Policy. 2020;16:8. https://doi.org/10.1186/s40504-020-00101-7.

38. Starkbaum J, Felt U. Negotiating the reuse of health-data: research, big data, and the European general data protection regulation. Big Data Soc. 2019:1–12. https://doi.org/10.1177/2053951719862594.

39. Bartoletti I. AI in healthcare: ethical and privacy challenges. Artif Intell Med. 2019:7–10. https://doi.org/10.1007/978-3-030-21642-9_2.

40. Chan S, Erikainen S. What's in a name? The politics of 'precision medicine' pages 50–52 | published online: 05 Apr 2018. https://doi.org/10.1080/15265161.2018.1431324

41. Lunshof J, Chadwick R, Vorhaus D, et al. From genetic privacy to open consent. Nat Rev Genet. 2008;9:406–11. https://doi.org/10.1038/nrg2360.

42. Caulfield T, Kaye J. Broad consent in biobanking: reflections on seemingly insurmountable dilemmas. Medical Law Int. 2009;10(2):85–100. https://doi.org/10.1177/096853320901000201.

# Personalized Medicine Literacy

**12**

Marius Geanta, Adriana Boata, Angela Brand,
Cosmina Cioroboiu, and Bianca Cucos

**What Will You Learn in This Chapter?**
The chapter provides an overview of the best practice model applied by the Center for Innovation in Medicine on reducing cancer fatalism in the Romanian population, as well as increasing the level of cancer literacy, including cancer innovations awareness, by periodical assessment of attitudes, perceptions, and behaviors, followed by personalized communication campaigns. Such a model can also be applied for a long-term sustainable increase in the level of personalized medicine literacy in any population.

M. Geanta (✉)
Center for Innovation in Medicine,
Bucharest, Romania

KOL Medical Media, Bucharest, Romania

United Nations University-Maastricht Economic and
Social Research Institute on Innovation and
Technology, Maastricht, The Netherlands
e-mail: marius.geanta@ino-med.ro

A. Boata · C. Cioroboiu · B. Cucos
Center for Innovation in Medicine,
Bucharest, Romania

A. Brand
United Nations University-Maastricht Economic and
Social Research Institute on Innovation and
Technology, Maastricht, The Netherlands

Department of Public Health Genomics, Manipal
School of Life Sciences, Manipal Academy of Higher
Education, Manipal, India

Faculty of Health, Medicine and Life Sciences,
Maastricht University, Maastricht, The Netherlands

Dr. TMA Pai Endowment Chair in Public Health
Genomics, Manipal School of Life Sciences, Manipal
Academy of Higher Education, Manipal, India

**Rationale and Importance**
Classical health strategies aiming to raise awareness around the theme of cancer innovations have proven ineffective because they do not take into account people's perceptions, attitudes, and behaviors. By not looking at these essential factors, communication campaigns are conducted on the one-size-fits all model. At the Center for Innovation in Medicine, we conducted research during 2016–2020 aiming to understand people's knowledge and attitudes toward cancer innovation. Based on this research, we performed personalized communication campaigns aiming to reduce the fatalism of cancer in the Romanian population. Implementing personalized health communication campaigns focused on citizens' needs is essential for any effective cancer control strategy, and it should be prioritized in national cancer control plans and aligned with the European initiatives. The COVID-19 vaccination campaigns all over the world were a real-time simulation of what delivering the right messages, by the right influencers, to the right population can do. When you intersect the two dimensions, you obtain a highly personalized approach in communicating health innovations to the citizens. This approach should constitute the first step in the efforts of increasing personalized medicine literacy.

## 12.1 Introduction

Health literacy is the degree to which individuals have the capacity to obtain, process, and understand health information needed to make health decisions that best suits their interest [1]. In time, the definition evolved to be more comprehensive—health literacy is not only about medical decisions, but it is also about healthcare and about health: one's health or a beloved one's health. As society evolved, the human's central meaning—*to live, to be alive*—transformed into *to be healthy*.

The COVID-19 pandemic showed us that the human race is now at the forefront of medical and technological advances. Even though humans are biased to perceive the negative side of events in a much more dramatic way [2], there was no better and safer time in human history to exist and be alive than now.

And this statement can be made in part because of the evolution of precision and personalized medicine, of health innovations in general. In order to be more holistic, we will refer to it as PHC—personalized healthcare. Because of PHC, the classical medical approach of one-size-fits-all is quickly disappearing—in some parts of the world faster than in others.

The omics sciences were first put in practice in oncology, but nowadays they are impacting nearly all dimensions of the response to the COVID-19 pandemic, including single-cell multi-omics analysis of the immune response at COVID-19 and multi-omics approach for the identification of potential therapeutic biomolecules [3].

Back to definitions. Although there is no universally accepted definition, the Horizon 2020 Advisory Group defines personalized medicine as "a medical model using the characterization of individuals' phenotypes and genotypes (e.g. molecular profiling, medical imaging, lifestyle data) for tailoring the right therapeutic strategy for the right person at the right time, and/or to determine the predisposition to disease and/or to deliver timely and targeted prevention" [4]. This definition was also used by EU health ministers in their council conclusions on personalized medicine from 2015, during the Luxembourg Presidency of the Council of Europe [5].

According to the 2012 definition of the European Consortium for Health Literacy (HL) [6], "Health literacy is linked to literacy and entails people's knowledge, motivation and competencies to access, understand, appraise, and apply health information in order to make judgments and take decisions in everyday life concerning healthcare, disease prevention and health promotion to maintain or improve quality of life during the life course." In other words, an adequate level of HL is defined by the ability of an individual to access health data, to sort and choose the appropriate sources of health-relevant information, to understand this information, to personalize it for his situation, and to apply the information in order to obtain a benefit for his own health.

HL is a complex concept that needs to be defined in a context. According to a 2017 analysis, there were more than 100 specific types of HL [7], but four of them are usually used when referring to HL: personal health literacy, organizational health literacy, digital health literacy, and quantitative literacy. In recent years, the 2012's definition of HL as it is above becomes more defined for the concept of personal health literacy [8].

Although no official definition of PHC exists, we can define it by integrating PM and the person, the citizen, and the individual as a social human being [9, 10].

Having this information in mind and the unprecedented advance in omics sciences, it is reasonable to argue that personalized medicine literacy, or more correctly, personalized health and care literacy, should become the fifth main part of the HL concept, at least for now. Though no PHCL official definition exists, it could be summed up:

> PHC literacy entails people's knowledge, motivation, and competencies to access, understand, appraise, and apply omics and other clinical and laboratory data and psychosocial and lifestyle information in order to make judgments and deci-

sions concerning the modifiable determinants of their health and prevention, healthcare, and health promotion, in order to maintain or improve quality of life during the life course.

And so, even though HL refers to the individual capacity for assessing and using health information mostly, PHC literacy is an essential catalyst for the responsible and effective translation of genome-based information for the benefit of population health [11].

## 12.2  What Does Being Healthy Mean? The Determinants of Health During the Time

In the 2006 Constitution of the World Health Organization, health was defined as the physical, mental, and social well-being [12], and not merely the absence of disease or infirmity. Achieving the highest standard of health is a fundamental human right, regardless of race, religion, political vision, or social and economic status.

As with health, "disease" is difficult to define. The first definition referred to a disturbance that occurs in an organism: "Organic or functional change in the normal balance of the organism; a pathological process that affects the body." A simple, but abstract and circular, definition could be "lack of health." Therefore, the two terms are closely related, and the current understanding of the human being determines the need to constantly redefine the terms, as the "state of health" becomes increasingly difficult to understand, and the "perception" of the person is what differentiates in fact between illness and health.

According to WHO 1998, the determinants of health are defined as "The range of behavioral, biological, socio-economic and environmental factors that influence the health status of individuals or populations" [13]. On average, 89% of our health occurs outside of the clinical space through our genetics, behavior, environment, and social circumstances [14], leaving only 11% for the clinical setting. Individual behavior shapes 36% of our health; social circumstances, 24%;

genetics and biology, 22%; and the environment, 7%. These are the main categories, based on the understanding of the state of health almost three decades ago.

While the percentages above show a bigger behavioral and social burden for the state of health when compared to biology and genetics, this is not the case for every individual. The disease is individual, even when referring to global pandemics.

Although access to healthcare services usually means access to better health outcomes, scientific evidence shows that healthcare, as it is today, with a focus on treatment, is only part of the problem. For example, the United Kingdom has begun to recognize social prescribing as a basic tool for public health [15].

Therefore, a personalized approach to determining the individual risks of developing diseases, and further early detection of those diseases, must take into account both the genetic and the environmental components (which refers to the conditions of development of the disease in an individual) integrating genetic, clinical, and lifestyle data and a person's developmental environment [16].

And although the determinants of health were refined over time, the challenge of understanding how they interrelate with each other is still a great topic of research. Nowadays, because of technological and scientific advances, the complexity of biological factors/determinants (including those related to omics) cannot be classified altogether. There are very few scientific papers that aim at classifying them, leading to a greater amount of information hard to understand not only by the general population but by the scientists and specialists themselves—leading to an infodemic.

Because of this complexity, a new emerging integrative field was born in the last years—social epigenomics: the study of how social experiences affect our genes and biology. Though social epigenomics is a relatively new area of research, studies exploring the individual and mutual influence of social, environmental, and genetic factors on health have become increasingly abundant. Social epigenomics is uniquely positioned at the intersection of population health and precision medicine, allowing us to understand how exposure to social and environmental stressors modifies the way in which genes are expressed and ultimately alters our risk for disease [17].

A recent study showed that the impact of genetic factors on the onset of disease decreases with age and other mechanisms are taking place as important, including those influenced by the environment in which the person lives [18].

Apart from the theoretical understanding of the determinants of health in the personalized medicine era that could lead to unimaginable long-term benefits for human health, there are more practical approaches that could be deployed, like meta-personalized public health interventions.

During the time, the majority of public health interventions were aimed at lifestyle, the modifiable factors—the classic understanding of primary prevention, without fully understanding the complex bond between these factors and the biological ones. It was acceptable 30 years ago, 20 years ago, but not in the last decade and not in the present, and not during our COVID-19 times. So the concept of personalized prevention emerged.

## 12.2.1  The Determinants of Health and Precision Cardiology

Let's look at cardiovascular diseases—the number one global killer. Cardiovascular diseases account for 36% of all deaths across the EU. Around 20% of all premature deaths (below the age of 65) in the EU are caused by CVD. CVDs are caused by so-called modifiable or non-modifiable (inherited, genetic) risk factors. The world's most common and non-modifiable CVD risk factor is familial hypercholesterolemia (FH). Less than 10% of those born with FH are diagnosed and adequately treated, leading to heart attacks, strokes, heart disease, and deaths, early in life, even as early as 4 years of age [19]. Interventions based on early screening of those at genetic risk might actually be the missing piece of the puzzle when it comes to CVD primary pre-

vention. And moreover, it would lead to a new understanding of the cardiovascular determinants and stimulate new therapies and preventive solutions development.

Although most of the cardiovascular medicine we see in Europe is mainly composed of classical prevention (targeting classical lifestyle factors) and treating the condition, cardiology benefits in practice from the personalized medicine approach in 3D modeling and simulation that can guide surgeons before cardiovascular surgery.

Artificial intelligence is also making its presence felt in CVD management—recently, a study showed that the predictive value of AI algorithms for determining the risk of developing cardiovascular disease using physiological data and laboratory results (such as blood pressure and cholesterol values) is much higher if data on the social determinants of health is added [20].

## 12.2.2  The Determinants of Health and Precision Oncology

On the other hand, in oncology, the other aspects of personalized medicine have been engaged: precision screening, personalized diagnosis, and personalized treatments. These are already a reality in many parts of the world. Genomics is already imposing major changes in cancer understanding and care, from redefining cancers to changing therapeutic standards.

When the sequencing projects for different types of tumors began, the aim was to create a "library" of mutations involved in cancer and to identify mechanisms that can be targeted therapeutically. Although this goal has been met, cancer is not a single disease; two tumors considered to be in the same category according to classical classifications may be completely different at the molecular level, and even cells in the same tumor may be different. Deciphering the human genome was only the first step. Over 300 different conditions are known as cancer nowadays.

Genomic tumor testing has evolved over time, from several biomarkers to extensive gene panels, which allow the analysis of all mutations that can be acted upon through targeted therapies.

Genomics also provides valuable insights into how the disease progresses and the response to a particular treatment can be anticipated.

Conventional oncological treatments, such as chemotherapy, involved the administration of cytotoxic agents that did not discriminate between a healthy and a pathological cell. New drugs appearing on the market are targeted at molecular alterations—at the level of DNA, and RNA, at the level of immune cells, etc. In the age of precision medicine, oncological therapies should be approached more and more from new perspectives, using molecular anomalies and not the organ in which the tumor appears for the choice of therapy.

Lung cancer is an important example of how genomic medicine has evolved. Up to 45% of patients with non-microcellular lung carcinoma have genetic mutations for which there are specific treatments already approved or under study. In recent years, several subgroups of patients with non-small cell lung cancer (NSCLC) have begun to be defined based on molecular abnormalities. There are already therapies approved by the authorities, targeting genetic abnormalities in the EGFR, ALK, ROS1, BRAF, NTRK, MET, and RET genes [21, 22].

### 12.2.3 The Determinants of Health and Precision Diabetology

Over the last two decades, many common diseases had to be rethought. Cancer has not only transformed and is continuously evolving in hundreds of distinct diseases but also forced the change from classification based on the primarily affected organ to classification by the mutations or biomarkers of the tumor (tumor-agnostic classification). Diabetes is another common disease that is in continuous change, from the glyco-centric approach to more complex mechanisms and new therapeutic approaches, putting high on the agenda the cardiovascular risk of the patient—these advances were so rapid in the last years that the American Diabetes Association approached them by a living guideline, updating it as new technologies emerged [23]. In 2018, the American

Diabetes Association and the European Association for the Study of Diabetes launched a consensus paper on the management of type 2 diabetes that underlined the patient-centered approach and evaluation of cardiovascular risk factors [24], and in 2021 a consensus paper on type 1 diabetes [25]. This deep understanding of diseases and health was possible through the personalized and precision medicine era we are in.

The major developments in the understanding of cancer, cardiovascular diseases, and other chronic diseases, based on the concept of personalized health and care, require a different approach as well as the doctor-patient relationship through the implementation of personalized communication and education as a part of the broader area of PHCL.

### 12.2.4 COVID-19 Pandemic and the Social and Behavioral Innovations

In a simple search on Google Scholar for "determinants of health," there are hundreds of papers from 2020 and 2021 analyzing the connection between *social determinants of health (including political and economical determinants) and COVID-19*. There are also hundreds of papers that estimate the lives lost because of people not having access to health services during the pandemic. On the other hand, researchers are trying to understand the complete biological burden of the COVID-19 disease—how long it will affect the body after the active phase, what kind of organs will be affected, and what is the genetic and biological predisposition to worse outcomes.

Being faced with such an emergency global state, the scientific world came up with more and more risk factors and new categories of determinants of health. In this complex context, adding the infodemic—too much information including false or misleading information in digital and physical environments during a disease outbreak [26]—it is impossible to continue to classify the determinants of health in an un-personalized manner. And arguably, it is even more impossible to make public health decisions based on the classical approach of classification and assessment of the health determinants, which most of the time exclude the assessment of psychosocial determinants of health. Depending on the population and individuals, certain determinants weigh more than others. Using only statistics is not enough to understand the complexity of features of an individual.

While we have described above the understanding of the determinants of health through the lenses of personalized medicine, PHC is a more complex concept—it involves health and care, and PM is only a part of it, although the terms are usually used to describe the same concept.

While health literacy is usually understood through a preventive attitude, healthy lifestyle behaviors, and the ability to navigate through the health system a person inhabits, these groundbreaking changes in how certain diseases are defined make the assessment and the increase the health literacy level one of the biggest challenges of our century.

COVID-19, a disease that will probably remain in the public and scientific focus for many years from now on, pushed the idea of prevention and health literacy further and showed us that while health literacy is important, during a global health emergency, it is more important and effective to be able to influence human behavior by understanding the attitudes and perceptions of certain populations. By doing this, you can indirectly increase the level of health literacy through practical experience and by indirectly targeting key beliefs and attitudes, using influencers of the community at three levels—micro, meso, and macro. This model is described in detail below, based on our experience.

To add another layer of evidence to the need to go beyond or rethink health literacy as it is now: a recent study showed that the countries that performed the best in the pandemic from the perspective of the number of infection cases were the countries in which citizens reported a high level of confidence in society and their governments and not those with the best plans of pandemic preparedness. The results also suggest that increasing health promotion for key modifiable risks is associated with a reduction of fatalities in countries where citizens trust the society and their leaders.

Overall, governments and communities can maintain or increase the public's trust by providing accurate, timely information about the pandemic, even when that information is still limited, and by clearly communicating the risk and relevant vulnerabilities [27]. The identity of the messenger in risk communication can also improve or damage trust.

The major point to be underlined here is that in countries with a very low level of trust in their leaders and a history of distrust in society, like ex-communist countries, timely, effective, and well-delivered communication might not still be enough. To take advantage of the full potential of personalized communication, a more sustainable approach is needed, based on citizens' perceptions and attitudes.

## 12.3 The Role of Attitudes and Perceptions Assessment for Influencing Pro-health Behavior of the Citizens: Two Case Studies on Cancer Literacy and Vaccination Literacy

**Abstract** In the following, we will present two case studies in the Romanian population—the seventh member state in the European Union in terms of population: one will focus on increasing the level of cancer literacy by assessing the attitudes and perceptions of the population on the subject and develop personalized communication and educational campaigns and the other on COVID-19 vaccination and HPV vaccination (two major vaccination campaigns in Romania from the public health perspective).

### 12.3.1 Cancer Literacy in Romania

Health literacy entails the knowledge, motivation, and competencies to access, appraise, understand, and apply information for making decisions concerning healthcare, disease prevention, and health promotion and to maintain and improve quality of life during the life course. In the context of cancer literacy, it refers to the knowledge and skills needed to find, understand, evaluate, and use the information and advice the health system has to offer with regard to prevention, diagnosing, and treatment [28]. A low level of cancer literacy has been shown to hinder patients at every stage of the disease journey. Improving cancer literacy in Europe can help save lives, time, and ultimately, costs.

As inequalities in cancer care are an international reality and a European reality, in the case of cancer literacy, we can observe the same trends and gaps. As innovations enter the clinical stage in countries with a low level of health literacy, health education, and literacy overall, the gaps between the EU Member States become more evident. Besides the influence that HL has on healthy behavior [29], a correlation can be drawn between the quality of healthcare services and the health literacy level of the population. The relationship between demand and supply is compromised in the healthcare system—healthcare providers will not be motivated to offer the best available quality of service if the patients are not empowered to request it and understand their rights.

The cancer domain, being the most positively impacted area by personalized medicine development, represents a key model for understanding how a high level of cancer literacy impacts and stimulates PHC literacy in a country at all levels. But as the pandemic highlighted, adding the attitudes and perceptions of a population on a certain subject, cancer in this particular example represents the missing essential piece of the puzzle for influencing human behavior in populations with a low level of HL and impacted by high inequalities in cancer care.

Moreover, in the current understanding of health, diseases, and the scientific advances in cancer, the actions that aim at increasing cancer literacy level should no longer address the patient, but the citizen, recognizing his role in society before, during, and after the cancer diagnosis.

### 12.3.1.1 Cancer Burden in Romania

Romania has some of the highest rates of avoidable deaths from both preventable and treatable causes in Europe [30]. Romania's cancer burden is high, with 83,461 newly diagnosed cases and roughly 50,902 total deaths occurring in 2018 [31]. Romania is also among the top ten European countries in terms of cancer mortality rates [32]. The lack of information and adequate screening and diagnosis services, together with the unstandardized cancer patient path, are some of the main causes of the late detection of cancer cases.

Although the access to new cancer treatments has improved in the last 6 years, the improvement was not reflected in the survival rate of the cancer patients. Romania provides public support and assistance to cancer patients through the National Programme for Cancer (NPC), operated by the National Health Insurance House. Over the years, the program has continuously evolved to include more patients and more types of cancer, but results are not published.

Besides the faulty healthcare services, at the macro level, the overall situation as shown by statistics is complex: half of the Romanian population live in the rural area, Romania has one of the lowest rates of education in Europe and some of the highest rates of school dropout, and Romania is an ex-communist country, with very conservative views, and many vulnerable populations living in poverty.

Going back to cancer statistics, Romania has also the biggest rate of mortality from cervical cancer in the European Union. Every year, in Romania, there are 1800 deaths from cervical cancer and 3400 new cases. At the European level, Romania ranks first in terms of incidence and mortality: the incidence is 2.5 times higher than the European average, and the mortality rate is over four times higher [33].

But seven out of ten cases of cervical cancer can be prevented with the HPV vaccine [34]. The HPV vaccination rate will be discussed in the next case study.

The guidelines of the European Society for Medical Oncology recommend mammography screening for breast cancer, annually or every 2 years, with priority for women in the 50–69 age group [35]. Moreover, in women with a family history of breast cancer, with or without knowledge of BRCA carrier status, annual MRIs and/or annual mammograms are recommended. The relative 5-year survival rate has increased by up to 90% due to the expansion of screening programs and therapeutic advances. Participation in screening programs is associated with a reduction in mortality of at least 30% and a reduction in the risk of severe disease by 40% [36].

Romania launched the first breast cancer screening pilot program in 2018 [37]. However, according to the latest Eurostat survey, only 9% of women in Romania aged between 50 and 69 reported in 2019 that they had a mammogram in the last 2 years [38]. Once again, Romania ranks last in the EU. For Bulgaria, which is on the penultimate place, the percentage is 36%—four times more than in Romania. In Sweden, the percentage is 95%, about 11 times higher.

### 12.3.2 Attitudes, Perceptions, and Behaviors on Cancer: National Survey (2016, 2018, 2020) in the Romanian Population

Increasing the overall health literacy level (including cancer literacy) has become more and more complex because of the unprecedented scientific development, at an unprecedented speed. But influencing health behavior through targeted interventions after assessing attitudes and perceptions seems to be a more sustainable approach.

In order to understand how the Romanian population relates to cancer and the degree of awareness of cancer innovations, the Center for Innovation in Medicine, a civil society organization with an interest in research, innovation, policy, personalized communication, and education at the European level, measured the level of citizens' awareness and their perception on prevention, diagnosis, and treatment of cancer, in 2016, 2018, and 2020 (pre-pandemic), through telephonic interviews (CATI—computer-assisted telephone interviewing): 1010 participants in

each study, sociologically relevant at the national level.

One of the major outputs of this study was to find that approximately 5% of Romanians had cancer at some point in life, and one in three people had a direct or indirect experience with cancer during their lifetime. These data are very valuable because there is no cancer registry at the national level and the IARC data on Romania is based on estimates from the Northern Region, where there is a functional cancer registry.

Another two major outputs consist of the fatalism rate in relation to cancer in the Romanian population and the drop observed from 2018 to 2020 in awareness of cancer innovation (personalized medicine and immuno-oncology), correlated with an actual increase of the information campaigns (but with poor and non-targeted messages), which led to a cancer infodemic.

#### 12.3.2.1 Fatalism

Measuring fatalism in relation to cancer is important because it can indicate people's willingness to take action in all the areas of the cancer continuum, from prevention to palliative care. In other words, the higher the fatalism in relation to the disease, the more people will resign and no longer participate in screening programs, will not adopt preventive measures, and will not try to find and access diagnosis and therapeutic options in case of a cancer diagnosis.

Despite the fact that they say, to a large extent, that they know that there are cancers that can be cured, when asked if a cancer diagnosis always leads to death, almost 48% (2020) agree. This indicates fatalism, a condition in which many citizens try to cope with the prospect of cancer, considering that the health system, for various reasons, cannot provide them with access to the means of screening, diagnosis, and treatment they may need (Fig. 12.1).

The rising rate of fatalism is also reflected in the knowledge about cancer innovation (personalized medicine, immuno-oncology, or biomarkers)—common terms in current cancer management. Another set of questions also assessed people's perceptions of access to medi-

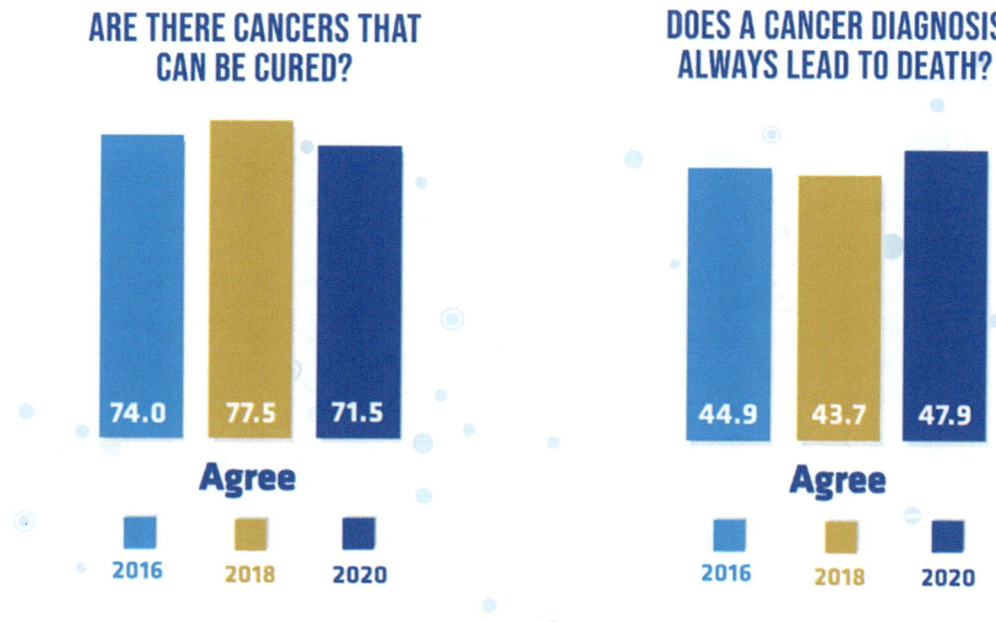

**Fig. 12.1** Fatalism results from the three studies. Property of Center for Innovation in Medicine

cines and therapies. The general conclusion is that most respondents did not know the price of therapies or how to access them.

The data show that although access to healthcare and new therapies has improved in Romania between 2016 and 2020, people's perceptions and fatalistic attitudes have worsened, despite multiple information campaigns. There has also been a decline in awareness of cancer innovation (immuno-oncology, biomarkers, and personalized medicine). One explanation could be the development of infodemia in relation to cancer—a multitude of incomplete, irrelevant, poor quality, and scientifically invalid information posted online and beyond, especially on social media and forums, which causes people to be confused and unable to identify the right message.

Independently, a recently published study in the United States had similar findings between 2016 and 2020, comparing two types of major populations in a state: *Rural Residents Tend to Hold Fatalistic Beliefs and Perceive More Cancer-related Information Overload Than Urban Residents* [39]. To assess whether cancer beliefs vary between rural and urban adults in the

United States, Jensen and colleagues analyzed the results of a survey conducted between 2016 and 2020 in 12 US National Cancer Institute-designated cancer centers.

Similarly, in the Romanian study conducted by the Center for Innovation in Medicine, the participants were asked to rate four statements related to:

- Prevention-focused cancer fatalism ("It seems like everything causes cancer" and "There's not much you can do to lower your chances of getting cancer").
- Cancer information overload ("There are so many different recommendations about preventing cancer, and it's hard to know which ones to follow").
- Treatment-focused cancer fatalism ("When I think about cancer, I automatically think about death").

The researchers found that, compared to urban participants, rural participants in the study exhibited higher levels of cancer fatalism and cancer information overload—a trait of cancer infod-

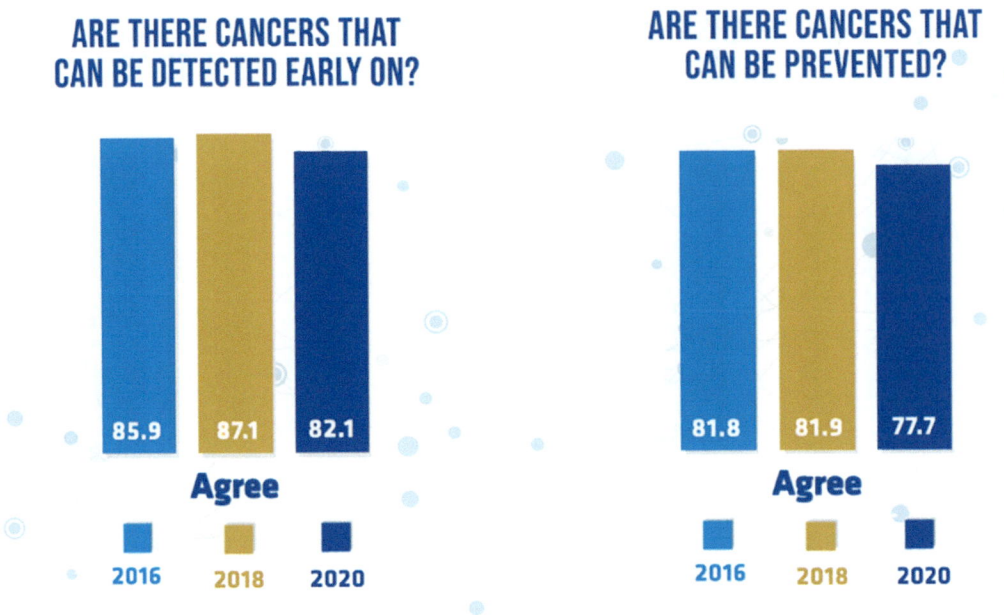

**Fig. 12.2** Prevention and diagnosis results from the three studies. Property of Center for Innovation in Medicine

emy. In particular, rural participants were 29% more likely to agree that everything causes cancer, 34% more likely to agree that prevention is not possible, 26% more likely to agree that there are too many different recommendations about cancer prevention, and 21% more likely to agree that cancer is always fatal (Fig. 12.2).

In 2020, 78% of the Romanians that participated in the study believed that cancer can be prevented, compared to 82% (in 2016 and 2018) (see Fig. 12.2). This belief was rather present in the segment that had no experience with the disease. Over 80% of respondents (85.9%, 2016; 87.1%, 2018; 82.1%, 2020) believed that the disease can be detected in early stages. Despite all this data, Romania has the lowest screening rates in the EU for cervical and breast cancer (the only types of cancer screening implemented so far).

### 12.3.2.2 Personalized Medicine and Cancer Innovation Awareness

Further data assessed levels of knowledge on immune-oncology and personalized medicine,

with varying trends being reported across the three studies (Fig. 12.3).

As of March 2020, approximately 42% of study participants said they have heard of the term "personalized medicine" (see Fig. 12.3). Comparably, in May 2018, there was a percentage of 44.0% when the notoriety of the term "personalized medicine" was evaluated among the adult population of Romania, and almost 40% of Romanians knew this term in 2016.

Regarding immuno-oncology, 39.1% of the respondents in March 2020 heard about this notion; the notoriety of the term "immunooncology" is also comparable with the data from previous waves (42.5% in 2018, respectively 37.1% in 2016).

The notoriety of the term "biomarkers" is also maintained at a constant value (42.8% in 2020, respectively 42.2% in 2018).

The 2020 study, similar to research conducted in 2016–2018, shows that new personalized therapies such as immuno-oncology and targeted therapies are associated with high prices/costs—only 20.9% of respondents estimate that they

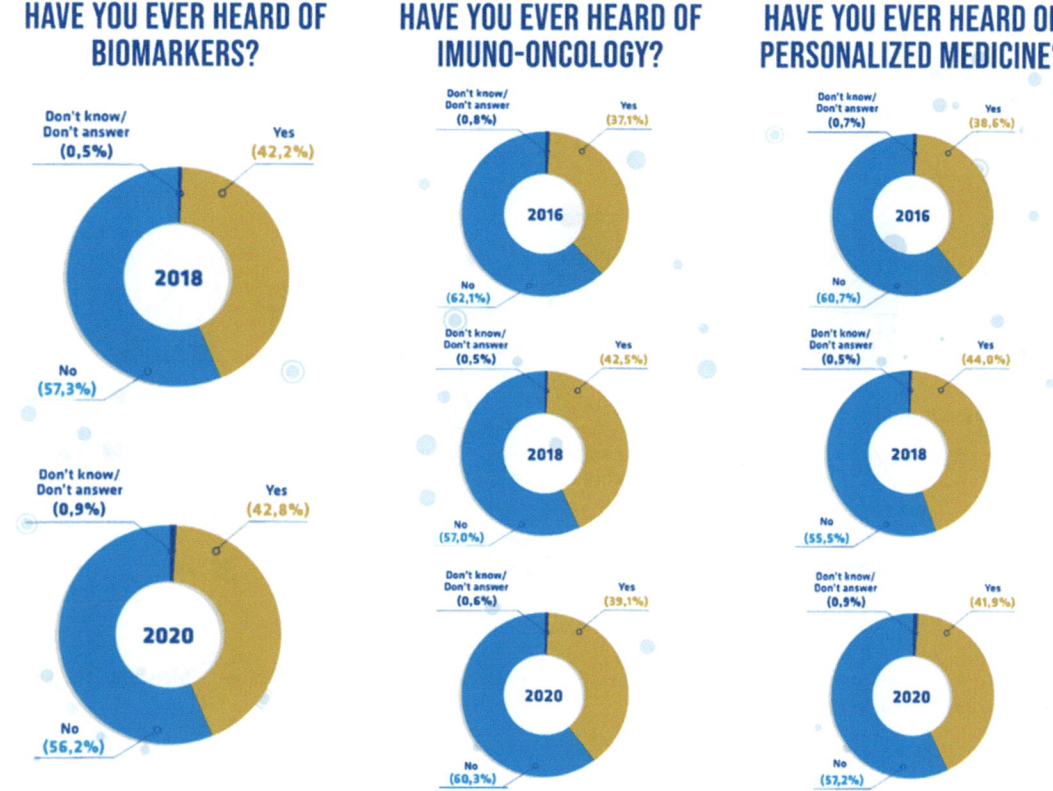

**Fig. 12.3** Health innovation awareness results. Property of Center for Innovation in Medicine

could afford it, and 33.3% of those surveyed consider that these treatments are not affordable at all (see Fig. 12.4). It should be noted that the percentage of those who consider that the new treatments are not accessible at all is still decreasing, from 44.1% in 2018, fueling the development of an increasing rate of fatalism.

### 12.3.2.3 Quality of the Cancer Information Campaigns: From One-Size-Fits-All to Personalized Communication

According to the *Digital News Report* [40], conducted by the Reuters Institute and Oxford University, in the last 3 years (2017–2020), television and the online environment have been and continue to be the main sources of information for Romanian citizens.

According to the data obtained from our surveys, the top five sources from which Romanian citizens are informed about cancer prevention, diagnosis, and treatment are represented by medical staff (doctors or nurses), the online environment blogs, medical forums, social networks, etc.), television, information materials (reports, posters, brochures, leaflets, etc.), and the written press. In 2020, the medical staff was responsible for informing a percentage of 68.2% of citizens, while the online environment reached a percentage of 61.3%, television 61%, news materials 53.5%, and print media 39.6%.

In 2017, after the results of the first round of the survey (2016) showed a high grade of fatalism (almost one in two Romanians believed that a cancer diagnosis always leads to death), and that the main source of information on cancer was the physician (usually itself being fatalistic

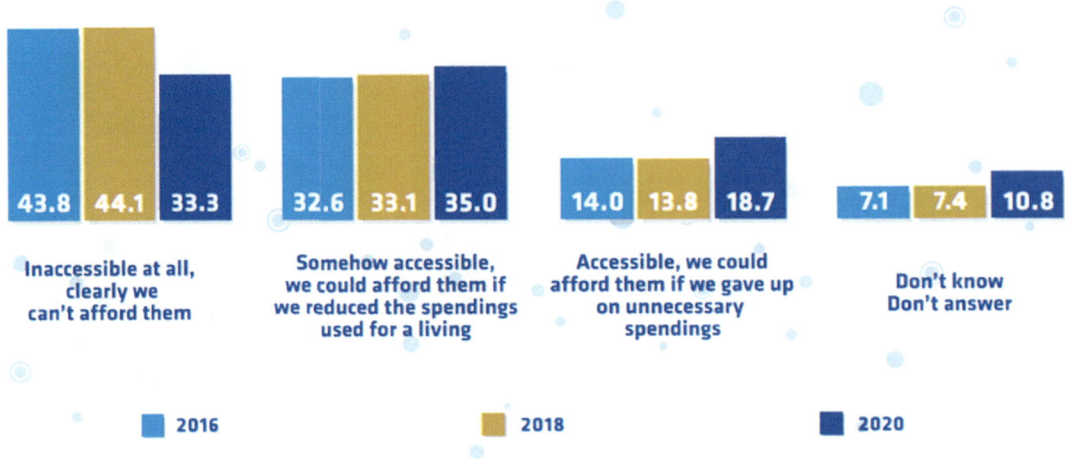

**Fig. 12.4** Associated costs of the new cancer therapies. Property of Center for Innovation in Medicine. Property of Center for Innovation in Medicine

when it comes to cancer topic), with the second place being held by the mainstream media (but no channel that actually referred to innovation in oncology, but rather presented the death cases and exclusively criticized the Romanian health-care system), the Center for Innovation in Medicine decided to implement a personalized multilayer communication and educational approach on cancer in Romania.

This study and the newly adopted definition of personalized medicine in Europe (2015) led to the first Personalized Medicine Conference organized in Bucharest, Romania, by the Center for Innovation in Medicine, in partnership with the presidential administration. The time for the conversation around cancer to change has come. The conference has been held annually since then. This was the first step at the macro level (politicians, decision-makers, mainstream health influencers, and mainstream media) to change perceptions around cancer by highlighting the benefits of innovations.

Less than a year later, at the beginning of 2017, the Center for Innovation in Medicine

launched the course *Innovation in Communication. Communicating the innovation* that aimed at training medical students, journalism students, and health engineering students to become health innovation communicators. The participants were thoroughly selected and 4 months later, the innovation communication platform, in Romanian, Raportuldegardă.ro [41] emerged, a platform that offered a different perspective for Romanian citizens that aimed at decreasing fatalism not only on cancer—another block was added to the macro level.

Based on the same findings from the 2016 national survey, at the end of 2017, the "Let's differently talk about cancer" campaign was launched, involving authorities, key opinion leaders, medical doctors, patients, cancer survivors, and citizens.

Adding another layer to the macro level of changing the perception of cancer and other fatal diseases in the Romanian population, the Center for Innovation in Medicine launched the initiative—science meets politicians—in partnership with the Romanian Parliament, consisting of the launch and debates around the *State of Innovation Annual Report of the Center for Innovation in Medicine* [22].

As a result, in 2018, the perceptions and attitudes were measured again, and we noticed that the fatalism rate experienced a decrease and the notoriety of cancer innovation-related terms "personalized medicine," "biomarkers," and "immuno-therapies" increased.

But between 2018 and 2020, another phenomenon happened in Romania. The media platform influencers became more vocal, and many health websites and platforms were launched, many of them engaging in mostly disease awareness campaigns mainly paid for by the industry. The cancer infodemic was reaching new heights. While "disease awareness campaigns" are not bad, the qualitative analysis we conducted showed us that these campaigns mostly fuel the fatalism and the infodemic around cancer. The messages followed the same pattern: *This percent of Romanians have died because of cancer in a year, get tested now.* While the conversation around death shouldn't be taboo, messages like these cannot be so bluntly delivered to a population in which one in two people believe that a cancer diagnosis always leads to death.

And so, the positive effect gained by the Center for Innovation in Medicine and partners between 2016 and 2018 at the macro level was neutralized and downgraded by the infodemic and indirectly negative effect at the macro and meso level.

On the other hand, as the social media platforms grew in popularity, many websites, pages, and health influencers promoted all sorts of wonder treatments, fueling the conspiracy theories around Big Pharma.

So in 2020, before the pandemic, when we measured the attitudes and perceptions again, the decrease in the fatalistic approach observed from 2016 to 2018 not only was not maintained but increased.

The work continues, with Raportuldegardă.ro being a source of information for approximately 2000 unique Romanian users daily, weekly covering the most important news on pandemic control and daily publishing the most relevant health innovations and trends at the international and European levels.

### 12.3.3 Vaccination Literacy in Romania

Two major vaccination campaigns can be identified in Romania in the last two decades: the HPV vaccination campaign that first started in 2008 and the SARS-CoV-2 vaccination campaign. The first one can briefly be described as a big international and national failure, and the last one can be described as a big European failure, but national relative success when compared to the previous experiences.

Both vaccination campaigns have the following key elements in common:

- The propagandistic approach and the alignment of the messages with the political agenda led to the misappropriation of public trust (a government trust already very low in public polls [42]).
- The incapacity of convincing the health professionals to deliver the pro-vaccination messages, as most of the people expected according to the national surveys.
- The failure of delivering the key messages: from the beginning of the COVID-19 vaccination campaign, the official message was that the vaccine will get you the normal life back and not that the vaccine protects you and your dear ones against complications and hospitalization.
- The oversimplified messages that did not refer to the unique genetic and biological traits of a person led to the false impression that everyone should react and get the same level of protection.
- The failure of complying with the people's needs, using the dissemination, channels, and influencers that reached people that had access and followed certain official pages, when the lowest rate of vaccination was among those who do not have access to these channels or are compliant to the fake news spreading channels.

#### 12.3.3.1    HPV Vaccination Literacy

The major problem when it comes to preventing disease by vaccination (particularly in Romania) is the inadequate communication that derives from a low level of understanding of the particularities of the population you are addressing. By adding the low level of health literacy and the fake news and conspiracy theories that arise from it to the low capacity of the authorities to communicate and engage with the citizens, a major gap in HPV vaccination rates and cancer survival rates was created between countries in Europe.

HPV vaccination is a very sensitive subject because it involves the prevention of a possible sexually transmitted disease from an early age. In societies with a strong traditional and religious background and in which around half of the population lives in rural areas, it is very difficult to communicate these messages properly. Romania has a strong communist background—a recent study showed that more than 60% of the respondents believe that the actual situation in Romania is worse than 30 years ago (7 years ago, only 40% of Romanians had that opinion). Over 60% of Romanians prefer traditional values to modern rights and freedoms [42].

At the EU level, Romania ranks first in terms of incidence and mortality for cervical cancer: the incidence is 2.5 times higher than the European average, and the mortality rate is over four times higher. When referring to HPV, this information can be explained partly by the following data, according to the national survey organized by the Center for Innovation in Medicine, Renaşterea Foundation, and National Institute for Public Health:

- (2018) 48% of women respondents said that in the last 3 years, they were not tested for HPV.
- (2018–2020) no less than 67% of women and girls from rural areas, aged 15–65, have genital infections.
- (2020) only 36% of Romanian women have heard of the HPV virus, and only 31% associate this infection with cervical cancer [43].

Currently, after the 2008's HPV vaccination failure (less than 2% of the target population was vaccinated at that time), Romania is trying to implement a new HPV vaccination program for girls aged 11–18 and a National Screening Program for Cervical Cancer, but these need to be strengthened by sustainable and highly personalized communication and training courses for the people involved in the process (from family doctors to school teachers).

In 2020, the Romanian Ministry of Health announced its intention to introduce free HPV vaccination also for boys. But as HPV infection is perceived as women's health issue exclusively in countries with a profile resembling Romania, the mere fact that there are free vaccines available for boys does not guarantee their vaccination (similar to COVID vaccination—doses available for the entire population, but the majority refused to get the vaccine).

Inequalities in vaccination in general and HPV vaccination rate, in particular, exist not only between countries but also within countries, communities, and groups. For example, a total of 41 counties, along with the municipality of Bucharest, constitute the official administrative divisions of Romania but based on Renaşterea Foundation's experience with their cancer screening projects, in the North and North-East of the country, the level of cervical cancer and HPV vaccination literacy is very low, being hard to get access to the people through classical methods. Different communication vectors and messages are needed.

### Two Sides of the Story About HPV Vaccination Efforts in Romania

The Center for Innovation in Medicine measured the level of citizens' awareness and their perception of prevention, diagnosis, and treatment of cancer, in 2016, 2018, and 2020 (before the pandemic). In 2018, 48% of women respondents said that in the last 3 years, they were not tested for HPV. Sixty-one percent stated that they had heard of the HPV vaccine. Fifty-four percent documented the subject but did not get the vaccine and only 2% have been informed and vaccinated.

According to data from the Renaşterea Foundation, no less than 67% of women and girls from rural areas, aged 15–65, have genital infections. In the urban areas, the percentage was 65%. Forty-one percent of girls aged 15–19 had a genital infection, and the percentage rose to 71% for girls aged 20–29. Moreover, only 17% of women in Romania took a Pap test, while at the EU level, the average is 70%.

Only 36% of Romanian women interviewed in an IRES (Romanian Institute for Evaluation and Strategy) survey conducted at the request of the National Institute of Public Health have heard of the HPV virus, and only 31% associate this infection with cervical cancer. Romania has the highest incidence and highest mortality rate from cervical cancer compared to European Union countries, although it is one of the few types of cancer that can be prevented by vaccination. Every 5 h, a Romanian woman in the 20–50 age group dies of cervical cancer.

In 2012, Romania launched its first cervical cancer screening program, targeting approximately six million women aged 25–63 (in a period of 5 years). By 2015, only 7% of the target population was tested [44].

In 2008, Romania was among the first countries to introduce HPV vaccination (for girls aged 10–11), simultaneously with the United Kingdom. However, the campaign was a total failure—only 2.6% of eligible girls were vaccinated and the program was suspended. In 2009, an information campaign was launched, followed by a second vaccination program, targeting girls aged 12–14. A catch-up program was also launched, where adult women were given the opportunity to get the vaccine free of charge through their health provider. Despite the accessibility of the vaccine, uptake remained low and the school-based program was discontinued at the end of 2011. The program was launched for the third time in April 2013 and for the fourth time in 2019.

In 2020 and 2021, according to preliminary data, less than 50,000 girls were vaccinated; no Romanian county has a vaccination rate higher than 5% of the target population, and the average vaccination rate in Romania in 2020/2021 is 2%.

The numbers speak for themselves—cervical cancer is not prevented by vaccination in Romania, and many women don't know about the virus and don't get tested for the infection. But in a parallel universe, a communication campaign ("Protect her wings") launched in 2017 to encourage HPV testing and to raise awareness of the disease caused by the infection with the virus, which, according to the jury that gave the award "contributed to the decision announced by the Ministry of Health to resume HPV vaccination," was awarded an international distinction for "Best in show" campaign in the entire CEE region.

Despite the fact that a pro-HPV awareness campaign implemented in Romania was awarded one of the most distinguished prizes for communication campaigns, the vaccination rate has not increased. One possible explanation was the one-size-fits-all approach of the campaign that did not include the assessment of behavior, perceptions, and attitudes in the Romanian population. Having a low HPV vaccination rate does not guarantee that if you just start communicating on the subject it will increase. It's that simple, but complex after all.

### 12.3.3.2  COVID-19 Vaccination Literacy

This kind of approach was also seen during the COVID-19 vaccination campaign. Again, the communication campaign was awarded a prize for its creativity, but Romania has one of the lowest COVID-19 vaccination rates in Europe. We can easily assume that the people that got the vaccine were not impressed by the campaign, since the campaign was carried on Facebook exclusively and sometimes on TV, mixed with political messages, but were convinced by the health emergency.

In 2020, COVID-19 was the cause of death for approximately 16,000 people in Romania—no less than 5% of the total number of registered deaths [30]. Approximately 18,500 COVID-19 deaths were recorded by the end of August 2021. COVID-19 mortality calculated by August 2021 was 12% higher in Romania than the EU average. Rates are calculated based on reported deaths, but the number may be much higher, not to mention the indirect burden.

The COVID-19 pandemic overlapped with an already overwhelmed health system and highlighted the need to increase the HL level of Romanian citizens. According to the results of the most recent Eurobarometer on the attitude of European citizens toward science and technology, only three out of ten Romanians know in 2021 that antibiotics have only an antibacterial effect and not an antiviral effect. Their percentage decreased compared to 2005 [45]. The data must be interpreted in the context in which the COVID-19 pandemic brought to the public discussion more than ever the subject of viruses and bacteria, but also considering the media coverage of nosocomial infections in Romanian hospitals and the antibiotic resistance crisis, which launched extensive communication campaigns 5–6 years ago.

Multiple sociological studies conducted by various institutions provide a number of information on the limited level of health literacy among the general population of Romania. Cancer is not the only disease impacted. In the Center for Innovation in Medicine's study from 2016 regarding another chronic disease, diabetes, patients did not seem to have long-term disease management skills, as 20% did not visit a diabetes specialist in the previous year. The results of another sociological study (2017) identified a low level of literacy in mental health and a negative attitude toward those diagnosed with mental illness among Romanians.

The low level of health literacy, the infodemic, and the impact of fake news were reflected in the results of the vaccination campaign against COVID-19 in Romania. According to the Public Health Barometer of October 2020 [46], 21% of Romanians would get vaccinated if a vaccine had been available, a third choice to be immunized only if they heard that there were no side effects, and 8% would have wanted to know more information to make the vaccination decision. In context, nine out of ten Romanians declared that they got a vaccine at some point in life.

Another study, conducted by IPSOS in September 2020 [47], showed that only 29% of Romanians in urban areas were determined to get vaccinated if a vaccine was available. The repetition of the study in early February 2021 showed that the intention to vaccinate increased by 7% among those over 16 years in urban areas.

An IRES (Romanian Institute for Evaluation and Strategy) study [48] conducted at the beginning of 2021 underlined that individuals who are pro-vaccination (declaring that they will certainly or probably be vaccinated) tend to come from urban areas (58%) and that anti-vaccination are mostly from rural areas (53%). Moreover, over 60% of the undecided respondents were willing to take the vaccination advice if it will come from their medic or a healthcare professional. The percentage was 20% in the case of those against vaccination. This finding is consistent in many sociological studies—the healthcare professionals are the main vectors that could recommend the vaccination.

Independently, this trend is observed in studies from other states and countries [49]. In the United States, the vaccination rate in people that received the recommendation of getting vaccinated against COVID-19 from their doctor was 15% higher than in people who did not receive such information from their doctor.

And although sociological studies show those rates, the gap between declared willingness and action is still important. One of the explanations can be drawn from the CDC study that showed that direct referring is very important. Even if doctors publicly recommended vaccination, the citizens expected that this would come from their physician or a physician they trust. In Romania, the vaccination rate of healthcare professionals had a slow increase when compared with other EU countries.

In May 2021, Romania was already vaccinating the general population, but only 3.5 million people were vaccinated with the two doses full scheme. From May to the end of October, another 2.4 million people were added, given that the indications for certain vaccines have been extended to children over 12.

In July 2021, only 16% of the 80+ population had the full scheme of vaccination, but Romania reported to ECDC that it had no difficulties with vaccinating the elderly population [50].

From October 2021 to February 2022, the number of people vaccinated with the full scheme (considered two doses in Romania) reached eight million (target population 5+), with less than

2000 people getting a first dose of the vaccine daily. Moreover, only two million of the eight million Romanians vaccinated got their booster dose. Almost 2 years since the vaccines were available, Romania recently reached the 50% vaccination rate against COVID-19 (two doses scheme). In the context of VOCs and Omicron variants, vaccine protection offered by the two doses vaccination scheme is low, and only two million Romanians got their booster doses since they became available. In this scenario, the 50% vaccination coverage doesn't have the same value as it would have 1 year ago.

The failure of the vaccination campaign and of the vaccination communication campaign can be partially explained by Eurobarometer data on Romanian's attitudes and perceptions about innovation and data from various national surveys. The pro-vaccination communication campaign failed to exceed the percentage of about 30% of people who constantly stated in opinion polls that they will be vaccinated.

On the other hand, the COVID-19 pandemic directed, in the first phase, public attention in Romania to the subject of health and health education and had a positive effect on discovering the communication and the applied importance of HL at the authorities' level. But after multiple waves of COVID-19, and after political messages of incertitude and measures that were not based and explained at the scientific standards required, the public opinion in Romania diverted, and the citizens turned their back to the individual protective measures in the face of the disease.

## 12.4 Conclusions and Discussions

Romania has the highest rate of preventable deaths among European countries (80.1% compared to 68% EU average) [51]. While these rates can briefly be explained by the limited access to health knowledge, healthcare, and technology, the extensive explanations are infinite.

While more and more therapies entered the Romanian market in the last years and were reimbursed, with cancer being one of the most active areas, the survival rates did not improve in accor-

dance. This can be partially explained by the high fatalistic attitudes, the low level of government trust, and the reticence in accepting innovations, the traditional and conservative beliefs of people, etc.

Despite multiple efforts to stimulate the adoption of a healthy lifestyle and increase the adherence to screening and treatment, in Romania and in other countries with a similar profile, the results did not meet the expectations. From our experience with HPV vaccination, COVID-19 vaccination, and cancer, we argue that there were three main general reasons for this: (1) lack of proper understanding of the behavior of individuals (based on micro, meso, and macro assessment) as the key determinant for pro-health behavior and implementation of the one-size-fits-all communication strategies, (2) the low level of health literacy within these countries, and (3) the failure of authorities in identifying the proper channels and personalize the messages for every group, as well as the failure of authorities to engage with relevant NGOs in order to improve the situation.

Overall, during more than 10 years of experience in the field of personalized health and care communication, we noticed that the lack of data and capacity to use data for public health decisions, as well as the lack of human touch and empathy (in some countries derived from the long history of communism), were the main drivers of this situation at the level of the decision-makers.

Time showed us that in populations with these characteristics, just offering better-structured information will not change the behaviors of the citizens, at least not fast enough for the speed of scientific advance of the moment. Many levels and actions need to be employed. We propose a concept of a personalized communication model based on the matrix with the social and behavioral determinants of health (see Fig. 12.5).

Based on the experience gained in our research studies done at the national level in Romania, we propose a new model, based on the individual behavioral health determinant matrix, for influencing pro-health behavior and indirectly increasing the overall health literacy and the faster

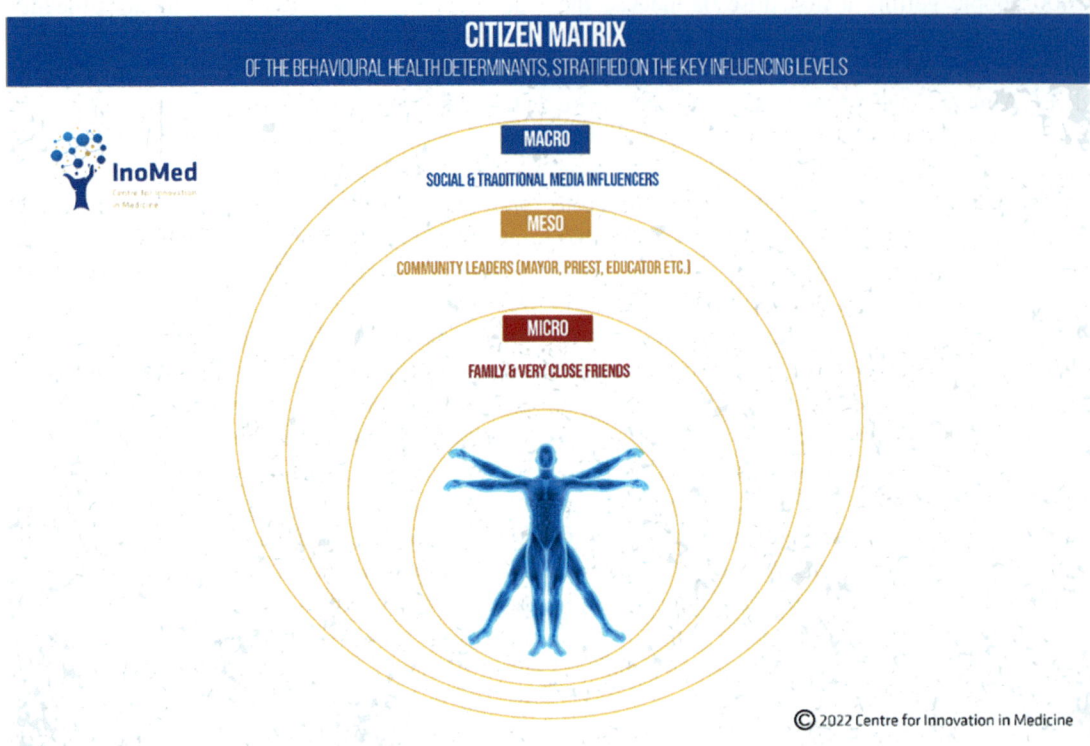

**Fig. 12.5** Citizen matrix. Property of Center for Innovation in Medicine

adoption of relevant and evidence-based health innovations.

The matrix has three layers:

- Micro-dimension—family members and inner circle of close friends who can influence behavior.
- Meso-dimension—community influencers who can influence pro-health behavior (e.g., religious leaders, family doctors, or mayors, especially in rural areas).
- Macro-dimension—(inter)national influencers driven by traditional media and social media who can influence behavior.

The citizen behavior matrix (micro, meso, macro) above represents an innovation in terms of health communication because it is based on a deep understanding of the high granularity of the reasons for the high level of fatalism. Another social innovation consists of the unique approach in using the citizens' perspectives and perceptions for influencing their behavior and not expecting that by only delivering the information, they can assimilate it and use it to make better decisions about their health (the one-size-fits-all approach) (Fig. 12.6).

The main motif of the personalized communication model revolves around health literacy, education, and communication based on personalized and sustained efforts in understanding the specific needs of the communities and individuals addressed. In countries with a medium-high literacy and education level, with at least a medium level of trust in the national and regional authorities, this level of detail might not be

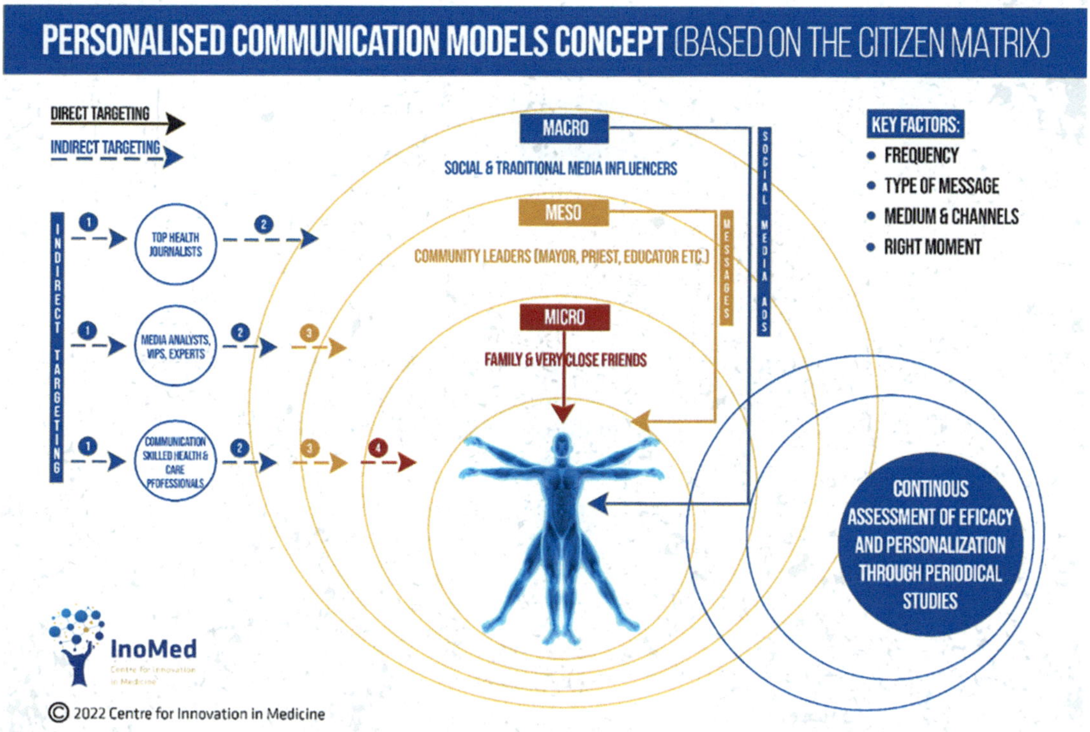

**Fig. 12.6** Personalized communication models concept. Property of Center for Innovation in Medicine

needed, but in countries like Romania, the experience and the time showed us that there is, unfortunately, no other way of doing it.

We have already validated the model of assessing the attitudes, perceptions, and behaviors of cancer every 2 years at the national level and take actions based on the survey results in our campaigns. We argue that this measurement could be a valuable tool for understanding the social determinants of health in a population and that by closing the community circle and assessing the level periodically, you can have a correct evaluation of the willingness of people to exercise a healthy behavior on main areas like getting vaccinated or participating in screening programs.

The same applies to the intake of innovations and personalized medicine understanding. If you are able to identify, target, and then sustainably address the meso influencers in the communities, their messages will penetrate and settle better at the citizen, individual level. This is not a one-time approach; you need to periodically reassess the quality of interventions—the 2-year span could become standard.

To sum up the concept:

- Assess attitudes, perceptions, and behaviors every 2 years.
- Conduct qualitative evaluation of health communication campaign at the national level.
- Take actions based on the information obtained from the surveys (launch a new platform, start a training course, construct a different communication campaign, etc.).
- Engage with the mainstream media and key macro stakeholders.
- Consult with communication and health literacy analysts (Fig. 12.7).

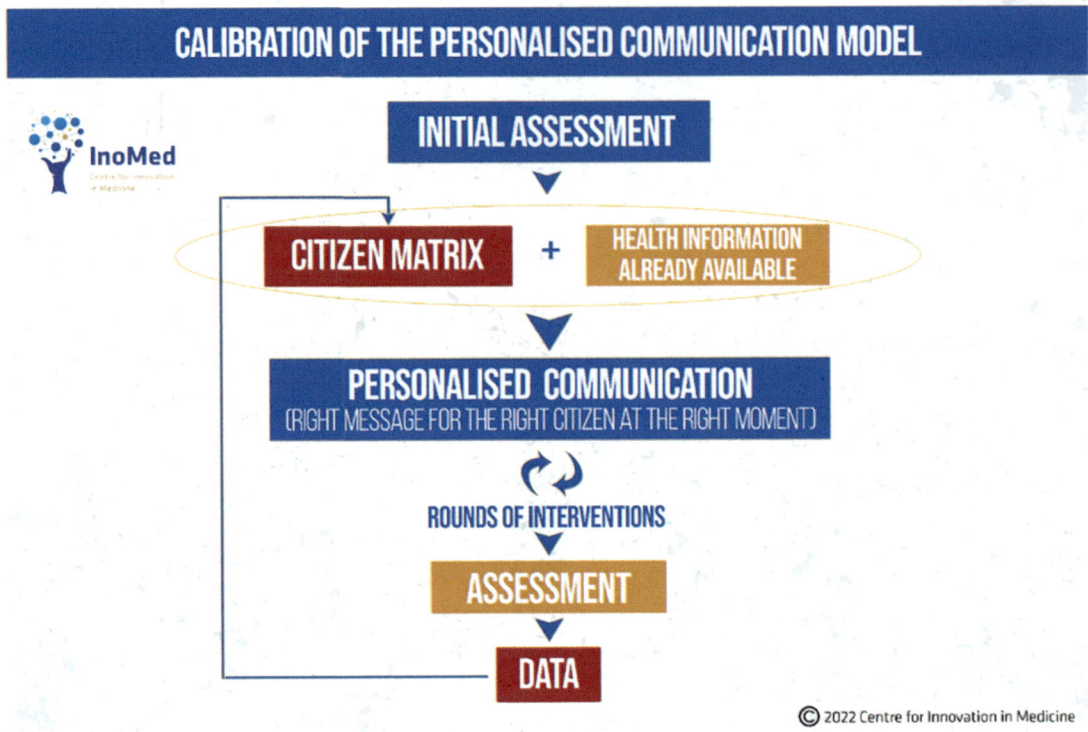

**Fig. 12.7** Calibration of the personalized communication model. Property of Center for Innovation in Medicine

### 12.4.1 Future Perspectives on Increasing the Individual Level of Health Literacy by Periodically Assessing Attitudes, Perceptions, and Behaviors, in Synergy with the European Opportunities

While following the concept and scheme presented above can bring great value to a population, collaboration at the national, European, and international levels is crucial. The good practice model needs to be shared, adapted, improved, and enriched with new data and scientific evidence.

Improving cancer literacy and changing health behavior following the model presented above can be a best practice model for improving or starting building on PHCL in certain populations. Europe's Beating Cancer Plan [52] and the Mission on Cancer [53] are key drivers in the fight against cancer at the European level. Though we refer to cancer literacy, the new vision in cancer battle has the citizens in the middle: it is no longer possible to fight cancer with what decision-makers perceive as being important. We need to take everything to the next level—identify the needs of patients and citizens and work with them to meet those needs (Table 12.1).

The "Missions" are a new tool in Horizon Europe—the European Union's Framework Program for Research and Innovation. Inspired by the Apollo 11 mission to send one man to the moon, EU missions are a commitment to addressing major social challenges. The five missions for the period 2021–2027 are fighting cancer, adapting to climate change, living in greener cities, ensuring healthy soils, and protecting the oceans.

The Cancer Mission, launched on September 29, 2021, together with Europe's Beating Cancer Plan (February 3, 2021) and the 2023 expected European Partnership for Personalised Medicine (EP PerMed), aims to improve the lives of more

**Table 12.1**  European initiative to support health literacy and citizens' engagement in cancer information extracted from the European Mission on Cancer and Europe's Beating Cancer Plan: Implementation Roadmap (updated version I January 2022) [52, 53]

| Initiatives and programs | Actions to support health literacy and citizens' engagement in cancer | When |
|---|---|---|
| Cancer Mission | UNCAN.eu – improve the understanding of cancer | 2021–2030 |
| | European Cancer Patient Digital Centre | 2021–2023 |
| | Support quality of life (living labs) | 2021–2025 |
| Europe's Beating Cancer Plan: Implementation Roadmap | Knowledge Centre on Cancer | 2021–2025 |
| | European Code Against Cancer | 2021–2025 |
| | "Health Literacy for Cancer Prevention and Care project" | 2021–2025 |
| | Propose mandatory front-of-pack nutrition labeling | 2021–2022 |
| | EU Clinical Trials Portal and Database | 2021–2025 |
| | Set up Partnership on Personalised Medicine | 2021–2025 |
| | Roadmap to personalized prevention | 2022–2022, 2024 |
| | Cancer Inequalities Registry | 2021–2025 |
| | EU Network of Youth Cancer Survivors | 2021–2025 |

than three million people affected by cancer by 2030. The four goals of the Cancer Mission are understanding cancer, preventing and detecting it early, optimizing diagnosis and treatment, and supporting quality of life.

Identifying the synergies and opportunities on how to better engage with citizens and let their voices be heard is crucial for a structured and sustainable approach over time. The table below sums up those actions and opportunities.

One first step is to adapt and include data on perceptions and attitudes in the Cancer Inequalities Registry, launched in February 2022. The initial framework of the Registry is based on the same classical approach to the disease and does not reflect the citizen's approach and inequalities' gap, but the change requires scientific evidence and new models of collaboration.

# References

1. Literacy I of M (US) C on H, Nielsen-Bohlman L, Panzer AM, Kindig DA. Introduction [Internet]. Health Literacy: A Prescription to End Confusion. National Academies Press (US); 2004 [cited 2022 Feb 12]. Available from: https://www.ncbi.nlm.nih.gov/books/NBK216033/
2. Buetti S, Lleras A. Perceiving Control Over Aversive and Fearful Events Can Alter How We Experience Those Events: An Investigation of Time Perception in Spider-Fearful Individuals. Front Psychol [Internet]. 2012 [cited 2022 Feb 12];3. Available from: https://www.frontiersin.org/article/10.3389/fpsyg.2012.00337
3. Geanta M, Tanwar AS, Lehrach H, Satyamoorthy K, Brand A. Horizon scanning: rise of planetary health genomics and digital twins for pandemic preparedness. OMICS J Integr Biol. 2022;26(2):93–100.
4. Personalised medicine I European Commission [Internet]. [cited 2022 Feb 14]. Available from: https://ec.europa.eu/info/research-and-innovation/research-area/health-research-and-innovation/personalised-medicine_en
5. EUR-Lex - C:2015:421:FULL - EN - EUR-Lex [Internet]. [cited 2022 Feb 14]. Available from: https://eur-lex.europa.eu/legal-content/EN/TXT/?uri=OJ%3AC%3A2015%3A421%3AFULL
6. Sørensen K, Van den Broucke S, Fullam J, Doyle G, Pelikan J, Slonska Z, et al. Health literacy and public health: a systematic review and integration of definitions and models. BMC Public Health. 2012;12(1):80.
7. Sørensen K, Pleasant A. Understanding the conceptual importance of the differences among health literacy definitions. Stud Health Technol Inform. 2017;240:3–14.
8. An Introduction to Health Literacy I NNLM [Internet]. [cited 2022 Feb 14]. Available from: https://nnlm.gov/guides/intro-health-literacy
9. Pires IM, Denysyuk HV, Villasana MV, Sá J, Lameski P, Chorbev I, et al. Mobile 5P-approach for cardiovascular patients. Sensors. 2021;21(21):6986.
10. The evolution of personalized healthcare and the pivotal role of European regions in its implementation I Personalized Medicine [Internet]. [cited 2022 Feb 14]. Available from: https://www.futuremedicine.com/doi/10.2217/pme-2020-0115
11. Sørensen K, Brand H. Health literacy: the essential catalyst for the responsible and effective translation of genome-based information for the ben-

efit of population health. Public Health Genomics. 2011;14(4–5):195–200.

12. Constitution of the World Health Organization [Internet]. [cited 2022 Feb 14]. Available from: https://www.who.int/publications/m/item/constitution-of-the-world-health-organization

13. Public Health Classifications Project - Determinants of Health - Final Report - HealthStats NSW [Internet]. [cited 2022 Feb 14]. Available from: https://www.health.nsw.gov.au/hsnsw/Pages/classifications-project.aspx

14. Determinants of Health Visualized [Internet]. [cited 2022 Feb 14]. Available from: https://www.goinvo.com/vision/determinants-of-health/

15. Social prescribing: applying All Our Health [Internet]. GOV.UK. [cited 2022 Feb 14]. Available from: https://www.gov.uk/government/publications/social-prescribing-applying-all-our-health/social-prescribing-applying-all-our-health

16. Personalized profiles for disease risk must capture all facets of health [Internet]. [cited 2022 Feb 14]. Available from: https://www.nature.com/articles/d41586-021-02401-0

17. Mancilla VJ, Peeri NC, Silzer T, Basha R, Felini M, Jones HP, et al. Understanding the interplay between health disparities and epigenomics. Front Genet [Internet]. 2020 [cited 2022 Feb 14];11. Available from: https://www.frontiersin.org/article/10.3389/fgene.2020.00903

18. Koirala R, Gurung N, Dhakal S, Karki S. Role of cancer literacy in cancer screening behaviour among adults of Kaski district, Nepal. PLoS One. 2021;16(7):e0254565.

19. Familial Hypercholesterolemia as a model for innovation in CVD prevention: evidence-based paediatric screening and early detection programs for FH at EU level [Internet]. Raportuldegardă.ro. [cited 2022 Feb 14]. Available from: https://raportuldegarda.ro/familial-hypercholesterolemia-model-innovation-cardiovascular-prevention-evidence-based-paediatric-screening-familial-hypercholesterolemia/

20. Zhao Y, Wood EP, Mirin N, Cook SH, Chunara R. Social determinants in machine learning cardiovascular disease prediction models: a systematic review. Am J Prev Med. 2021;61(4):596–605.

21. ESMO. ESMO Clinical Practice Guidelines: Lung and Chest Tumours [Internet]. [cited 2022 Feb 14]. Available from: https://www.esmo.org/guidelines/lung-and-chest-tumours

22. Publications - Center for Innovation in Medicine [Internet]. [cited 2022 Feb 14]. Available from: https://ino-med.ro/publications.html

23. Introduction: Standards of Medical Care in Diabetes—2021. Diabetes Care. 2020;44(Supplement_1):S1–2.

24. Buse JB, Wexler DJ, Tsapas A, Rossing P, Mingrone G, Mathieu C, et al. 2019 Update to: Management of Hyperglycemia in Type 2 Diabetes, 2018. A Consensus Report by the American Diabetes Association (ADA) and the European Association for the Study of Diabetes (EASD). Diabetes Care. 2020;43(2):487–93.

25. Holt RIG, DeVries JH, Hess-Fischl A, Hirsch IB, Kirkman MS, Klupa T, et al. The management of type 1 diabetes in adults. A Consensus Report by the American Diabetes Association (ADA) and the European Association for the Study of Diabetes (EASD). Diabetes Care. 2021;44(11):2589–625.

26. Infodemic [Internet]. [cited 2022 Feb 14]. Available from: https://www.who.int/westernpacific/health-topics/infodemic

27. Pandemic preparedness and COVID-19: an exploratory analysis of infection and fatality rates, and contextual factors associated with preparedness in 177 countries, from Jan 1, 2020, to Sept 30, 2021 - The Lancet [Internet]. [cited 2022 Feb 14]. Available from: https://www.thelancet.com/journals/lancet/article/PIIS0140-6736(22)00172-6/fulltext

28. Improving cancer literacy in Europe to save time, costs and lives (Guest blog) [Internet]. [cited 2022 Feb 14]. Available from: https://efpia.eu/news-events/the-efpia-view/blog-articles/improving-cancer-literacy-in-europe-to-save-time-costs-and-lives-guest-blog/

29. Vandenbosch J, Van den Broucke S, Vancorenland S, Avalosse H, Verniest R, Callens M. Health literacy and the use of healthcare services in Belgium. J Epidemiol Community Health. 2016;70(10):1032–8.

30. Romania: Country Health Profile 2021 | en | OECD [Internet]. [cited 2022 Feb 14]. Available from: https://www.oecd.org/publications/romania-country-health-profile-2021-74ad9999-en.htm

31. Analiza-de-Situatie-Cancer-2018 – Institutul Național de Sănătate Publică [Internet]. [cited 2022 Feb 14]. Available from: https://insp.gov.ro/wpfb-file/analiza-de-situatie-cancer-2018-pdf/

32. Breathing in a new era: a comparative analysis of lung cancer policies across the Asia-Pacific region [Internet]. Economist Intelligence Unit. [cited 2022 Feb 14]. Available from: https://www.eiu.com/n/campaigns/breathing-in-a-new-era-a-comparative-analysis-of-lung-cancer-policies-across-the-asia-pacific-region/

33. Todor RD, Bratucu G, Moga MA, Candrea AN, Marceanu LG, Anastasiu CV. Challenges in the prevention of cervical cancer in romania. Int J Environ Res Public Health [Internet]. 2021 Feb [cited 2022 Feb 14];18(4). Available from: https://www.ncbi.nlm.nih.gov/labs/pmc/articles/PMC7916723/

34. Virus Papiloma Uman (HPV) [Internet]. [cited 2022 Feb 14]. Available from: https://ec.europa.eu/health/vaccination/hpv_ro

35. Early breast cancer: ESMO Clinical Practice Guidelines for diagnosis, treatment and follow-up† - Annals of Oncology [Internet]. [cited 2022 Feb 14]. Available from: https://www.annalsofoncology.org/article/S0923-7534(19)31287-6/fulltext

36. Trimboli RM, Giorgi Rossi P, Battisti NML, Cozzi A, Magni V, Zanardo M, et al. Do we still need breast cancer screening in the era of targeted therapies and precision medicine? Insights Imaging. 2020;11(1):105.

37. Sănătății M. Ministerul Sănătății va demara primul program de screening al cancerului de sân – Ministerul Sănătății [Internet]. [cited 2022 Feb 14]. Available from: http://www.ms.ro/2017/11/06/ministerul-sanatatii-va-demara-primul-program-de-screening-al-cancerului-de-san/

38. 66% of women in the EU aged 50–69 got a mammogram [Internet]. [cited 2022 Feb 14]. Available from: https://ec.europa.eu/eurostat/web/products-eurostat-news/-/edn-20211025-1

39. Jensen JD, Shannon J, Iachan R, Deng Y, Kim SJ, Demark-Wahnefried W, et al. Examining rural-urban differences in fatalism and information overload: data from 12 NCI-designated cancer centers. Cancer Epidemiol Biomarkers. 2022;31(2):393–403.

40. Digital News Report 2021 [Internet]. Reuters Institute for the Study of Journalism. [cited 2022 Feb 14]. Available from: https://reutersinstitute.politics.ox.ac.uk/digital-news-report/2021

41. Raportuldegardă.ro [Internet]. Raportuldegardă.ro. [cited 2022 Feb 14]. Available from: https://raportuldegarda.ro/

42. Sondaj Inscop. Unul din patru români consideră că Rusia răspândeşte ştiri false pe teritoriul ţării noastre. Care sunt cele mai de încredere surse de informaţii | adevarul.ro [Internet]. [cited 2022 Feb 14]. Available from: https://adevarul.ro/news/politica/sondaj-inscop-unul-patru-romani-considera-rusia-raspandeste-stiri-false-teritoriul-tarii-noastre-cele-mai-incredere-surse-informatii-1_606433e85163ec42717a1ca0/index.html

43. Două din trei românce nu au auzit de virusul HPV. La fiecare 5 ore, o româncă din grupa de vârstă 20–50 de ani moare de cancer de col uterin [Internet]. [cited 2022 Feb 14]. Available from: https://www.hotnews.ro/stiri-sanatate-25242029-doua-din-trei-romance-nu-auzit-virusul-hpv-fiecare-5-ore-romanca-din-grupa-varsta-20-50-ani-moare-cancer-col-uterin.htm

44. Colzani E, Johansen K, Johnson H, Celentano LP. Human papillomavirus vaccination in the European Union/European Economic Area and globally: a moral dilemma. Eur Secur. 2021;26(50):2001659.

45. European citizens' knowledge and attitudes towards science and technology - September 2021 - - Eurobarometer survey [Internet]. [cited 2022 Feb 14]. Available from: https://europa.eu/eurobarometer/surveys/detail/2237

46. mara.stroescu. Barometrul de Sănătate Publică – Octombrie 2020 | ISPRI Ion I. C. Bratianu [Internet]. [cited 2022 Feb 14]. Available from: https://ispri.ro/barometrul-de-sanatate-publica-octombrie-2020/

47. Controverse şi convingeri despre vaccinarea anti-COVID-19 în România [Internet]. Ipsos. [cited 2022 Feb 14]. Available from: https://www.ipsos.com/ro-ro/controverse-si-convingeri-despre-vaccinarea-anti-covid-19-romania

48. ANALIZĂ DE PROFIL ÎN FUNCŢIE DE INTENŢIA DE VACCINARE ÎMPOTRIVA COVID-19 [Internet]. [cited 2022 Feb 14]. Available from: https://ires.ro/articol/416/analiza-de-profil-in-func-ie-de-inten%C8%9Bia%2D%2Dde-vaccinare-%C3%AEmpotriva%2D%2Dcovid-19

49. Nguyen KH. Report of Health Care Provider Recommendation for COVID-19 Vaccination Among Adults, by Recipient COVID-19 Vaccination Status and Attitudes — United States, April–September 2021. MMWR Morb Mortal Wkly Rep [Internet]. 2021 [cited 2022 Feb 14];70. Available from: https://www.cdc.gov/mmwr/volumes/70/wr/mm7050a1.htm

50. Overview of the implementation of COVID-19 vaccination strategies and deployment plans in the EU/EEA [Internet]. European Centre for Disease Prevention and Control. 2022 [cited 2022 Feb 14]. Available from: https://www.ecdc.europa.eu/en/publications-data/overview-implementation-covid-19-vaccination-strategies-and-deployment-plans

51. says VF. Two-thirds of deaths of under 75-year-olds could have been prevented [Internet]. www.euractiv.com. 2019 [cited 2022 Feb 14]. Available from: https://www.euractiv.com/section/active-ageing/news/two-thirds-of-deaths-of-under-75-year-olds-could-have-been-prevented/

52. Europe's Beating Cancer Plan [Internet]. European Commission - European Commission. [cited 2022 Feb 14]. Available from: https://ec.europa.eu/commission/presscorner/detail/en/IP_21_342

53. EU Mission: Cancer [Internet]. European Commission - European Commission. [cited 2022 Feb 14]. Available from: https://ec.europa.eu/info/research-and-innovation/funding/funding-opportunities/funding-programmes-and-open-calls/horizon-europe/eu-missions-horizon-europe/cancer_en

# Precision Medicine in Infectious Disease

13

Maria Josefina Ruiz Alvarez ⓘ, Mandana Hasanzad ⓘ,
Hamid Reza Aghaei Meybodi ⓘ,
and Negar Sarhangi ⓘ

**What Will You Learn in This Chapter?**
Personalized medicine follows the concept that people all harbor unique biological variables (genomic information) that drive their response to disease and leverages these differences for improved diagnostics and therapeutics. Besides this, the identification of pathogen-specific factors, in combination with clinical data, can further target patient management. So, addressing personalized medicine (PM) approaches on infectious diseases could optimize their diagnostic, treatment, and also disease prevention, mainly for the (rapid) identification of a disease-causing microbe and determination of its antimicrobial resistance profile, to guide appropriate antimicrobial treatment for the proper management of the patient. Furthermore, these PM tools can help to the appropriate recognition of infectious outbreaks.

This chapter focuses on providing a comprehensive understanding of the personalized medicine approaches in selected infectious diseases. The pathogens described in this chapter have been selected taking into account their role as an example of personalized infectious disease.

**Rationale and Importance**
This chapter will answer the question: Is there scope to develop a personalized approach for the diagnosis and treatment of infectious diseases?

The management of infectious diseases is consistent with the goals of the PM, following the identification of the causative organism and creating data repositories to direct specific treatment for infectious diseases. New technology has been incorporated to increase the knowledge on resistance and to protect populations. The main scope is to understand if the personalized approach can fit the infectious disease field. And following this concept, try to answer why, reasons, and determinants of the different outcomes when someone is exposed to a pathogen (Fig. 13.1). After the exposition, a person can be infected or not and

M. J. R. Alvarez (✉)
Research Coordination and Support Service, Istituto Superiore di Sanità, Rome, Italy
e-mail: mariajose.ruizalvarez@iss.it

M. Hasanzad
Medical Genomics Research Center, Tehran Medical Sciences, Islamic Azad University, Tehran, Iran

Personalized Medicine Research Center, Endocrinology and Metabolism Clinical Sciences Institute, Tehran University of Medical Sciences, Tehran, Iran

H. R. A. Meybodi
Endocrinology and Metabolism Research Center, Endocrinology and Metabolism Clinical Sciences Institute, Tehran University of Medical Sciences, Tehran, Iran

N. Sarhangi
Personalized Medicine Research Center, Endocrinology and Metabolism Clinical Sciences Institute, Tehran University of Medical Sciences, Tehran, Iran

© The Author(s), under exclusive license to Springer Nature Singapore Pte Ltd. 2022
M. Hasanzad (ed.), *Precision Medicine in Clinical Practice*,
https://doi.org/10.1007/978-981-19-5082-7_13

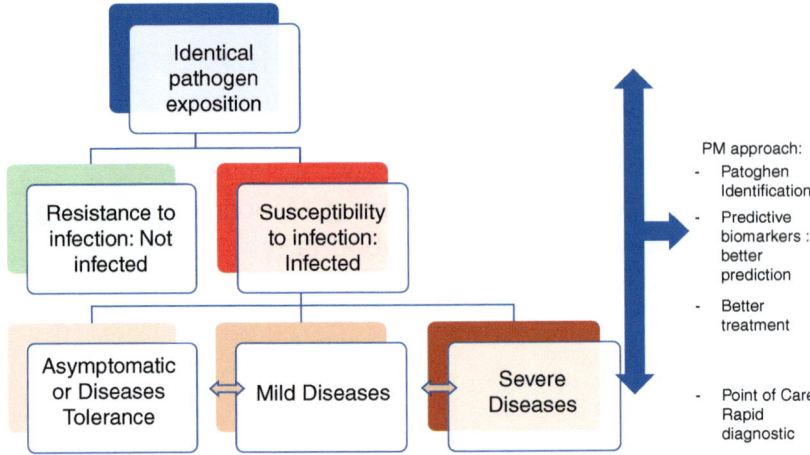

**Fig. 13.1** Pathogen exposition and infectious disease variability. After the exposition, a person can be infected or not and can develop a different level of diseases (from tolerance or asymptomatic to severe diseases) or just not be infected if someone is resistant to this pathogen. Even the immunocompromised patient population suffers from different levels of clinical infectious complications. The PM approach will facilitate pathogen identification, thanks also to the point-of-care rapid diagnostic and consequently, improve the treatment. Likewise, it facilitates the prediction through predictive biomarkers

can develop a different level of diseases (from tolerance or asymptomatic to severe diseases) or just not be infected if someone is resistant to this pathogen. Even the immunocompromised patient population suffers from different levels of clinical infectious complications.

The PM approach will facilitate pathogen identification, thanks also to the point-of-care rapid diagnostic, and consequently, improve the treatment [1]. Likewise, it facilitates the prediction through predictive biomarkers. It is evident the role of host immunity on infectious disease manifestation. So, going deeper into these studies, it will be feasible to identify persons with higher risk for infection, predict individuals who will have poorer outcomes with therapy, and determine who is destined for therapeutic failure, therefore allowing healthcare providers to adjust therapy in time, but we may also see the rise of immune therapeutics tailored specifically for infectious diseases.

If individuals can be identified as potentially more vulnerable or resistant to an infection disease, they may require different prevention strategies such as vaccination and/or prophylactic treatment.

This chapter will drive the concept of personalized infectious disease approach looking at the host face through different infectious diseases considered as a global health issue and trying to better understand the possibility of identifying determinants of clinical outcomes. These data will be used for predicting diseases, improved treatment, and also prevention strategies specific for infectious pathogens. Additionally, new technologies are supporting the rapid identification of the infective agents and targeted approaches based on the genetic resistance of pathogens to antibiotics.

## 13.1 Addressing Personalized Medicine Approaches on Infectious Diseases: Prevention, Diagnostic, and Treatment

*What can PM do for infectious diseases?* Personalized medicine (PM) applied to infectious diseases pursues the identification of pathogen biomarkers or molecular markers and/or their interaction with the human individual characteristics (immune response, infectious disease susceptibility, host-microbiota interactions, or hypersensitivity to antimicrobial drug treatment). Both lines of these study aspects have an impact on the prevention, epidemiological surveillance, diagnosis, treatment, or prognosis of the diseases [2].

Deeper studies focus on clinical application and the designs of models are requested. Further studies focus on clinical application and the design of PM models are being requested. These clinical studies should be able to facilitate the identification and genomic characterization of microorganisms that cause infectious diseases, distinguishing colonization from infection, and understanding the full spectrum of mixed infections.

As a general recommendation, the support for a personalized approach on infectious diseases' clinical studies is really needed, as well as other epidemiological or observational studies, including the integration of associated social science research. Until now, most medical treatments have been designed for the "average patient," following the "one-size-fits-all" approach. However, treatments can be very successful for some patients but not for others. The personalized approach and its research strategy want to explore the personal management of disease approach and consider individual differences in people's genes, environments, and lifestyles.

Often overlooked, advances in genomics are already changing both infectious diseases' medical practice and public health. PM doesn't want to pay off for a few people. It focuses on all the population, so PM research has to follow in parallel the research on social determinants of health and population-level interventions. To sum up, the currently personalized approach to infectious disease can be divided for easier understanding on two lines of application, the effect on patient care and the effect on population/public health. Both of them are distributed following three main pillars as the study of vaccines, for the prevention and consequent reduction of burden diseases; the pathogen identification, for rapid diagnosis and rapid outbreak detection; and, finally the antimicrobial selection, for adequate treatment and the resultant reduction of antimicrobial resistance (Fig. 13.2).

*Personalized innovation technologies:* Increasingly automated, standardized, and affordable molecular technologies are being integrated into the diagnosis, treatment, and control of infections. Indeed, the deeper knowledge of the infections and the causative pathogens are supporting the development of innovative specific diagnostics. The advances in genomics, such as whole-genome sequencing (WGS), and computing are being applied to infectious diseases. They can enhance infectious disease management in

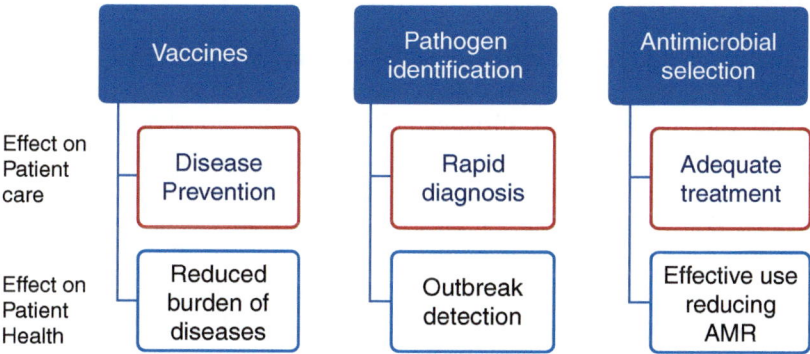

**Fig. 13.2** Personalized medicine approach on infectious diseases. To sum up, the currently personalized approach to infectious disease can be divided for easier understanding on two lines of application, the effect on patient care and the effect on population/public health. Both of them are distributed following three main pillars as the study of

vaccines, for the prevention and consequent reduction of burden diseases; the pathogen identification, for rapid diagnosis and rapid outbreak detection; and, finally the antimicrobial selection, for adequate treatment and the resultant reduction of antimicrobial resistance

both patient care and public health by delivering more timely and precise identification of pathogens. So, we can apply whole-genome sequencing in clinical microbiology for the identification of genotypes that predict phenotypes and direct therapeutic strategies and for surveillance and identification of genetic relatedness in outbreaks. Indeed, the adoption of routine use of whole-genome sequencing can be used to predict drug resistance not only in bacteria but also in viruses, fungi, and eukaryotic parasites. In the same way, new developments in molecular diagnostic tests combined with bioinformatics and epidemiology enhance public health surveillance.

Identifying biomarkers (genetics, metabolomics, proteomics, and others) related to susceptibility or resistance to infections and the study of pharmacogenomics can improve the rational drug use for infectious disease management.

Fulfilling the new technologies, collaboration and coordination among of stakeholders involved are needed, including funding and regulatory bodies, public health agencies, the diagnostics industry, healthcare systems, professional societies, and individual clinicians [3].

## 13.2 Rapid Point-of-Care Diagnostics (POC-Ds) and Their Integration to Decision-Making in General Practice

The time-consuming and complex laboratory-based conventional diagnostic tools are a challenge to medical care on infectious diseases. Currently, the development of point-of-care testing (POC-Ds) is needed for fast diagnosis of infectious diseases along with "on-site" results that are helpful in timely and early action for the best treatment. This fast diagnosis is also key in hampering the transmission of this pathogen by offering real-time testing and lab-quality microbial diagnosis within minutes.

Point-of-care diagnostic testing can be defined as "patient specimens assayed at or near the patient with the assumption that test results will be available instantly or in a very short timeframe to assist caregivers with immediate diagnosis and/ or clinical intervention" [4]. If it is effective, the result of a microbial identification test performed would be available within an hour. It required a new technological and communication platform together with economic support to be applied to the largest number of patients.

The development of rapid diagnostics suitable for each practice setting or for a particular type of infection can be quite possible and adopted easily. However, it must be affordable, sensitive, specific, user-friendly, and rapid. In this way, patients benefit from more immediate use of effective antimicrobials, and society takes advantage of less indiscriminate use of antimicrobials, a significant determinant of the emergence and spread of resistance.

The ASSURED criteria for diagnostic tests set by WHO [5] are instrumental in diagnostic testing at or near the site of patient care for an actionable decision for initiating the proper treatment for the right disease at the right time. At first, the use of rapid diagnostics that can offer real-time identification of infecting pathogens, whether it is viral or bacterial, can be a great step in initiating the appropriate therapy. Several factors need to be improved before their further adoption in daily general practice, such as user-friendliness, the affordable device cost, the appropriate system's connectivity to medical records, storage, and shelf life of reagents/cartridges, the time (not more than 1 h), and the uncertainty about data quality. In general, it is associated with a lack of specificity and sensitivity (false-positive/false-negative results) in comparison with the prevailing gold standard laboratory tests.

For this reason, the investment in research for the implementation of rapid point-of-care on microbiology is needed to decrease the duration of the diagnostic cycle in order to accelerate infectious disease management.

Real-time PCR (rt-PCR) was the first technology approved for clinical microbiology testing that offers a diagnosis in less than 1 h directly from a clinical sample [6]. Nucleic acid-based tests and molecular microbiology methods in different configurations are the basis of the POC for infectious diseases. In addition, due to the reduction in the number of technical steps required to perform it, new instruments and technologies have been applied. More recently, other technologies are being applied, such as mass spectrometry [7] or electrospray ionization mass spectrometry capable of delivering post-blood culture microbial identification in a matter of minutes [8]. An example is the detection of *Escherichia coli O104* outbreak associated with contaminated foodstuff and the second next-generation sequencing that demonstrated its usefulness during the last pandemic situation [9].

POC-Ds include also rapid microscopy or immunological diagnostic tests that can be realized outside clinical laboratories with less sensitivity and/or specificity. One of them is the serum procalcitonin level [10]. Procalcitonin measurement is able to guide clinical decisions to diagnose bacterial infections earlier and reduce unnecessary tests, procedures, and length of hospital stay. In fact, it is considered a biomarker that can be used in clinical practice as a surrogate marker for the diagnosis of bacterial infection in suspected cases of septicemia. It includes also indications of the severity of bacterial infections such as community-acquired pneumonia and sepsis. For fungal meningitis, CSF (1,3)-beta-D-glucan nonspecific marker is described as useful for their detection and should be used in conjunction with organism-specific testing [11].

Another priority is to add the susceptibility profile on the rapid test for bacterial detection with the scope of having the appropriate treatment for most of the infectious diseases. For both, to address this combined test and to reduce the assay time, innovative phenotypic assays (imaging, microfluidic culture) and the evaluated molecular methods (PCR, nanoparticle-based assays, microfluidic-based capture and enrichment, electrochemical sensors, CRISPR, sequencing, etc.) are being applied in different clinical studies. AST and rapid bacterial detection methods have been developed in whole blood using digital PCR one-step. It has been conducted as a proof-of-concept study (pCasCure) to demonstrate that CRISPR-Cas9-mediated (clustered regularly interspaced short palindromic repeats-CRISPR-associated protein-9 nuclease) resistance gene and plasmid curing can effectively re-sensibilize CR *Enterobacteriaceae* to carbapenems. It creates site-precise double-strand breaks for removing carbapenem-resistant (CR) genes and plasmids [12].

A new ongoing technology to assess antibiotic susceptibility is based on bacterial nanomotion (characteristic bacterial oscillations that cease when the organism dies). The aim is to reduce time to results without any apparent loss in quality, so far. It is being studied in subjects with sepsis, by reducing the time to tailored antibiotic treatment. These rapid AST techniques are based on the monitoring of the oscillations of bacteria upon exposure to antibiotics, due to the fact that all living organisms oscillate in the range of nanometers [13].

In conclusion, under the PM approach, these new rapid tests should be improved by taking into account the next three factors: first, the sensibility for detecting bacteria at low concentrations (<1 to 100 CFU/ml); second, the adjustment or diminution of current cost with expensive requested equipment and lengthy and, lastly, the need for high-quality diagnostic test (also considering complex samples). Finally, the adequate correlation of the genotyping of several immunogenetic targets will offer a deep understanding of human susceptibility to infection and disease severity [1].

## 13.3   Bacterial Infections

There are several examples of integrated personalized medicine strategies on bacterial infectious diseases, most of them addressed to the rapid diagnosis. It has been shown that these tests can save lives, decreased the spread of these infections, and reduced healthcare costs.

Group B streptococcal infections are an important cause of neonatal morbidity and mortality, and it makes necessary a rapid method for the detection of this organism in pregnant women at the time of delivery to allow early treatment of neonates. One example of this method is the prevention of neonatal infections with the rapid detection of *Streptococcus agalactiae* in parturient women. Furthermore, the positive test indicates the appropriate antibiotic regimen before baby delivery [14].

Other several PCR assays for the detection of *group B streptococci* have been shown to be specific and sensitive, from the specific amplification of the "cfb" gene to the novel molecular detection approach, multiple cross displacement amplification (MCDA) coupled with polymer nanoparticle-based lateral flow biosensor (MCDA-LFB), that offer an accurate result on around 50 min, including sample collection, template, preparation, multiple cross displacement amplification reactions, and result reporting [15]. Similar technologies are followed by several other groups, for the rapid diagnostics of hospital-acquired infections associated with *methicillin-resistant Staphylococcus aureus* (MRSA), *vancomycin-resistant enterococci* (VRE), and *Clostridium difficile*, among others. The multiplex PCR assay for the detection of MRSA colonies and MRSA-positive blood culture bottles is a rapid, sensitive, and specific method that requires a total time of 3 h only [16].

In the case of difficult-to-grow organisms such as *Bordetella pertussis*, *Bordetella parapertussis*, and *Helicobacter pylori*, these rapid diagnostics are a valid alternative, to PCR or other molecular methods, for their detection and identification. There are several commercialized PCR-based kits for their detection in nasopharyngeal swab specimens [17]. Recently, a commercialized real-time PCR-based assay to detect the *H. pylori* has been published after evaluating their performance "glmM "gene and mutations in the 23S rRNA genes conferring clarithromycin resistance [18].

The catheter is one of the causes of nosocomial bloodstream infections, being crucial not only for the early detection with the opportunity of diagnosis without requiring the catheter removal. One tentative has been done to the culture of catheter blood which can offer the right diagnosis more than 2 h before that peripheral blood cultures.

### 13.3.1 Climate Change-Related Infectious Disease

The distribution and severity of certain infectious diseases are being affected by the climate crisis, as there are links between climatic conditions and infectious diseases. To fight against this situation at a personalized public health level, measures responding to the climate crisis have additional benefits, and the strategy needs to be adopted. These include public health training, more personalized models of surveillance, and rapid emergency response systems with sustainable prevention and control programs, with the citizen's collaboration.[1]

### 13.3.2 Tuberculosis Diseases (TB)

It is understood that the gene-gene interactions in an individual's susceptibility to a complex disease seem more explicative than single polymorphisms would on their own. In addition, it has been hypothesized that the interaction between the genotypes of the human host, and the bacterial strain, could determine both the progression to disease and the type of disease and host genotype conferring susceptibility.

TB can manifest under several forms linked to the host susceptibility, such as latent infection, drug-susceptible and drug-resistant TB disease, childhood TB, extrapulmonary TB, subclinical TB, TB comorbidity with other communicable diseases and noncommunicable diseases, and TB-related long-term pulmonary functional disability. About one-third of the world's population is estimated to have been exposed to TB bacteria and potentially infected.[2] Only a small proportion of infected will become sick with TB.

This complex disease needs to implement novel approaches for the following [19]:

1. Rapid diagnosis of active TB: Cheap, rapid, and accurate point-of-care diagnostic tests able to characterize drug resistance are urgently needed.

---

[1]Meiro-Lorenzo M, et al. Climate change and health approach and action plan. Investing in climate change and health series. World Bank Group, 2017. http://documents.worldbank.org/curated/en/421451495428198858/Climate-change-and-healthapproach-and-action-plan

[2]WHO. 2013a. Global TB report https://apps.who.int/iris/handle/10665/91355

2. Improve treatment protocols with the diminution of therapy duration and better strategies or standard treatments for both drug-sensitive and drug-resistant TB: Host-directed therapies are better tolerated, can shorten the duration of therapy, and improve treatment outcomes with a reduction of compliances [20].

3. At the level of public health, to prevent the relapse, decrease the drug resistance, prevent long-term lung damage, and prevent latent TB infection progressing to active TB, potential drug interactions with HIV treatment should be examined, given high levels of HIV-TB coinfection.

4. Vaccine development [21] adjunct host-directed therapies based on repurposed drugs, cellular therapies, and other immunomodulators.

The adaptation of six lineages of *Mycobacterium tuberculosis* to specific populations has been described [22]. It showed that strains from a "defined sub-lineage might be selected by a human population" in a defined geographical setting. The broader variation of HLA allele frequency between human populations supported this theory [23] and that the HLA genotype has been associated with susceptibility to *M. tuberculosis*. One example of this association was found between the C allele of the TLR2 597 T/C SNP of the host and the bacterial genotype in *M. tuberculosis* disease.

Regarding the vulnerability, people with weakened immune systems caused by the prolonged use of medicines such as steroids or TNF-α inhibitors, HIV-infected patients, and patients with chronic diseases such as diabetes, renal insufficiency, and pulmonary diseases have a high risk of developing serious TB diseases. However, it has been described the variable outcome of disease by the inadvertent immunization of 251children with the same dose of a virulent strain of *M. tuberculosis* in Lübeck, Germany, in 1926. Of these children, 77 died, 127 had radiological signs of disease, and 47 showed no evidence of tuberculosis [23]. Other several studies have evidenced those host genes are involved in tuberculosis susceptibility, for example, mutations in genes related to immunity against intracellular pathogens in the interleukin IL-12/IL-23/ interferon (IFN) axis [24].

Nowadays, genome-wide approaches, which have no prior hypothesis, have been studied and have already identified several host regions containing potential susceptibility genes. This approach includes linkage studies to trace chromosomal regions containing putative susceptibility genes. These outcomes support the idea of host susceptibility due to different ethnic populations living in diverse environments. In this direction, population-based studies are focused on different potential susceptibility genes. As introduced before, several genes have been associated with TB [25], such as human leukocyte antigen (HLA), mainly class II region, NRAMP1 (divalent transporter localized to the late endosomal membrane that regulates cytoplasmic cation levels by specifically regulating the iron metabolism in the macrophages), IFNG (IFN-g receptor), NOS2A nitric oxide synthase 2A gene, SP110 (SP110 nuclear body protein gene), CCL2 (C-C chemokine ligand-2 gene), MBL (gene encoding the collectin mannose-binding lectin), CD209 (dendritic cell-specific intracellular adhesion molecule (ICAM)-grabbing nonintegrin), VDR (vitamin D receptor gene), and TLR (family of mammalian Toll-like receptors). Studies of genome-wide association, epigenetics, copy number, and rare mutations have also been involved.

For the prevention of the identification of TB morbidity and mortality, the study of biomarkers related to the development of TB can be more effective using multiomic approaches. These omics [26] correlate metabolites, miRNA (micro RNA mapping), and cytokines/chemokines.

For surveillance and identification of genetic relatedness in outbreaks, a combination of PM tools with previous epidemiologic methods may rise to disease mapping and lead to health policy decisions. The study [27] of extensively drug-resistant tuberculosis (XDRTB), analyzing targeted and whole-genome sequencing to account for the geographic distribution of XDR-TB strains and the strains sequenced, was divided into those attributed to acquired resistance (due to poor adherence, subtherapeutic drug levels, or inappropriate treatment) or those attributed to transmitted resistance. A higher prevalence of transmitted resistance was correlated with a high frequent population of entertainment community areas.

For the rapid *Mycobacterium tuberculosis* identification and primary assessment of the drug multiresistance profile, there are tests based on the N-acetyltransferase 2 genotype of the patient. It is used to determine her/his pharmacogenetic profile, to guide the isoniazid dosage, and to limit drug hepatotoxicity [28].

TB latent infection: People with latent TB infection do not have symptoms, and they cannot spread TB bacteria to others. However, the myco-bacteria can replicate and induce inflammation, being sick at every time. For this reason, diagno-sis and treatment are really needed.[3]

Comorbidities: HIV infection is associated with a TB atypical clinical progression [29]. The probability of developing TB among people liv-ing with HIV is 16–27 times higher than those who are HIV-negative [30]. Based on multiomics approach, there are several ongoing studies to identify biomarkers with high accuracy for pre-dicting the evolution of TB and HIV. There have been [31] six biomarker signatures identified that can be used to identify those at highest risk of TB after ART initiation and who may benefit from additional monitoring and intensified or immune-modulatory treatment [32].

Novel studies are working on the more accu-rate relationship [33] as the involvement of interleukin (IL)-17A-mediated inflammatory responses in HIV-tuberculosis coinfection [34].

## 13.4   Viral Diseases: Zika and HIV Diseases

CD4 and CD8 T-cell's right function is a crucial part of the host's capacity to defend itself not only against cancer diseases but also against viral infections. Most of the studies on these heterog-enous populations and also the association with the CD4/CD8 ratio provide a valuable tool for highly personalized therapy[4].

### 13.4.1   Zika Diseases

The ZIKV is an arbovirus and belongs to a com-mon family *Flaviviridae*, which emerged in Brazil, and spread rapidly in the whole of America due to human activity and travel. ZIKV is mainly propagated through arthropods like mosquitoes, most commonly from the *Aedes* spe-cies [35], but can also be transmitted by blood transfusion and sexual intercourse. Several com-plex host factors that interact with viral proteins have been identified [36]. Nevertheless, the mechanisms of the interplay between virus and host factors during the diverse evolution of the virus infection need further investigation [37].

Currently, there is no specific antiviral agent or vaccine against *Zika*,[5] and following the first isolate in 1947, this virus has developed epidemic capacity due to a change in phenotype.

Because Zika virus spread is continually evolving and adapting, more deep research on virus-host interactions is needed. Host determi-nants can facilitate antiviral drug development and the understanding of viral maintenance and emergence [38]. Up to 90% of ZIKV-infected individuals develop clinical symptoms charac-terized by headache, high fever, maculopapular rash, myalgia, arthralgia, and severe asthenia that might last for months or years. However, about 20–25% manifest a wide range of symp-toms (headaches, fever, maculopapular rashes, arthralgia, conjunctivitis, and swelling at the extremities). Additionally, ZIKV has been the

---

[3]https://www.cdc.gov/tb/topic/treatment/decideltbi.htm

[4]https://www.immunopaedia.org.za/online-courses/previous-iuis-courses/immunocolombia/overview-of-t-cell-subsets/section-2-cytokines-determine-subsets-of-cd4-and-cd8-t-cells/

[5]WHO Director-General Summarizes the Outcome of the Emergency Committee Regarding Clusters of Microcephaly and Guillain-Barré Syndrome. 2016. Available online: https://www.who.int/news/item/01-02-2016-who-director-general-summarizes-the-outcome-of-the-emergency-committee-regarding-clusters-ofmicrocephaly-and-guillain-barr%C3%A9-syndrome

first of its kind that shows neurological complications such as autoimmune ascending paralysis or Guillain-Barré syndrome and neurological birth defects such as microcephaly due to the infection of the placenta after sexual intercourse.

Currently, studies have indicated that ZIKV induces robust T-cell activation, suggesting a significant role for CD4 and CD8 T cells in the immune response to ZIKV. Briefly, T cell responses have pathogenic consequences for the host. However, there is described as a protective role linked to CD4 T cell response against non-structural proteins; In the same way, the CD8 T cell response is driven against structural proteins and shows a cross-reactivity -protective immunity. In this line, cross-reactive epitopes need to be studied for their application in vaccination and targeted therapy. CD8 T cells seem also to have a role in enhancing ZIKV pathogenesis in the mice model [39]. Another question often addressed by human studies relates to how T-cell responses to ZIKV change over time.

Research on the virus protein interaction with host proteins can help in the diagnosis and the design of specific drugs to break these interactions and reduce the health damage to the host.

For example, viral infection activates the type I interferon (IFN-I) signaling leading to STAT1/2 (signal transducer and activator of transcription1/2) activation. STAT1 was under strong purifying selection when populations shifted from hunting and gathering to farming because this went along with a change in the pathogen spectrum.

The occurrence of mutations in the STAT1 molecule can modify the function, in positive or in negative, causing different phenotypes and symptoms [40]. It has been demonstrated that ZIKV5 protein acts as an antagonist of the IFN-I pathway by stimulating STAT2 (but not STAT1) degradation. This STAT1 activity is modulated by PIAS1, interacting indirectly as the partner of NS5.

With the CRISPR/Cas9 methodology, PIAS1-depleted cells seem more sensitive to ZIKV infection-dependent lethality. So, STAT1 interaction with PIAS1 might play an important role in ZIKV biology by modulating NS5 protein levels [41].

Following the CDC guidelines,[6] the diagnostic testing for Zika virus infection can be accomplished using both molecular and serologic methods. However, serological testing is not recommended for diagnosis since antibodies against Zika persist for years and cross-react with other similar viruses, including dengue.

## 13.4.2   Human Immunodeficiency Virus (HIV)

The human immunodeficiency virus (HIV) is grouped to genus *Lentivirus* within family *Retroviridae* and subfamily *Orthoretrovirinae*. HIV is classified into types 1 and 2 (HIV-1, HIV-2) based on genetic characteristics and differences in the viral antigens. Epidemiologic and phylogenetic analyses currently available imply that HIV was introduced into the human population around 1920–1940 [42]. The clinical manifestation of HIV follows an acute viral syndrome, with subsequent resolution of symptoms and an 8–10-year asymptomatic period until the development of AIDS, defined in adults by a decline in CD4+ T-cell number to less than 200 cells/mm³. Among these long-term survivors (LTS) or long-term non-progressors (clinically asymptomatic for ≥10 years; no antiviral therapy), there is a subgroup of "elite controllers": people, for at least 2 and some for over 10 years, have not shown replication of the virus in the plasma, although they remain infected.

The research of the characterization of infected individuals that spontaneously control infection to very low viral loads such as elite controllers and the long-term progressors has characterized the investigation for a long time and also for the vaccines' development.

The personalized medicine approach and the HIV diagnosis and treatment have been developing in parallel, with the nucleotide sequencing tests used for the phylogenesis studies and the resistance detection. These tests have supported the antiviral treatment according to the genotype

---

[6]https://www.cdc.gov/zika/laboratories/types-of-tests.html

of the virus circulating at the time of testing, upon interrogation of a database of known antiviral drug resistance mutations [43].

There is considerable scope for novel approaches and the harnessing of innovative technologies to improve engagement with populations.

Nowadays, 95% of people diagnosed with HIV infection receive antiretroviral therapy, and 95% of people receiving antiretroviral therapy have effective viral suppression. Current demographic trends foresee increasing numbers of young people at risk of HIV exposure and increasing numbers of adults living with chronic HIV disease at risk of comorbidities as they age. To enhance coverage and effective viral suppression, special attention will need to be paid to vulnerable populations and the barriers to access to care.

Long-term HIV management requires an increasing focus on coinfections and comorbidities, associated with polypharmacy, with the risk for drug-drug interactions [44]. Furthermore, challenges with antiretroviral drug resistance are ongoing. Lastly, there is an imperative need to assess novel delivery mechanisms for innovative biomedical methods of prevention, including antiretroviral-based interventions, broadly neutralizing antibodies, and the continuous effort to HIV vaccines. Effective implementation of such innovations will require innovative people-centered and community-oriented approaches.

While studies of biological susceptibility remain important, understanding human behavior, including uptake and adherence of novel HIV prevention methods, is paramount[7] for stopping the dissemination of HIV.

Some natural intracellular resistance to HIV replication, due to host cell proteins, has been described and is being studied deeply. The most recent are the APOBEC3G and F, Trim5a (Lv-1/Ref-1), Lv-2, Tetherin, and Murr-1, with different types of protection [45].

Several factors influence the host response to HIV (virus' subtype infecting with different levels of virulence, intracellular resistance factors or host proteins, and C-C chemokine receptor type 5 (CCR5)). In addition, the human leukocyte antigen (HLA) type of the individual can correlate with the extent of HIV disease progression. HLA is setting up the protection and progression, and the HIV-1 adaptation to HLA is part of the immune response, both for protection and for disease progression. The virus is continually adapting to the immune response pressures through HLA-associated immune escape mutations [46]. Using univariable and multivariable analyses, HLA associations have been studied among different populations, describing protective and risk HLA alleles in the population in study [47].

Abacavir hypersensitivity: A potentially life-threatening hypersensitive reaction occurs in association with the initiation of HIV nucleoside analog abacavir therapy in 4–8% of patients. Preliminary studies appear to confirm the role of the immune system in abacavir hypersensitivity. Hypersensitivity reaction to abacavir is strongly associated with the presence of the HLA-B*5701 allele and is related to the result of the presentation of drug peptides into HLA, which may induce a pathogenic T-cell response. Prospective HLA-B*5701 genetic screening has now been instituted in clinical practice to reduce the risk of a hypersensitivity reaction [48].

Due to the effective antiretroviral therapy stopping the viral replication, HIV can be considered as a "chronic infection disease," so some studies are focusing on the characterization of the principal aspect of the chronic inflammation in persons living with HIV to predict adverse and unfavorable treatment outcomes or to identify clinical biomarkers to target and to improve the quality of life.

An example of the real implementation of PM in the health system is Botswana: a recent change in the HIV management policy in Botswana where the country opted out of efavirenz-based therapies as first-line anti-retroviral therapy (ART), in favor of dolutegravir (in 2016). Genomics studies had showed that about 13.5% of the Botswana

---

[7] Joint United Nations Programme on HIV/AIDS (UNAIDS). (2014). Fast-Track: ending the AIDS epidemic by 2030. https://www.unaids.org/sites/default/files/media_asset/JC2686_WAD2014report_en.pdf

population are unable to effectively metabolize efavirenz-based therapies [47].

HIV vaccination: So far, all trials (for preventive and for the therapeutic vaccine) focusing on combined and diverse viral proteins have been unsuccessful. It seems that new technologies bring some hope to develop an effective and preventive vaccine.

## 13.5 Innovation Technologies: CRISPR/Cas9 and by CART (Chimeric Antigen Receptor T Cell)

CRISPR/Cas9 is an "adaptive immune system where bacteria and archaea have evolved to resist the invading viruses and plasmid DNA by creating site-specific double-strand breaks in DNA." Innovative studies are applying this gene-editing system in inhibiting human immunodeficiency virus type 1 (HIV-1) infection by targeting the viral long terminal repeat and the gene coding sequences [48]. Based on it, Excision BioTherapeutics has launched the first clinical trial to see if a one-time IV infusion could effectively cure HIV[8].

Immune modulation by CART (chimeric antigen receptor T cell). This technology is based on the engineering hematopoietic and T stem cells that attack and kill cells infected with HIV. Providing a self-renewing population of both CD8+ and CD4+ HIV targeted T cells resistant to direct HIV infection [49].

## 13.5.1 Neglected Infectious Diseases (NIDs)

Neglected ("forgotten") diseases are a set of infectious diseases, many of them parasitic, that primarily affect the most vulnerable populations: "the poorest of the poor, the most marginalized, and those with the least access to health services."[9]

The personalized approach can help this population with a more integrated and multi-disease approach to reduce the negative impacts that these diseases have on the health, social, and economic well-being. Many of these diseases are avoidable or treatable. However, more precise tools for diagnosis, better treatment regimens or combination therapies, novel drugs, and enhanced awareness are needed to make progress in the control and elimination.

One priority is their coinfection with malaria, TB, or HIV in the context of noncommunicable diseases. The development of drugs, diagnostics, and vaccines will be a priority, along with the improved understanding of the consequences of coinfection and comorbidity.

*Protozoan infections*: Effective and safe chemotherapies and vaccines against protozoan infections are generally absent. Currently, drug resistance and drug toxicity are increasing, such as the antimonial resistance, with the direct inter-human transmission.

New AI diagnostics, biochemistry methodologies, and parasite genomics can support the understanding of parasite biology and the existence of infection without clinical symptomatology. A few clinical studies are currently ongoing to evaluate the use of mAbs for the treatment or prophylaxis for *leishmaniasis*, *Chagas disease*, *malaria*, and *toxoplasmosis* [50].

*Helminth infections*: There is still no effective human vaccine against helminths, and with numerous infections worldwide, many publications have revealed increasing anthelmintic resistance (benzimidazoles, imidazothiazoles (levamisole), and macrocyclic lactones (avermectins and milbemycins)) families. Some studies focused on the stimulation of dendritic cells to elicit a type 2 or regulatory immune responses on the host immune system may contribute to controlling gastrointestinal helminth infections, for example, with the administration of probiotics [51].

Unfortunately, there are not so many clinical studies evaluating the use of mAbs for the

---

[8] https://www.globenewswire.com/news-release/2021/09/15/2297456/0/en/Excision-Receives-FDA-Clearance-of-IND-for-Phase-1-2-Trial-of-EBT-101-CRISPR-Based-Therapeutic-for-Treatment-of-HIV.html

[9] https://www.who.int/neglected_diseases/zoonoses/en/

treatment or prophylaxis for parasitic infections. One of the reasons is the failure of many candidates at the preclinical level. Other factors are the complexity of the mechanisms of host-pathogen interaction, some of them still pending being fully deciphered and the indirect effect that mAb might have on host susceptibility to infections or on the reactivation of latent ones. In this way, consulting the clinical trial website, one can see that few studies are withdrawn, in the recruitment of phase 1 or completed but with results not published yet.

## 13.6   Malaria

Malaria is an acute febrile illness caused by *Plasmodium* parasites, transmitted by infected female *Anopheles* mosquitoes. Five parasite species cause malaria in humans. The deadliest malaria parasite is. *P. falciparum*, most prevalent on the African continent. *P. vivax* is the dominant one outside of sub-Saharan Africa.[10]

Priority populations for malaria interventions include children, adolescents, and pregnant women living in high-transmission regions, people with genetic hemoglobinopathies, immuno-compromised individuals, migrants, and mobile populations. However, there is a serious gap in full access to preventive interventions, diagnostic testing, and treatment [52]. Needed diagnostic tools to differentiate *Plasmodium falciparum* and *Plasmodium vivax* and sensitive methods for the rapid diagnosis of asymptomatic malaria infections are urgently required. Also, field testing of diagnostics to identify low-level infections and resistance mutations is crucial [53, 54].

Most of the mutations of the human malaria parasite *Plasmodium falciparum* are single nucleotide polymorphism (SNP), a high frequency of this SNP origin resistance to the chemotherapeutic agents, vaccines, and vector control strategies. Several markers have been identified in the genome, for example, pfcrt, pfmdr1, pfdhps, pfdhfr, pfkelch13, pfatpase6, and pfmrp1, which open the field to new diagnosis technologies [55].

As one of the portable sequencers, MinION nanopore sequencing for genotyping the malaria parasite *Plasmodium falciparum* of nine representative genes causing resistance to anti-malaria drugs is diagnosed. This approach will change the standard methodology for the sequencing diagnosis of malaria parasites, especially in developing countries [56].

Vaccine: The development and evaluation of new and improved malaria vaccine candidates.

Will be key to both improved malaria control and malaria elimination, with the most promising vaccine candidates including sporozoite, blood stage, and transmission-blocking vaccines [57].

## 13.7   Antimicrobial Resistance Diseases

The rise in antibiotic resistance is clearly one of the greatest challenges to public health. Starting from the simple Darwinian evolution of microbes, our society has incorporated the misuse of antibiotics in clinical and agricultural settings and the lack of interest of pharmaceutical companies in antibiotic development. In this way, ethical dilemmas such as balancing restrictions on individual liberty for the protection of public health and the well-being of future generations need to be acknowledged and confronted (see ELSA chapter in this book).

The novel resistant strains can be considered new pathogens since their diagnostic and treatment need to be identified.

Nowadays, antimicrobial resistance (AMR) is compromising the use of not only multiple antibiotics but also antimalarial, antiviral, and antifungal therapeutics with global impact and different landscapes on the different regions worldwide [58] where it is influenced by specific environmental and epidemiological factors as poverty-related and neglected infectious diseases.

The personalized key is the development of point-of-care diagnostics, to determine susceptibility/resistance to antibiotics and to distinguish

---

[10]World Health Organization. (2018). World malaria report 2018.
https://www.who.int/malaria/publications/world_malaria_report/en/

bacterial and viral infections so that unnecessary use of antibiotics can be avoided. Antibiotic stewardship and promoting the use of new e-health technologies will help reduce antibiotic consumption. The development of new vaccines and optimized host immune interventions will reduce the burden of diseases treated with antibiotics but may also drive changes in disease etiology [59].

Phenotypic methods supported the gold standard or accurate determination of antimicrobial susceptibility patterns, while the genotypic determination of the AMR potential of Gram-positive pathogens (e.g., methicillin or vancomycin) may offer more opportunities for yielding fast results. Nevertheless, the strategy applicable to Gram-negative bacteria is complicated by the broad variety of resistance mechanisms and the genetic drift of resistance gene alleles that limit the spectrum of antibiotic options.

Lastly, the need for rapid antimicrobial susceptibility testing (RAST) in bloodstream infections is important for adjustment of therapy, and many attempts have been made to shorten the time required for reporting antimicrobial susceptibility testing results. EUCAST has developed and recently validated a disc diffusion RAST method directly from positive blood cultures delivering reliable AST results within 4–8 h of positivity of blood culture bottles [60].

In parallel, the determination of the patient's pharmacogenetic profile will provide an additional assessment of the drug metabolizer phenotype and/or of the risk of potential adverse drug interactions, while the determination of the immunogenetic profile of the patient could be used to evaluate her/his susceptibility to infection.

Progress in whole-genome sequencing (WGS) technologies and appropriated software to interpret the results can simultaneously provide fast pathogen identification, epidemiological typing, and detection of drug susceptibility genes [61]; third-generation systems can provide fairly long reads at high speed.

The European Committee on Antimicrobial Susceptibility Testing reviewed in 2017 the development status of WGS for AST [62] and suggested the need for implantation of the database of all known resistance genes/mutations improving the interoperability between different systems and bioinformatics tools. In this way, GWAS in bacteria is being driven forward by advances in genome sequencing, associated with high-quality metadata, being a reliable method to identify loci related to a phenotype and to understand the expression of clinically important bacterial traits.

Whole-genome sequencing technologies have improved the understanding of resistance and allowed rapid identification of resistance mechanisms in multiple organisms. Innovative tools applying artificial intelligence (machine learning application for risk definition, decision support systems for personalized therapies) need to be explored and connected with rapid diagnostics.

While recognizing that presence of a resistance gene does not always equate with clinically relevant drug resistance, the introduction of resistance mutation databases would aid patient management, enabling personalized treatment. There is an urgent need for personalized management through drug resistance screening in certain patient groups. Clinical trials addressing the implementation of personalized therapies according to geographical and care appropriate settings, local epidemiology, age, gender, metabolism, and clinical characteristics are needed.

Furthermore, it is essential to address social and behavioral factors that contribute to AMR. Systems biology can shed light on mechanisms of action, identify targets for rational clinically useful combination chemotherapy to suppress or minimize resistance, drive drug repurposing, and provide a framework for the discovery and development of novel antimicrobial interventions and therapeutics. Findings should be translated into implementation strategies for use in clinical medicine. Therapeutic use should be linked to PM, where accurate and rapid diagnostic tests directly connect with effective therapy.

## 13.8 Personalized Epidemiology

The general response to infectious diseases is being remodeled by the personalized approach, from the genome-based approaches for diagnosis and individual-level treatment to trace and detect pathogen transmission, resulting in potential enhancements in the implementation of population-level public health interventions (Fig. 13.3). The control of infectious disease relied on incidence data and interview-based contact tracing to estimate key epidemic parameters and to reconstruct transmission chains with the difficulty to do an exhaustive case reporting due to the complex work of contact tracing due to the use of clinical symptoms to identify cases [63]. Technological advances in communication methods and their easier use have also impacted our ability to respond to infectious diseases.

Infectious Outbreaks

It is vital to strengthen the capacity for preparedness to address emerging/re-emerging infectious diseases and to undertake the rapid evaluation of interventions in clinical trials when outbreaks occur, and treatments are inadequate or lacking entirely [64]. Throughout history, outbreaks caused major loss of life, severe social upheaval, and great financial losses, which drove medical research to the development of vaccines against the serious infections which have caused major outbreaks in the past. There are three different related definitions:

1. An outbreak is a simple detection of more cases of a disease than expected, in a specific place or population, over a specific time period. It also includes a single case it is unexpected.
2. Epidemic means larger outbreaks, involving multiple cases.

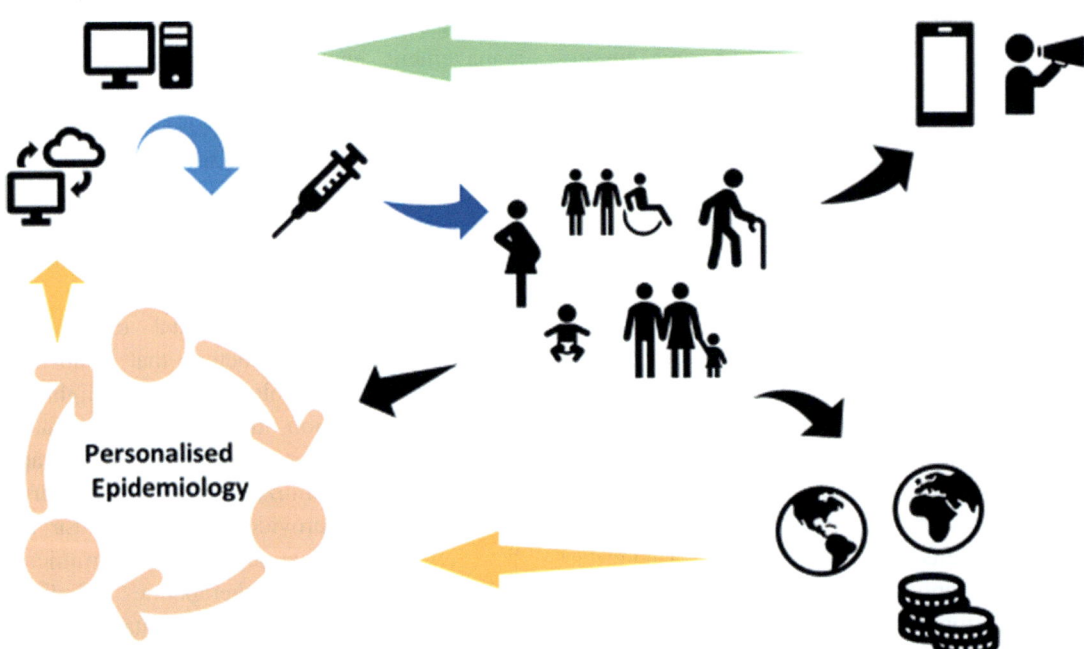

**Fig. 13.3** Personalized epidemiology: new sequencing and data analysis facilitate the broad application of whole-genome to pathogens for quicker reaction to outbreaks of infectious disease. Data has been achieved without delay from clinical samples and in near real time during an outbreak. With the combination of these genomes and sequences generated from the same outbreak with their metadata, as well as previously characterized variants, researchers can inform individual- and population-level intervention strategies to minimize the burden of infectious diseases. The communication technologies are helping communities to disseminate and advise the population about the outbreak and quick prevention measure that can be individualized following the clinical history of patients

3. A pandemic is a large epidemic, where it has spread into several countries or continents.

Outbreaks of infectious disease are still occurring notwithstanding all the medical and technological progress, and the current pandemic has taught us how infection control and healthcare equity intersect and how complex it is to apply healthcare equity under infectious outbreaks. For an effective epidemic/outbreak preparedness and response, multidisciplinary teams (experts in the investigation, prevention, and control of infectious diseases) must work in collaboration and coordination to avoid a socioeconomic and health impact.

The International Health Regulations (IHR) 2005 provide an overarching legal framework for countries' rights and obligations to deal with public health events and cross-border emergencies. This regulation provides health security to populations and is a legally binding global health security framework agreed.

The WHO has defined infection prevention and control (IPC) as a practical, evidence-based approach that prevents patients and health workers from being harmed by avoidable infection and as a result of antimicrobial resistance. A list of tools and resources have been included in a toolkit following five categories: guidance materials, implementation manuals and resources, communications and advocacy tools, measurement tools, and training and education resources.

On the WHO Disease Outbreak News (DONs) web page, confirmed acute public health events or potential events of concern are listed showing us the necessity to invest in new tools for the right diagnostic and surveillance. For example, during the last period of 2021, there were reported different outbreaks as already known diseases: yellow fever in West and Central Africa and Ghana; *hepatitis E virus* in the Republic of South Sudan, *Ebola virus* disease in the Democratic Republic of the Congo, cholera in Cameroon, dengue fever in Pakistan; the Middle East respiratory syndrome coronavirus (MERS-CoV) in the United Arab Emirates, circulating vaccine-derived poliovirus type 2 (cVDPV2) in Yemen and Ukraine; monkeypox in the USA, and Zika virus disease in India.

Artificial intelligence and big data approaches have become necessary to join local, national, and international public health efforts to reveal, report, and control emerging outbreaks. One example is the adoption of big data for the initial identification of emerging infections through the development of the Global Public Health Intelligence Network (GPHIN), a cooperative effort between (at the time) Health Canada and the World Health Organization (WHO) [63].

Briefly, it is based on an automated web-based system that scans newspapers and other communications worldwide such as social media tools for potential indicators of outbreaks. All the information is analyzed and rapidly assessed by a multilingual, multidisciplinary team at the agency. The system identifies a risk and analysts check and alert for further decision-making.

The use of smartphones and health apps (application programs that offer health-related services for smartphones and tablet PC) is also acquiring an important position in this field, such as the detection of one Ebola outbreak in Nigeria earlier than the WHO announcement [65].

Other smartphone applications such as Flu Near You [66] and DoctorMe [67] encompass crowdsourcing systems that capture voluntarily submitted symptoms. These data are rapidly aggregated and offer feedback close to real time.

Research needs to work in parallel with the surveillance system, so the R&D Blueprint framework (WHO) works to facilitate a rapid research response focused on therapeutics, vaccines, and diagnostics for a list of priority diseases.

Moreover, it is important to consider in this emergence situation the role of ethics in research. A guideline can be consulted on the WHO Guidance for Managing Ethical Issues in an Infectious Disease Outbreak. The key point is to follow decisions supported by evidence, share data rapidly, and start the research under the approval of ethics committee and national authorities, with the appropriate informed consent.

## 13.9 Vaccination Strategy

New genomic technologies are improving the identification of pathogen strategies on immune response evasion and mechanisms of virulence. These can be applied not only for the development of immunotherapy as the monoclonal add vaccines.

In this section, "personalizing" vaccination recommendations is described, without letting down the universal vaccination recommendations. The objective is to integrate vaccination as part of health interventions targeting an individual's needs, to improve coverage, and to reduce the total burden of disease in addition to current public health interventions.

Universal vaccination, implemented at a population level, is recommended and administered, with very similar products to all individuals, regardless of their individual risk. However, the personalized vaccination strategy adds the possibility to make the risks and benefits personal, without forgetting the unavoidable effects of being a range of age and residing in a particular geographical situation.

For expanding the adult immunization programs, broad risk groups should be identified by age, preexisting comorbidity, immune deficiency, pregnancy, etc. [68] Personalized tools will offer the capability to better interpret those individuals at risk for certain conditions (including NCDs), who will derive the advantage from vaccination and, also essential for the public trust, who are most at risk for adverse events from vaccination [69].

In this way, health record analyses and machine learning applications will allow to identify high-risk patients and target them for increasing the vaccination coverage, including those who the "universal concept" didn't identify as at risk. A second step can be the use of digital outreach with the scope to improve vaccine acceptance.

Among the policy recommendations, the WHO enclosed the access to vaccines as part of the "older person-centered and integrated care" due to the observation that an individual's functional capacity will represent a more precise indicator of the need for vaccination than their chronological age [70].

One example is the role of vaccination as preventative activity against cardiovascular disease, from childhood and continuing throughout life, taking into account factors such as familial risk, personal exposure history, comorbidities, and obesity. It is clinically demonstrated the need to prevent influenza and herpes zoster in patients with serious cardiovascular disease.

Independently of the healthcare system resources, the first step in this direction is the instruction on the vaccination recommendations to increase the vaccine uptake, by identifying and personalizing the risk of disease and the benefits of vaccination for citizens. The public need to be involved in the discussion (e.g., by including a more detailed analysis of lifestyle and familial factors) to be motivated to comply.

Several initiatives are ongoing on how to deal with responses linked to genome sequencing and analysis capabilities and training the different actors involved. For example, the Association of Public Health Laboratories (APHL)-CDC bioinformatics fellowship program and the H3Africa initiative [71], which is backed by the US National Institutes of Health and the UK Wellcome Trust.

## 13.10 Microbiome

With the explosion of genetic tools and omics, over the years, the human microbiome has been deeply studied. It is the ecosystem of microorganisms that live in the human body and it is being considered an emerging health determinant. With a higher population in the gut, the ratio of bacterial and human host genesis is 200:1. The human microbiota starts their formation in the placenta, conditioned by the delivery mode. During the postnatal period, it changes following the environment changes and food transition completing the development during the first 3 years of life. Accordingly, the gut microbiota can affect the host physiology due to the effect on the host phenotype, having a potential role in human health and in some diseases, such as cardiometabolic disorders, inflammatory bowel diseases, neuropsychiatric diseases, and cancer [72]. For example, chronic inflammation in the elderly is associated with an alteration of microbiota composition or dysbiosis.

With the omics technologies, the composition of the microbiota is better understood, and the key metabolites produced are being identified with even the isolation of novel bacteria.

However, to clarify better their therapeutic or prevention applicability, several local factors that influence the composition of the microbiota should be contemplated, such as alimentation, drug treatments, intestinal motility, and stool regularity and consistency.

Wrapping up, the personalized vision is already being practiced in infectious diseases, and the use of different genomic tools has been enhanced in patient care. However, the integration of genomics into the infectious clinic needs to be standardized and streamlined to reduce the cost.

## 13.11 Coronaviruses

### 13.11.1 Introduction

*Coronaviruses* (CoVs) are a group of enveloped single-stranded RNA viruses, belonging to order *Nidovirales*, family *Coronaviridae*, and subfamily *Orthocoronavirinae* [69, 70]. The subfamily of coronavirus is divided into four genera, α, β, π, and γ, according to the genome of the serotype. After the appearance of SARS-coronavirus (SARS-CoV) in Southern China, scientific interest in coronaviruses increased exponentially [71, 72].

Coronaviruses are primarily responsible for enzootic infections in birds and mammals and have shown that they can also infect humans in recent decades [73]. A novel coronavirus (CoV) has been called "2019 novel coronavirus" or "2019-nCoV" by the World Health Organization (WHO) which is responsible for recent pneumonia in Wuhan City, Hubei Province, China, starting in early December 2019, a disease now officially called "the coronavirus disease 2019 (COVID-19)" [74–76]. The International

Committee on Taxonomy of Viruses (ICTV) has designated the virus name as SARS-CoV-2 because of its similarity of its symptoms to those caused by severe acute respiratory syndrome [77]. SARS-CoV-2 belongs to genus β and subgenus sarbe [72, 77].

In recent years, human coronaviruses have emerged rapidly due to mutation, high nucleotide substitution rates, and their ability to invade a new host and transmission of cross-species. Since it harbors error-prone RNA-dependent RNA polymerases (RdRp), mutations and recombination events often happen [78]. The mortality rates of SARS and MERS are expected to be about 10% and 35%, respectively [79, 80]. SARS-COV-2 case fatality ratio (CFR), which estimates this proportion of deaths among identified confirmed cases, is widely variable from less than 0.1% to over 25% [13, 81].

Moreover, it was reported that the asymptomatic incubation duration for 2019-nCov-infected individuals ranged from 1 to 14 days (most likely 3–10 days), longer than that of SARS-CoV [82].

Older age, particularly >65 years of age, and people with comorbidity are more likely to develop infection and severe symptoms and are at risk of death [83]. Children appear to be less symptomatic of infection and less susceptible to severe illness, and children under 18 years of age account for 2.4% of all reported cases [84–86].

The most common symptoms are fever (44–98%), the range of fever may be lower at initial hospital presentation or in the outpatient setting, cough (46–82%, usually dry), shortness of breath at onset (31%), myalgia or fatigue (11–44%), and loss of taste or smell in which the potential sign is seen in early infection, but not unique to COVID-19 as may be seen with other viral infections. The less common symptoms are pharyngitis, headache, productive cough, GI symptoms, and hemoptysis.

Critical illness (respiratory failure, septic shock, and/or multiple organ dysfunction/failure) is reported in only fewer than 6% of cases [85].

Three scenarios have been described for SARS-CoV-2 kinetics and immunopathogenesis: paucisymptomatic patient with nasopharyngeal high viral titer (and virus in feces), symptoms and then decompensation (~day 10, respiratory decompensation) with low viral titer compared to earlier in nasopharyngeal samples, and lastly progression/death with high viral titers in upper and lower respiratory samples plus persisting viremia [87]. Interventions of SARS-CoV-2 are non-pharmaceutical and pharmaceutical.

The quick expansion of SARS-CoV-2 is worrying, and also it causes both mortality and financial damage, which presents the global concern of this emerging disease [88].

## 13.12 SARS-CoV-2 Genome

Coronaviruses are the most common type of positive-stranded RNA viruses (26–32 kb) with variable numbers [74–79] of open reading frames (ORFs). SARS-CoV and MERS-CoV genomic organizations include an enveloped, single, positive-stranded RNA genome with one non-structural coding gene (*rep*) and encoding four major viral structural proteins, namely, spike (S), envelope (E), membrane (M), and nucleocapsid (N) proteins 3–5 [74]. A typical CoV includes a minimum of six ORFs in its genome. The genome is surrounded by a helical capsid and an envelope; the spike protein forms large crown-shaped protrusions in the envelope, giving the virus a coronal appearance [89, 90]. The spike protein (S proteins) which is located on the surface of coronaviruses plays a crucial role in viral infection and pathogenesis, and it is the main target for typing [91]. Four main structural proteins are encoded by ORFs 10 and 11 on the one-third of the genome near the 3' terminus [92].

The coronavirus "SARS-CoV-2" genome is fully sequenced, showing similar but distinct genome composition of SARS-CoV and MERS-CoV [74, 93]. Genomic organization of "WH-Human 1" coronavirus named as SARS-CoV-2 is a positive-sense single-stranded RNA 29,903 nucleotides in length with the gene order 5' to 3' replicase orf1ab, spike (S), envelope (E), membrane (M), and nucleocapsid (N) [82]. The replicase orf1ab gene of SARS-CoV-2 contained at least 16 nonstructural proteins followed by at least 13 downstream ORFs [74, 82].

A phylogenetic analysis of the Wuhan-Hu-1 viral genome showed that the virus was most closely related (96.2% nucleotide similarity) to a group of bat SARS-related coronaviruses (SARSr-CoV, RaTG13) (genus *Betacoronavirus*, subgenus *Sarbecovirus*) that had previously been found in bats in China (bat-SL-CoVZC45 and bat-SL-CoVZXC21) indicating that they share a common ancestor [74, 94–96].

Whole-genome sequence identity of the novel virus shows roughly 79% and 50% similarity to SARS-CoV and MERS-CoV, respectively [97, 98]. The phylogenic tree analysis between different isolates from China and the USA indicated that all isolates are nearly identical across the S gene. The range of S gene amino acid similarity of 2019-nCoV isolates with Bat_SARS-like CoVs and other SARS-like CoVs is between the mean of 81.5% and 77.55%, respectively [99, 100]. The amino acid length of "S" protein of nCoV is longer than SARS and BatSL. The "S" protein of nCoV showed some differences with SARS-CoV including three short insertions at the N-terminal region, along with four changes in the receptor-binding motif within the receptor-binding domain [99]. Five of the six essential amino acid (AA) residues in RBD differed between SARS-CoV-2 and SARS-CoV which leads to a higher binding affinity to angiotensin-converting enzyme 2 (ACE2) in contrast to SARS-CoV [99, 101].

## 13.13 Variations in SARS-CoV-2 Genome

We have considered 205,965 sequences' data analysis available in January 2022 [102], which have investigated the reported genetic variants in 7456 locations.

The genome sequence released data from the Global Initiative on Sharing All Influenza Data (GISAID) until mid-January 2022 includes 6,713,752 of 2019-n CoVgenome sequences (complete coronavirus genome sequences and partial coronavirus genome sequences),194,343 complete genome sequence (complete coronavirus genome sequences), 205,573 human genome sequences (coronavirus genome sequences that are from human (i.e., isolation source) and genome sequences include complete genome sequences and partial genome sequences), and 193,986 human complete genome sequences (coronavirus genome sequences that are complete genome sequences and are from human (i.e., isolation source)) [95, 102].

Totally, 2499 single nucleotide polymorphisms (SNP), 4566 insertions, 10,893 deletions, and 1022 Indel mutations have been reported across the genome of SARS-CoV-2 until January 11, 2022. Genomic regions of nCov 2019 which harbor mutations include 5'UTR, ORF1ab, S, ORF3a, Intergenic, E, M, ORF7a, ORF8, N, ORF10, and 3'UTR (Fig. 13.4) [102].

Mutation analyses of the reported mutations indicated the highest rate of variation in the orf1ab region. It seems that orf1ab is a hot spot region in the genome of nCov19 with 21,074 reported nucleic acid mutation (until January 11, 2022), and variation types in this region include SNPs, deletion, insertion, and Indel. The annotation types of these variations were missense-variant, synonymous-variant, coding-sequence-variant, inframe-deletion, inframe-insertion, frameshift-variant, and stop-gained.

The most important second region with significant variation in nucleic acid and amino acid sequences is the spike region with 3801 nucleic acid mutations and 3042 amino acid mutations.

The genomic regions of SARS-CoV-2 including 5'UTR, ORF1ab, S, ORF3a, E, M, ORF6, ORF7a, ORF8, N, ORF10, and 3'UTR have 45,980 nucleic acid mutations which are reported in China National Center for Bioinformation until January 11, 2022 (Table 13.1) [102–104].

The dynamic pattern of the mutation sites is provided over time in the form of a heatmap by calculation of mutation sites' frequency. In this pattern, the population frequency of the 3037 position and 14,408 located in orf1ab region gradually increased from 0 to 0.9. A similar trend is seen in the 23,403 position located in S from 0 to 0.9. The population frequency of mutation site 241 located in the 5'UTR region increased from 0 to 0.9. The annotation type of each region is indicated in [102].

The difference in COVID-19 mortality rate across countries may be due to diverse factors, including demographic measurements and genetic factors [105]. Genetic studies are very essential during the pandemic in the case of the genetics of virus or host genetic susceptibility [106, 107].

Since the COVID-19 outbreak in December 2019, SARS-CoV-2 has quickly been prone to several mutations; therefore, it has been found to have many different clades. These clades can provide scientists with information on where certain strains of the virus are concentrated and how

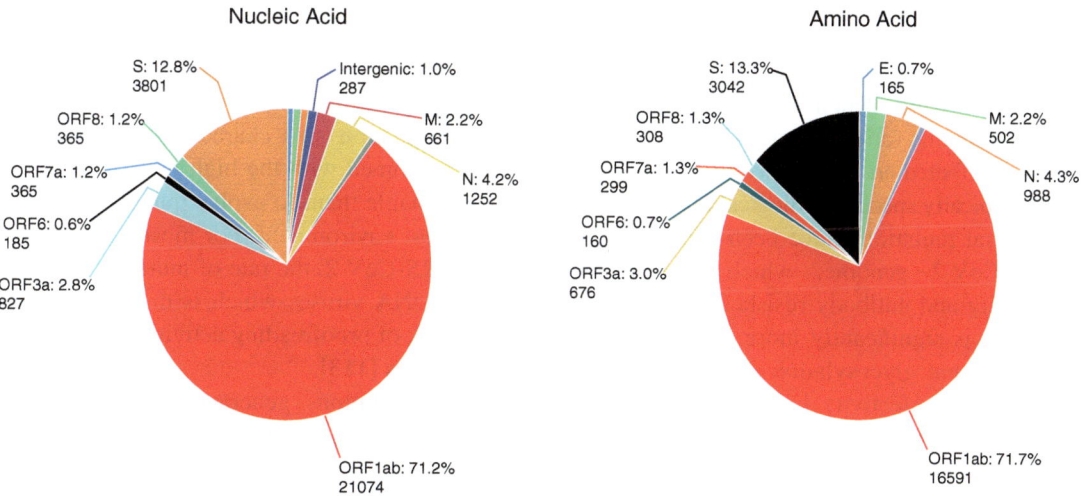

**Fig. 13.4** Variation annotation in 2019-nCoV genomic regions [102]

**Table 13.1** Variation annotation frequency in 2019-nCoV genome (last updated mid-November) [38]

| Location | Number of nucleic acid mutation | Number of amino acid mutation | Frequency of nucleic acid mutation | Frequency of amino acid mutation |
|---|---|---|---|---|
| 5′UTR | 245 | 0 | 0.924 | 0 |
| orf1ab | 21,083 | 16,623 | 0.990 | 2.34 |
| S | 3804 | 3052 | 0.995 | 2.39 |
| ORF3a | 827 | 676 | 0.998 | 2.45 |
| E | 227 | 165 | 0.995 | 2.2 |
| M | 661 | 502 | 0.988 | 2.26 |
| ORF6 | 185 | 160 | 0.994 | 2.62 |
| ORF7a | 365 | 299 | 0.997 | 2.47 |
| ORF8 | 365 | 308 | 0.997 | 2.45 |
| N | 1252 | 988 | 0.993 | 2.35 |
| ORF10 | 116 | 93 | 0.991 | 2.44 |
| 3′UTR | 189 | 0 | 0.825 | 0 |

these different clades can affect SARS-CoV-2 virulence, disease progression speed, and drug or vaccine resistance [108, 109].

SARS-CoV-2 whole-genome sequencing (WGS) can help to find tracing patterns spread across countries and a patient who becomes infected with a certain SARS-CoV-2 clade [109, 110].

Five clades of SARS-CoV-2 that were characterized by 11 major mutations worldwide have been shown by analyzing the public database of the GISAID [95]. These mutations are dominant in one or two clades in each geographical region. These clades were defined as G614, S84, V251, I378, and D392. The mutation in these clads is focused on Orf8, Orf3a, Orf1ab, and S [109].

According to several studies, the most common clade, which is widely spread globally, is G614 which is most prevalent in Europe [107, 111].

The genomic variability of SARS-CoV-2 specimens distributed around the world may implicate geographically specific etiological effects [111].

Several mutations have been reported in the SARS-CoV-2 genome, which confers viral infectivity and antibody resistance. The D614G mutation is significantly more infectious, but it is shown that glycosylation deletion leads to a reduction in infectivity in the N331 and N343 mutations. Other mutations such as the F490L are significantly resistant to some neutralizing antibodies [107].

According to the population frequency study, which is reported in Novel Coronavirus Resource [104], similar trend is seen in the other genome regions of SARS-CoV-2 which is important in future studies for assessing antibody resistance or infectivity degree.

The exceptionally high mutation rates of viruses are not represented by any other organism in the evolution of life. In a report by B Korber et al., D614G mutation as a dominant SARS-CoV-2 lineage in Europe, the USA, Canada, and Australia (the amino acid aspartate (D) at the 614th amino acid position of the spike protein is replaced by glycine (G)) is now present as more transmissible form of SARS-CoV-2 [112].

Mutation rates may be different between as high as $10^{-3}$ and $10^{-11}$ per incorporated nucleotide in different viruses. Short generation times and large population size characteristics in viruses when combined with the high mutation rates of viruses enable them to evolve quickly and adapt to the host environment. For most RNA viruses like SARS-CoV-2, the rate of mutation is higher than in DNA viruses which is the consequence of the lack of proofreading activity of RNA virus polymerases [113].

Virus mutations generate genetic diversity, which is the result of opposing selection and natural genetic drift, which are both directly influenced by the size of the population of the virus.

In the case of SARS-CoV-2 as the population size is large, the selection will be prevalent and random drift will occur with less frequency. It ensures that deleterious alleles will be completely eliminated from the population, while adaptive alleles will have the ability to take over the population. In contrast to the small population of viruses that deleterious alleles are at a high level in the population, the adaptive alleles may be lost by chance [113].

Therefore, high mutation rates develop many viral variants. A variant of the original virus is one that has one or more additional mutations. The viruses with a large number of variant genomes have been called viral quasispecies. The rich cloud of mutants shows the potential to encode viruses with enhanced resistance to a drug.

Some SARS-CoV-2 variants are thought to be a concern because they preserve (or even improve) their replication capability in the face of increased population immunity, either through infection recovery or vaccination.

An "emerging variant" is a term used to describe a novel variant that appears to be spreading in a population.

Some of the potential impacts of emerging variants are increased transmissibility and mortality and the ability to evade detection by diagnostic tests and to evade natural immunity and to infect vaccinated individuals [114].

Variants that appear to meet one or more of these criteria may be labeled "variants under investigation" or "variants of interest" (VOI). The variant of concern (VOC) for SARS-CoV-2 is a category where mutations in their spike protein receptor-binding domain (RBD) significantly increase the binding affinity in the RBD-hACE2 complex and are also related to widespread in human populations. This variant is recognized by WHO [115].

## 13.14  Genomics of Host Susceptibility to COVID-19

The genetics of an individual can influence their risk of infection as well as the severity of disease symptoms. A huge worldwide investigation has discovered parts of the human genome that can influence the risk of COVID-19 severity. One of the most prominent features of SARS-CoV-2 infection is the wide range of symptoms, from asymptomatic to life-threatening. While recognizing host characteristics correlate with disease severity, these risk factors do not account for all disease variables. COVID-19 susceptibility and severity genetic factors may provide new biological insights into illness pathogenesis and propose molecular targets for therapeutic development or drug repurposing [116].

The COVID-19 Host Genetics Initiative (COVID-19 HGI) (https://www.covid19hg.org) is a bottom-up collaborative effort. The COVID-19 HGI tried to find genetic variations that account for individual vulnerability to COVID-19 as well as disease severity.

COVID-19 HGI follows three main aims:

1. Provide an environment to foster the sharing of resources to facilitate COVID-19 host genetics research (e.g., protocols, questionnaires).
2. Organize analytical activities across studies to identify genetic determinants of COVID-19 susceptibility and severity.
3. Provide a platform to share the results from such activities, as well as the individual-level data where possible, to benefit the broader scientific community.

The results of three genome-wide association meta-analyses involving up to 49,562 COVID-19 participants (including 13,641 individuals who were hospitalized with the infection, and of those, 6179 who were critically ill with the disease) from 46 studies in 19 countries have been compared with the genomes of around two million control individuals without known infection. Thirteen genetic loci in the human DNA have found that influence on COVID-19 susceptibility and severity. Variants in nine loci are associated with critically ill COVID-19, whereas variants in four other loci are associated with susceptibility to COVID-19 [116].

More than 40 candidate genes were found in the proximity of each locus, which was previously known with function in immune function or function in the lungs. It was also shown that some genetic variants, most notably at the *ABO* and *PPP1R15A* loci, in addition to *SLC6A20*, can impact COVID-19 susceptibility rather than progress to severe COVID-19 once infected [52]. Several genes have been introduced such as *SLC6A20, LZTFL1, CCR9, ABO, IFNAR2, OAS3, OAS1, DPP9, ICAM5/TYK2, HLA-G, CCHCR1, HLA-DPB1, FYCO1, CXCR6*, and *XCR1* which are working in different ethnicities [117].

SARS-CoV-2 entry depends on two factors: the first one is the host receptor named ACE2 and the second one is TMPRSS2 which is host cell protease for spike protein modification [118]. The ACE2 gene encodes the angiotensin-converting enzyme-2, the receptor of both SARS-coronavirus (SARS-CoV) and recently SARS-CoV-2 [97, 119]. ACE2 is the principal host cell receptor of 2019-nCoV through the RBD of spike protein and plays a critical role in the entry of the virus into the cell to induce the final infection [120]. Single-cell transcriptome analysis confirms high expression of ACE2 in epithelial cells of the tongue [121]. It is reported that a number of variants of *ACE2* may reduce the connection between ACE2 and S-protein in SARS-CoV [122].

Variations in both the viral spike protein and the host ACE2 sequences have also been shown to act as a barrier to viral infection across species. The expression level and expression pattern of human ACE2 in various tissues could be critical for the susceptibility symptoms, and outcome of 2019-nCoV/SARS-CoV-2 infection. In a systematic analysis, the allele frequency (AF) of the candidate functional coding variants in ACE2 between different populations has been investigated. Totally, it was shown that 32 variants potentially affect the amino acid sequence of *ACE2* in databases [101].

Molecular modeling of human *ACE2* allelic variants showed significant variations in the intermolecular interactions between *ACE2* alleles, rs73635825 (S19P), and rs143936283 (E329G) with SARS-CoV-2 spike protein. These alleles have a low binding affinity and lack of some of the key residues in the complex formation with SARS-CoV-2 spike protein [123].

A systematic analysis of coding region variants in *ACE2* by Yanan Cao et al. [36] revealed that no strong evidence has been provided genetically supporting the presence of coronavirus S-protein binding-resistant ACE2 mutants in different populations. In the expression quantitative trait loci (eQTL) variants, the East Asian populations have much higher AFs associated with higher ACE2 expression in tissues. The top six variants including rs2158082, rs4646127, rs6629110, rs5936011, rs4830983, and rs5936029 have shown very high AF in the East Asian population (>95%), while the AFs of these variants were significantly lower in European populations (52%–65%) [101]. All of these eQTL variants are associated with high expression of ACE2 in tissues which may consider different susceptibility or responses of different populations under similar conditions to 2019-nCoV/SARS-CoV-2 [101].

Our knowledge of genetic susceptibility and disease vulnerability of SARS-CoV-2 infection and COVID-19 severity remains in its infancy. Several COVID-19 host genetics initiatives have been launched to encourage studies to generate and analyze data to illustrate the genetic determinants of COVID-19 susceptibility, severity, and outcomes. COVID-19 host genetics initiatives follow three aims including [1] providing an environment to foster the sharing of resources to facilitate COVID-19 host genetics research in these initiatives, [2] organizing analytical activities across studies to identify genetic determinants of COVID-19 susceptibility and severity, and [3] providing a platform to share the results from meta-analytical activities to benefit the broader scientific community. So far, 220 COVID-19 initiatives have been registered worldwide [1].

Genome-wide association studies (GWASs) may help identify potential host genetic factors contributing in the pathogenesis of severe COVID-19 with associated respiratory failure. In a GWAS involving 1980 COVID-19 patients with respiratory failure, at chromosome locus 3p21.31, a cluster of six genes, including *SLC6A20, LZTFL1, CCR9, FYCO1, CXCR6*, and

*XCR1*, showed the peak association signal, and several of them have potential functions relevant to COVID-19 and the other signal was seen at locus 9q34.2 which is matched to ABO blood group system [124].

Genome-scale CRISPR loss-of-function screening technology has been used to identify host factors needed for SARSCoV-2 viral infection of human alveolar epithelial cells. Top-ranked genes were identified which involved several pathways. By considering the crucial role of the ACE2 receptor in the early stages of the viral entrance, Rab7 is introduced as a regulator of ACE2 called surface expression. It was indicated that RAB7A loss reduces viral entry [125].

## 13.15 Proposed Targets of Repurposed and Investigational Therapies for COVID-19

SARS-CoV-2 infects either lung or cardiac cells through the viral structural spike (S) protein that attaches to the ACE2 receptor [118]. This receptor binding allows the anchorage and the entry of the virus particles by the endocytic pathway. In addition, this entry requires S protein priming that is facilitated by host type 2 transmembrane serine protease called TMPRSS2 [118]. Once the virus enters the cell, not only the RNA will be translated into a number of viral polypeptides, but also it will be translated into a replicase-transcriptase complex. Nonstructural proteins are synthesized after the proteolysis process of viral polypeptides. RdRp of virus synthesizes RNA. Then the structural proteins are translated resulting in the formation and release of additional viral particles. The virus then synthesizes RNA via its RdRp. Then the structural proteins are translated leading to the formation and release of additional viral particles by exocytosis that will infect other cells [118, 126, 127]. This particular phenomenon is associated with an excessive immune response, which is called a cytokine storm, an over-inflammatory process, and can be mediated by interleukin-6. Cytokine storm is known to be one of the main causes of ARDS and multiple organ failures [128].

This first phase of entry and anchoring can be repressed by a molecule called camostat mesylate through TMPRSS2 inhibition. The membrane fusion of the viral envelope through the interaction of spike protein and ACE2 can be inhibited by Arbidol. The endocytosis step can be interrupted through the immunomodulators' effects of chloroquine and hydroxychloroquine. Lopinavir and darunavir prevent the proteolysis of the viral polypeptide by inhibiting 3chymotrypsinlike protease. Remdesivir, favipiravir, and ribavirin inhibit viral RdRp [129]. Two monoclonal antibodies, tocilizumab and sarilumab, which have previously been used for autoimmune disorders can inhibit IL-6 signaling through IL-6 receptor binding and prevention of IL-6 receptor activation. In addition to all of these molecules, antihypertensive medication such as losartan has been shown to have the ability to upregulate ACE2 protein that facilitates the entry of the virus [130].

## 13.16 Precision Medicine and Pharmacogenomics Approach for COVID-19

Despite the historical vision and recommendations, medical therapy used a very wide strategy based on clinical and genetic/genomic data from various groups rather than focusing on each patient [131].

To choose a therapeutic regimen, clinicians followed guidelines based on evidence, experience, and knowledge of previous patient/disease which is called the evidence-based medicine approach.

In clinical trials, about 20% of the population did not respond to a specific treatment or even experience adverse drug reactions due to genetic variations [132]. The mapping of the human genome was a breakthrough that provided a better knowledge of an individual's genetic makeup [133].

The aim of precision medicine is to combine current medicine with advances in genomic medicine to target patients individually and increase the efficacy and effectiveness of the therapeutic approach [134]. Moreover, the unique identity of each individual's DNA provides critical information about disease development and progression, as well as response to various therapy regimens [135].

The actual vision of precision medicine is the combination of the human DNA, environmental factors, disease assessments, and medication in order to generate a stronger therapeutic outcome.

Although definitions of precision medicine (PM) vary, it is broadly introduced to be the use of individual variability in genomics, transcriptomics, proteomics, metabolomics, and also other omics with the implication of prediction, prevention, and personalized treatment for the need of the individual patient. The recent development of large-scale biological databases leads to applying more broadly this new concept of medicine [136].

Every expert scientist believes that a new virus epidemic like SARS-CoV-2 is inevitable, and a pandemic like this will cause global economic and social disruption. Although most of the reported death occurred in older age (ʼ60 years), healthy people younger than 60 are affected by this virus too [75]. For years such pandemics remained an essential cause of intensive care unit (ICU) admission, and current therapies will probably only be effective in constrained subpopulations during specific disease periods [119].

The SARS-CoV-2 caused significant pulmonary damage, partly due to the elevation of robust pro-inflammatory cytokine responses resulting in acute respiratory distress syndrome (ARDS) [137–139]. It is currently unknown if SARS-CoV-2 and SARS-CoV sequence similarities translate into similar behavioral characteristics, including pandemic potential. Since it will take months until we will fully understand the whole picture of SARS-CoV-2 characteristics and the host immune responses in overcoming the infection, interest in the use of biomarkers for evaluating host response and determining severity and complications may be growing. It is well recognized that these biomarkers especially based on genomics data can be powerful tools for predicting severity and mortality if used properly [140]. It seems that the precision medicine and pharmacogenomics concept which are previously introduced in the context of noncommunicable diseases will be applicable in infectious diseases [141, 142].

The personalized (precision) medicine approach considers the seven Ps including preventive, predictive, protective, participatory, pervasive, personalized, and precise (treatments) pillars.

A revolution in disease taxonomy is forming in medicine. Researchers have created a myriad of new disease subclassifications as a result of high-throughput genomic technology, molecular diagnostic assays, and developments in disease biology. More than 60 COVID-19-related subtypes are proposed in 2020 [143]. In the precision medicine approach for COVID-19, the most necessary factors for attention are the substantial mutation rate in different parts of the SARS-CoV-2 genome and host genetic susceptibility.

Infectious disease subclassification needs consideration of both the host and the pathogen. SARS-CoV-2 strains are not the same. COVID-19 is caused by an emerging virus with various mutations, lineages, and variants, rather than a single, genetically identical RNA virus [144]. Antigenicity, transmission, and virulence of SARS-CoV-2 are influenced by the corresponding variants. SARS-CoV-2 variant of concern (VOC) has an increased risk-adjusted odds of mortality compared with patients infected with original SARS-CoV-2 [145]. Therefore, a better understanding of COVID-19 prognosis and spread will be achieved through pathogen subclassification.

Variations in disease "tolerance" can also be explained by host variability [146]. Various predisposing host factors influence disease susceptibility and progression. Moreover, the host shows a variable response to SARS-CoV-2.

According to host-pathogen complexity, thousands of potential subtypes of COVID-19 may exist. Five essential subtype criteria could help in the classification of COVID-19 subtypes including biologically feasible, reproducibility, and treatment response [143].

Clinical big data, pathogen sequencing, and multiomic readouts should all be considered when subtyping COVID-19 to inform not only prognosis but also treatment regimens. The pandemic of the SARS-CoV-2 virus provides the ideal opportunity for both researchers and healthcare professionals to change toward a more individualized approach that considers patients'

needs. These variations in different positions of the genomes are very diverse in each country as shown in Fig. 13.5.

Personalized treatment or pharmacogenetics approach helps to choose the best treatment with the highest efficacy and lowest adverse drug reactions (ADRs), because in COVID-19 patients, minimizing the risk of toxicity is very crucial [147]. A list of pharmacogenomics information including the clinical annotation and drug label annotation is shown in Table 13.2 [148].

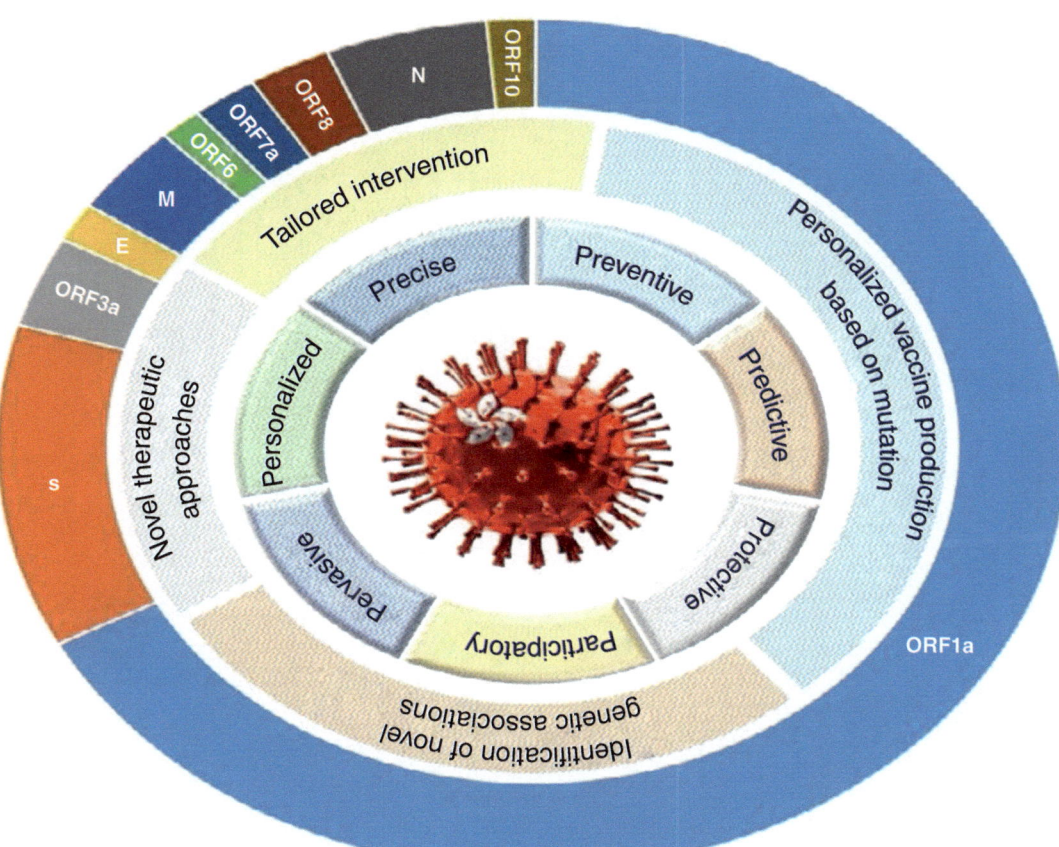

**Fig. 13.5** Precision (personalized) medicine approach in COVID-19. Three targets will be achievable by considering the seven pillars of precision medicine; these targets include tailored treatment, personalized vaccine production by considering the mutation spectrum, and identification of predisposing genetic factors which makes some affected individuals more susceptible to some symptoms like acute respiratory distress syndrome (ARDS) or mortality. The variation was seen in the SARS-Cov2 genome position which can determine the access to abovementioned goals

**Table 13.2** Pharmacogenetics (genes) implicated in COVID-19 involving drug treatments [84]

| Category | Drugs | Clinical annotation | | | Drug label annotation | | |
|---|---|---|---|---|---|---|---|
| | | Gene | Type | Level | PGX level | Gene | Source |
| Viral entry and endocytosis inhibitors | Hydroxychloroquine | – | – | – | Actionable PGX / Actionable PGX | G6PD / G6PD | FDA / Swissmedic |
| | Chloroquine | – | – | – | Actionable PGX / Actionable PGX | G6PD / G6PD | FDA / Swissmedic |
| TMPRSS2 inhibitors | Camostat Mesylate | – | – | – | – | – | – |
| | Bromhexine | – | – | – | – | – | – |
| Viral protease inhibitors | Lopinavir | ABCB1 | Efficacy | 3 | – | – | – |
| | | ABCC1 | Efficacy | 3 | – | – | – |
| | | ABCC2 | Toxicity/ADR | 3 | – | – | – |
| | Ritonavir | ABCB1 | Efficacy | 3 | Informative PGX | IFNL3 | FDA/EMA |
| | | CYP3A5 | Metabolism/PK | 3 | Informative PGX | CY2D6 | EMA |
| | | UGT1A7 | Toxicity/ADR | 3 | Informative PGX | CYP3A4 | EMA |
| | | APOE | Toxicity/ADR | 3 | Actionable PGX | IFNL3/ IFNL4 | Swissmedic |
| | | APOC3 | Toxicity/ADR | 3 | – | – | – |
| | | UGT1A1 | Toxicity/ADR | 3 | – | – | – |
| | | UGT1A3 | Toxicity/ADR | 3 | – | – | – |
| | Atazanavir | UGT1A1 | Toxicity/ADR | 1A, 3 | Testing recommended | CYP2C19 | EMA |
| | | ABCB1 | Toxicity/ADR | 3 | Actionable PGX | CYP2C19 | PMDA |
| | | CYP3A5 | Metabolism/PK | 3 | Informative PGX | CYP2C19 | HCSC |
| | | UGT1A7 | Toxicity/ADR | 3 | – | – | – |
| | | CYP3A4 | Metabolism/PK | 3 | – | – | – |
| | | SORCS2 | Metabolism/PK | 3 | – | – | – |
| | | UGT1A3 | Toxicity/ADR | 3 | – | – | – |
| | | ABCC2 | Toxicity/ADR | 3 | – | – | – |
| | | NR1I2 | Metabolism/PK | 3 | – | – | – |

| Drug class | Drug | Gene | Effect | Level | | | |
|---|---|---|---|---|---|---|---|
| RNA-dependent RNA polymerase (RdRp) | Ribavirin | IFNL3 | Efficacy | 1A | – | – | – |
| | | IFNL4 | Efficacy | 1A | – | – | – |
| | | ITPA | Toxicity/ADR, efficacy, and dosage | 1B | – | – | – |
| | | SCL29A1 | Efficacy | 3 | – | – | – |
| | | HLA-B | Efficacy and toxicity/ADR | 3 | – | – | – |
| | | VDR | Efficacy | 3 | – | – | – |
| | | IL18 | Efficacy | 3 | – | – | – |
| | | LDLR | Efficacy | 3 | – | – | – |
| | | HLA-C | Efficacy | 3 | – | – | – |
| | | EGFR | Efficacy | 3 | – | – | – |
| | | CY27B1 | Efficacy | 3 | – | – | – |
| | | CYP2R1 | Efficacy | 3 | – | – | – |
| | | SCARB1 | Efficacy | 3 | – | – | – |
| | | BCL2 | Efficacy | 3 | – | – | – |
| | | FTO | Efficacy and toxicity/ADR | 3 | – | – | – |
| | | IL6 | Toxicity/ADR | 3 | – | – | – |
| | | SLC28A2 | Toxicity/ADR | 3 | – | – | – |
| | | OASL | Efficacy | 3 | – | – | – |
| | | SLC6A4 | Toxicity/ADR | 3 | – | – | – |
| | | M1CB | Efficacy | 3 | – | – | – |
| | Remdesivir | CYP2C9, CYP2D6, CYP3A4 | Efficacy | – | – | – | – |
| | Favipiravir | – | – | – | – | – | – |
| Angiotensin receptor inhibitors | Losartan | CAMK1D | Toxicity/ADR | 3 | – | – | – |
| | | AGTR1 | Efficacy | 3 | – | – | – |
| | | CYP2CP | Efficacy and toxicity/ADR | 3 | – | – | – |
| | | STK39 | Efficacy | 3 | – | – | – |
| | | ABCB1 | Efficacy | 3 | – | – | – |
| | | NPHS1 | Efficacy | 3 | – | – | – |
| | | AGTR1 | Efficacy | 3 | – | – | – |
| Antibiotics | Azithromycin | ABCB1, MRP2, PGP | – | – | – | – | – |
| Janus kinase 1 and 2 inhibitors | Ruxolitinib | – | – | – | Informative PGX | CYP3A4 | EMA |
| | Baricitinib | – | – | – | – | – | – |

*FDA* US Food and Drug Administration, *EMA* European Medicines Agency, *Swissmedic* Swiss Agency of Therapeutic Products, *PMDA* Pharmaceuticals and Medical Devices Agency, Japan, *HCSC* Health Canada (Santé Canada)

## 13.17 Pharmacogenomics Aspect in COVID-19 Treatment Approach

The evidence for the efficacy of COVID-19 treatments and recommendations for medication remain controversial, and clinical trial data is continuously being produced. However, in certain patients, there is a lack of therapeutic response or the onset of adverse drug reactions. Based on the genetics of COVID-19 patients, pharmacogenetics could address the interindividual variability in treatment response.

Because proposed drugs for COVID-19 treatment including antivirals, antibiotics, antiparasitics, and/or anti-inflammatory have been previously used for other infectious and noninfectious diseases, a background framework of pharmacogenetic studies exists [149].

In the pharmacogenomics context, there are several genetic polymorphisms in *CYP3A4*, *CYP3A5*, *CYP2D6*, *CYP2C8*, *ABCB1*, *SLCO2B1*, *ABCC2*, *G6PD*, and *CES1* which can help for the improvement of COVID-19 clinical outcome [150]. Most of the drugs used to treat COVID-19 are metabolized by CYP3A4 and are P-gp and OATPB substrates [150]. Genetic polymorphism in drug-metabolizing enzymes, receptors, and transporters can help personalized therapy in COVID-19 patients.

### 13.17.1 Hydroxychloroquine and Chloroquine

Hydroxychloroquine and chloroquine phosphate have been used for the prevention and treatment of malaria for several years [151]. The mechanism of action is to block the endocytosis step through inhibition of host receptor glycosylation and proteolytic process. They indicate immunomodulatory effects by attenuating cytokine production and inhibiting autophagy [152]. The clinical trial and outcome data have not yet shown evidence for the efficacy of chloroquine/hydroxychloroquine treatment of COVID-19 [153]. QT prolongation adverse effects have been reported for chloroquine and hydroxychloroquine [85].

No antiviral therapy exists for COVID-19 in humans. Several RCTs are ongoing. The data from the retrospective study on SARS suggests that earlier treatment with antivirals may be more effective than receiving medications after severe organ failure [154].

Cytochrome P450 (CYP) enzymes play an important role in drug metabolism. CYP enzyme activity is strongly related to specific SNPs in CYP genes. Variation in the CYP genes leads to the difference in hydroxychloroquine hepatic metabolism, which consequently influences the whole-blood level of hydroxychloroquine. The clinical response of hydroxychloroquine is directly correlated to the whole-blood level of hydroxychloroquine. Hydroxychloroquine (HCQ) is metabolized to N-desethyl HCQ (DHCQ) in the liver by CYP2C8, CYP3A4, and CYP2D6 cytochrome P450 (CYP) enzymes [155, 156]. Associations of rs1065852 (CYP2D6*10), rs1135840 (CYP2D6*10), rs776746 (CYP3A5*3), and rs28371759 (CYP3A4*18B) with blood concentrations of hydroxychloroquine (HCQ) and its metabolite, N-desethyl HCQ (DHCQ), have been investigated in 194 SLE patients taking HCQ for >3 months. CYP2D6*10 (rs1065852 and rs1135840) allelic variants were significantly associated with the [DHCQ]/[HCQ] ratio. Therefore, CYP polymorphisms could demonstrate why there is a wide variation in HCQ blood concentrations [155]. A multicenter observational and pharmacogenetic study of 200 patients with discoid lupus erythematosus (DLE) treated with hydroxychloroquine was conducted. Clinical response to hydroxychloroquine was the primary outcome. Although the majority of the patients responded to hydroxychloroquine treatment, a significant proportion (39%) failed to respond to the drug or developed toxicity, requiring withdrawal. This study did not demonstrate a significant association between CYP2D6*3, *4, and *5 and for CYP2C8*3 and *4 genotypes and response [156].

The association between CYP2D6 genotypes, drug exposure to artemisinin-based therapy (chloroquine and primaquine), and tolerance and efficacy has been investigated in 58 patients with malaria due to *Plasmodium vivax* mono-infection. Eight SNPs in CYP2D6 gene including (G-1584C [rs1080985], C1023T [rs28371706], G1846A [rs3892097], C100T [rs1065852], G2988A [rs28371725], C2850T [rs16947], G3183A [rs59421388], and G4180C [rs1135840]) and one deletion (2615-2617delAAG [rs5030656]) were genotyped. In the primaquine group, impaired CYP2D6 metabolism was associated with treatment failure. It was shown that patients with a longer chloroquine half-life are more likely to report mild pruritus and there is no evidence of the impact of CYP26 genetic variations on chloroquine response.

Glucose-6-phosphate dehydrogenase (G6PD) facilitates the synthesis of NADPH and ribose-5-phosphate and is one of the first genes found to be involved with variable drug reactions. Individuals with G6PD deficiency may have increased chances of adverse reactions to many medications, including primaquine and chloroquine. More than 400 variations of the G6PD enzyme have now been reported, based on clinical manifestations and biochemical properties, and G6PD deficiency is the most prevalent enzyme deficiency in the world [157–161].

Most of the genetic variants appear to be single-point mutations, and the lack of large or out-of-frame deletions may suggest that the complete absence of enzyme activity is fatal [158, 159, 161–163].

Chloroquine- and hydroxychloroquine-related cardiac disorder is reported as a rare but severe adverse drug reaction that can result in death. KCNH2 gene or human ether-a-go-go-related gene (hERG) encodes a potassium voltage-gated ion channel. It seems that most of the drugs that cause drug-induced QT prolongation such as chloroquine and hydroxychloroquine are KCNH2/I Kr blockers. However, the syndrome of drug-induced LQTS is most commonly caused by the block of the hERG channels encoded by the KCNH2 gene [164–168].

## 13.18  Remdesivir

Remdesivir (GS-441524) is a monophosphoramidate nucleoside analog prodrug that was initially designed to treat Ebola virus infection. It can inhibit viral replication through binding to the viral RNA-dependent RNA polymerase. Remdesivir is metabolized by hydrolases, CYP2C8, CYP2D6, and CYP3A4 [169]. Several CYP gene variations have been reported that have a significant impact on enzyme activity.

## 13.19  Viral Protease Inhibitors (Lopinavir, Ritonavir, Atazanavir, Oseltamivir)

The mechanism of action is through inhibition of 3-chymotrypsinlike protease. Lopinavir/ritonavir reduces in vitro replication by 50% in the MERS coronavirus. Lopinavir/ritonavir (LPV/r) is a fixed-dose combination FDA-approved medication for the treatment and prevention of HIV/AIDS. The treatment combination with interferon-alpha or ribavirin has been recommended. Timing of administration during the early peak viral replication period seems to be critical since delayed treatment with lopinavir/ritonavir showed no influence on clinical outcomes [170]. Lopinavir/ritonavir (LPV/RTV) has been commonly used in China and elsewhere; however, several RCTs of lopinavir/ritonavir are ongoing, and the exact therapeutic effect of lopinavir/ritonavir on COVID-19 is difficult to determine [171].

Lopinavir/ritonavir is metabolized by CYP3A4 and is mainly contraindicated with drugs that their clearance is extremely dependent on the CYP3A family. CYP3A4 is also involved in the metabolism of darunavir [172].

In addition to CYP3A4, the impact of variations in ABCB1, CYP3A5, ABCC2, and SLCO1B1 on lopinavir plasma concentrations has been investigated in various pharmacogenetic analyses that conflicting results have been reported [150].

Oseltamivir as an ethyl ester prodrug is used to treat influenza A and B infections. It has been used for the COVID-19 treatment but its effectiveness has been demonstrated [150].

CES1 polymorphism, rs71647871 p.Gly143Glu, is related to oseltamivir pharmacokinetic variations [173].

## 13.20 Antimicrobial (Azithromycin)

Azithromycin as an antimicrobial agent prevents bacterial protein synthesis by binding to 50S component of the 70S ribosomal subunit. It does not interact with cytochrome P450 enzymes but it can be a substrate of P-gp and MRP2 transporters. *ABCB1* genetic variation is reported in pharmacogenetic studies of azithromycin [150].

## 13.21 Corticosteroids

Dexamethasone is used to inhibit cytokine release and suppress lung infiltration. It is widely metabolized by CYP3A4 into 6-hydroxydexamethasone in the liver. It is also a substrate for P-gp. In addition to CYP3A4, several genetic variations in *ABCB1, DOK5, SERPINE1, LINC00251, BMP7, PYGL, DOK5*, and *CTNNB1* have been reported in association with the efficacy or toxicity of dexamethasone [148, 150, 174].

## 13.22 RNA-Dependent RNA Polymerase

Favipiravir is an anti-influenza drug available in China and Japan that is investigated for COVID-19 in ongoing clinical trials. Most of the previous clinical data is derived from influenza and Ebola RNA viruses. It can inhibit RdRp and avoid viral replication [175].

Remdesivir is a nucleotide analog that blocks viral nucleotide synthesis to stop viral replication through inhibition of RdRp. It is likely the most promising drug for COVID-19. No pharmacogenetic label is recommended for RdRp drugs. For

Janus kinase inhibitors including ruxolitinib, baricitinib, and CYP3A4, drug label annotation is recommended by European Medicines Agency (EMA).

## 13.23 Conclusion

The emerging novel coronavirus has become a global concern within a short time. SARS-CoV-2 tends to infect a large number of human populations and the epidemic can cause severe economically social impacts. The rapid development of vaccines is needed as the virus became a major global concern. Based on the experiences from managing SARS-CoV and MERS-CoV, rapid improvements in our knowledge of the genetic architecture action, epidemiology, and pathogenesis of SARS-CoV-2 can be achieved at an unprecedented speed. By integrating novel technology such as next-generation sequencing (NGS) into the key findings on the clinical features of the infected patients and immunological responses, the complex SARS-CoV-2 puzzle of the pandemic can be solved piece by piece. In this period of the SARS-CoV-2 pandemic, it is important to remember the importance of functional studies in confirmation of the reported mutations in association with phenotype, severity, and mortality of COVID-19.

The new approaches which connect science, medicine, and technology will be quite valuable in the face of major problems in COVID-19 management.

Accordingly, the collaborative efforts of medical communities are required in order to eradicate the outbreak.

## References

1. Bissonnette LBM. Infectious disease management through point-of-care personalized medicine molecular diagnostic technologies. J Pers Med. 2012;2(2):50–70. https://doi.org/10.3390/jpm2020050.
2. Chan IS, Ginsburg GS. Personalized medicine: progress and promise. Annu Rev Genomics Hum Genet. 2011;12:217–44. https://doi.org/10.1146/annurev-genom-082410-101446.

3. Al-Mozaini MA, Mansour MK. Personalized medicine: is it time for infectious diseases? Saudi Med J. 2016;37(12):1309–11. https://doi.org/10.15537/smj.2016.12.16837.

4. Ehrmeyer SS, Laessig RH. Point-of-care testing, medical error, and patient safety: a 2007 assessment. Clin Chem Lab Med. 2007;45(6):766–73. https://doi.org/10.1515/CCLM.2007.164.

5. Kosack CS, Pageb AL, Klatser PR. A guide to aid the selection of diagnostic tests. Bull World Health Organ. 2017;95:639–45. https://doi.org/10.2471/BLT.16.187468.

6. Yang S, Rothman RE. PCR-based diagnostics for infectious diseases: uses, limitations, and future applications in acute-care settings. Lancet Infect Dis. 2004;4(6):337–48. https://doi.org/10.1016/S1473-3099(04)01044-8.

7. Mahmud I, Garrett TJ. Mass spectrometry techniques in emerging pathogens studies: COVID-19 perspectives. J Am Soc Mass Spectrom. 2020;31(10):2013–24. https://doi.org/10.1021/jasms.0c00238.

8. Shuren JST. COVID-19 molecular diagnostic testing-lessons learned Covid-19. N Engl J Med. 2020;383:e97. https://doi.org/10.1056/NEJMp2023830.

9. Pierce S, Bell R, Hallberg R, Cheng CM, Chen KS, Williams-Hill D, Martin W, Allard M. Detection and Identification of Salmonella enterica, Escherichia coli, and Shigella spp. via PCR-electrospray ionization mass spectrometry: isolate testing and analysis of food samples. Appl Environ Microbiol. 2012;78 https://doi.org/10.1128/AEM.02272-12.

10. Bibi A, Basharat N. Procalcitonin as a biomarker of bacterial infection in critically ill patients admitted with suspected Sepsis in Intensive Care Unit of a tertiary care hospital. Pak J Med Sci. 2021;37(7):1999–2003. https://doi.org/10.12669/pjmss.37.7.4183.

11. Poplin V, Boulware DR, Bahr NC. Methods for rapid diagnosis of meningitis etiology in adults. Biomark Med. 2020;14(6):459–79. https://doi.org/10.2217/bmm-2019-0333.

12. Hao MHY. CRISPR-Cas9-mediated carbapenemase gene and plasmid curing in carbapenem resistant. Antimicrob Agents Chemother. 2020:e00843–20. https://doi.org/10.1128/AAC.00843-20.

13. Kasas S, Malovichko A. Nanomotion detection-based rapid antibiotic susceptibility testing. Antibiotics. 2021;10(3):287. https://doi.org/10.3390/antibiotics10030287.

14. Bergeron M, Ke D, Ménard C, Picard F, Gagnon M, Bernier M, Ouellette M, Roy PH, Marcoux S, Fraser W. Rapid detection of group B streptococci in pregnant women at delivery. N Engl J Med. 2000;343(3):175–9. https://doi.org/10.1056/NEJM200007203430303.

15. Cheng X, Dou Z. Highly sensitive and rapid identification of streptococcus agalactiae based on multiple cross displacement amplification coupled with lateral flow biosensor assay. Front

Microbiol. 2020;11:1926. https://doi.org/10.3389/fmicb.2020.01926.

16. Palavecino EL. Rapid methods for detection of MRSA in clinical specimens. Methods Mol Biol. 2014;1085:71–83. https://doi.org/10.1007/978-1-62703-664-1_3.

17. Chow SK, Arbefeville S. Multicenter performance evaluation of the Simplexa Bordetella Direct kit in nasopharyngeal swab specimens. J Clin Microbiol. 2021;59:e01041–20. https://doi.org/10.1128/JCM.01041-20.

18. Pichon M, Pichard B. Diagnostic accuracy of a non-invasive test for detection of Helicobacter pylori and resistance to clarithromycin in stool by the Amplidiag H. pylori+ClariR real-time PCR assay. J Clin Microbiol. 2020;58:e01787–19. https://doi.org/10.1128/JCM.01787-19.

19. Walzl G, McNerney R, du Plessis N, et al. Tuberculosis: advances and challenges in development of new diagnostics and biomarkers. Lancet Infect Dis. 2018;18(7):e199–210. https://doi.org/10.1016/S1473-3099(18)30111-7.

20. Tiberi S, du Plessis N, Walzl G, et al. Tuberculosis: progress and advances in development of new drugs, treatment regimens, and host-directed therapies [published correction appears in Lancet Infect Dis. 2018 Apr 27]. Lancet Infect Dis. 2018;18(7):e183–98. https://doi.org/10.1016/S1473-3099(18)30110-5.

21. Voss G, Casimiro D, Neyrolles O, et al. Progress and challenges in TB vaccine development. F1000Res. 2018;7:199. https://doi.org/10.12688/f1000research.13588.1. Published 2018 Feb 16

22. Krishnan N, Malaga W, Constant P, Caws M, Thi Hoang Chau T, Salmons J, et al. Mycobacterium tuberculosis lineage influences innate immune response and virulence and is associated with distinct cell envelope lipid profiles. PLoS One. 2011;6(9):e23870. https://doi.org/10.1371/journal.pone.0023870.

23. Alcaïs A, Abel L. Application of genetic epidemiology to dissecting host susceptibility/resistance to infection illustrated with the study of common mycobacterial infection. 2004 In Bellam, & Cambridge University Press., Susceptibility to Infectious Diseases: The Importance of Host Genetics (p. 7–44). Cambridge: Bellamy R. https://doi.org/10.1017/CBO9780511546235.002

24. Al-Muhsen S, Casanova JL. The genetic heterogeneity of mendelian susceptibility to mycobacterial diseases. J Allergy Clin Immunol. 2008;122(6):1043–53. https://doi.org/10.1016/j.jaci.2008.10.037.

25. Aravindan PP. Host genetics and tuberculosis: Theory of genetic polymorphism and tuberculosis. Lung India. 2019;36(3):244–52. https://doi.org/10.4103/lungindia.lungindia_146_15.

26. Krishnan S, Queiroz ATL, Gupta A, Gupte N, Bisson GP, Kumwenda J, Naidoo K, Mohapi L, Mave V, Mngqibisa R, Lama JR, Hosseinipour MC, Andrade

BB, Karakousis PC. Integrative multi-omics reveals serum markers of tuberculosis in advanced HIV. Front Immunol. 2021;12:676980. https://doi.org/10.3389/fimmu.2021.676980.

27. Shah NS, Auld SC, Brust JCM, Mathema B, Ismail N, Moodley P, et al. Transmission of extensively drug-resistant tuberculosis in South Africa. N Engl J Med. 2017 Jan 19;376(3):243–53. https://doi.org/10.1056/NEJMoa1604544.

28. Gausi K, Wiesner L. Pharmacokinetics and drug-drug interactions of isoniazid and efavirenz in pregnant women living with HIV in high TB incidence settings: importance of genotyping. Clin Pharmacol Ther. 2021;109(4):1034–44. https://doi.org/10.1002/cpt.2044.

29. Glaziou PSC. Global epidemiology of tuberculosis. Cold Spring Harb Perspect Med. 2014;5(2):a017798. https://doi.org/10.1101/cshperspect.a017798.

30. Fauci AS, Morens DM. Zika virus in the Americas — yet another arbovirus threat. N Engl J Med. 2016;374:601–4. https://doi.org/10.1056/NEJMp1600297.

31. Pardy RD, Richer MJ. Protective to a T: the role of T cells during Zika Virus infection. Cells. 2019;8(8):820. https://doi.org/10.3390/cells8080820. Published 2019 Aug 3

32. Manabe YC, Andrade BB, Gupte N, et al. A parsimonious host inflammatory biomarker signature predicts incident tuberculosis and mortality in advanced human immunodeficiency virus. Clin Infect Dis. 2020;71(10):2645–54. https://doi.org/10.1093/cid/ciz1147.

33. Du Bruyn E, Fukutani FK. Inflammatory profile of patients with tuberculosis with or without HIV-1 co-infection: a prospective cohort study and immunological network analysis. Lancet Microbe. 2021;2(8):375–E385. https://doi.org/10.1016/S2666-5247(21)00037-9.

34. Deschamps M, Laval G, Fagny M, Itan Y, Abel L, Casanova JL, et al. Genomic signatures of selective pressures and introgression from archaic hominins at human innate immunity genes. Am J Hum Genet. 2016;98(1):5–21. https://doi.org/10.1016/j.ajhg.2015.11.014.

35. Golubeva VA, Nepomuceno TC, de Gregoriis G, Mesquita RD, Li X, Dash S, Garcez PP, Suarez-Kurtz G, Izumi V, Koomen J, Carvalho MA, Monteiro ANA. Network of interactions between ZIKA virus non-structural proteins and human host proteins. Cell. 2020;9(1):153. https://doi.org/10.3390/cells9010153.

36. German Advisory Committee Blood (Arbeitskreis Blut), Subgroup 'Assessment of Pathogens Transmissible by Blood'. Human Immunodeficiency Virus (HIV). Transfus Med Hemother. 2016;43(3):203–22. https://doi.org/10.1159/000445852.

37. Wichit S, Gumpangseth N, Hamel R, et al. Chikungunya and zika viruses: co-circulation and the interplay between viral proteins and host factors.

Pathogens. 2021;10(4):448. https://doi.org/10.3390/pathogens10040448. Published 2021 Apr 9

38. Ávila-Ríos S, Parkin N. Next-generation sequencing for HIV drug resistance testing: laboratory, clinical, and implementation considerations. Viruses. 2020;12(6):617. https://doi.org/10.3390/v12060617.

39. Joint United Nations Programme on HIV/AIDS (UNAIDS). The youth bulge and HIV; 2018. https://www.unaids.org/sites/default/files/media_asset/the-youth-bulge-and-hiv_en.pdf

40. Levy JA. HIV pathogenesis: 25 years of progress and persistent challenges. AIDS. 2009;23(2):147–60. https://doi.org/10.1097/QAD.0b013e3283217f9f.

41. Gingras SN, Tang D, Tuff J, McLaren PJ. Minding the gap in HIV host genetics: opportunities and challenges. Hum Genet. 2020;139(67):865–75. https://doi.org/10.1007/s00439-020-02177-9.

42. Goulder PJ, Walker BD. HIV and HLA class I: an evolving relationship. Immunity. 2012;37(3):426–40. https://doi.org/10.1016/j.immuni.2012.09.005.

43. Munung NS, Mayosi BM, de Vries J. Genomics research in Africa and its impact on global health: insights from African researchers. Glob Health Epidemiol Genom. 2018;3:e12. https://doi.org/10.1017/gheg.2018.3. Published 2018 Jun 8

44. Yin L, Hu S, Mei S, et al. CRISPR/Cas9 inhibits multiple steps of HIV-1 infection. Hum Gene Ther. 2018;29(11):1264–76. https://doi.org/10.1089/hum.2018.018.

45. Qi J, Ding C, Jiang X, Gao Y. Advances in developing CAR T-cell therapy for HIV cure. Front Immunol. 2020;11:361. https://doi.org/10.3389/fimmu.2020.00361. Published 2020 Mar 10

46. Longoni SS, Tiberti N. Monoclonal antibodies for protozoan infections: a future reality or a utopic idea? Front Med (Lausanne). 2021;8:745665. https://doi.org/10.3389/fmed.2021.74566.

47. Saracino MP, Vila CC. Searching for the one(s): using probiotics as anthelmintic treatments. Front Pharmacol. 2021;12:714198. https://doi.org/10.3389/fphar.2021.714198.

48. González R, Sevene E, Jagoe G, Slutsker L, Menéndez C. A public health paradox: the women most vulnerable to malaria are the least protected. PLoS Med. 2016;13(5):e1002014. https://doi.org/10.1371/journal.pmed.1002014. Published 2016 May 3

49. WHO Strategic Advisory Group on Malaria Eradication. Malaria eradication: benefits, future scenarios and feasibility. Geneva: World Health Organization; 2019. https://malariaworld.org/sites/default/files/WHOCDS-GMP-2019.10-eng.pdf

50. Feachem RGA, Chen I, Akbari O, et al. Malaria eradication within a generation: ambitious, achievable, and necessary. Lancet. 2019;394(10203):1056–112. https://doi.org/10.1016/S0140-6736(19)31139-0.

51. Murmu LK, S. A. Diagnosing the drug resistance signature in Plasmodium falciparum: a review from contemporary methods to novel approaches. J Parasit

Dis. 2021;45(3):869–76. https://doi.org/10.1007/s12639-020-01333-2.

52. Runtuwene LR, Tuda JSB. Nanopore sequencing of drug-resistance-associated genes in malaria parasites, Plasmodium falciparum. Sci Rep. 2018;8:8286. https://doi.org/10.1038/s41598-018-26334-3.

53. Rabinovich RN, Drakeley C, Djimde AA, Hall BF, Hay SI, Hemingway J, et al. malERA: an updated research agenda for malaria elimination and eradication. PLoS Med. 2017;14(11):e1002456. https://doi.org/10.1371/journal.pmed.100245.

54. O'Neill. Tackling drug-resistant infections globally: final report and recommendations. Government of the United Kingdom, 2016. https://apo.org.au/sites/default/files/resource-files/2016/05/apo-nid63983-1203771.pdf

55. Ndihokubwayo JB, et al. Antimicrobial resistance in the African Region: Issues, challenges and actions proposed. Brazzaville: WHO, Regional Office for Africa; 2013. http://apps.who.int/medicinedocs/documents/s22169en/s22169en.pdf

56. Akerlund A, Jonasson E. EUCAST rapid antimicrobial susceptibility testing (RAST) in blood cultures:validation in 55 European laboratories. J Antimicrob Chemother. 2020;75:3230–8. https://doi.org/10.1093/jac/dkaa333.

57. Van Belkum A, Bachmann T. Developmental roadmap for antimicrobial susceptibility testing systems. Nat Rev Microbiol. 2019;17:51–62. https://doi.org/10.1038/s41579-018-0098-9.

58. Ellington MJ. E. O. The role of whole genome sequencing in antimicrobial susceptibility testing of bacteria: report from the EUCAST subcommittee. Clin. Microbiol. Infect. 2017;23(1):2–22. https://doi.org/10.1016/j.cmi.2016.11.012.

59. Ladner JT, Grubaugh ND, Pybus OG, et al. Precision epidemiology for infectious disease control. Nat Med. 2019;25:206–11. https://doi.org/10.1038/s41591-019-0345-2.

60. Odlum M, Yoon S. What can we learn about the Ebola outbreak from tweets? Am J Infect Control. 2015;3(6):563–71. https://doi.org/10.1016/j.ajic.2015.02.023.

61. Wójcik OP, Brownstein JS. Public health for the people: participatory infectious diseases surveillance in the digital age. Emerg Themes Epidemiol. 2014;11:7. https://doi.org/10.1186/1742-7622-11-7.

62. Dion M, AbdelMalik P. Big data and the global public health intelligence network (GPHIN). Can Commun Dis Rep. 2015;41(9):209–14. https://doi.org/10.14745/ccdr.v41i09a02.

63. Susumpow P. Participatory disease detection through digital volunteerism: how the doctorme application aims to capture data for faster disease detection in Thailand. In: I. W. (IW3C2). 2014; pp. 663–666. https://doi.org/10.1145/2567948.2579273.

64. Bonanni PBG. Focusing on the implementation of 21st century vaccines for adults. Vaccine. 2018;36(36):5358–65. https://doi.org/10.1016/j.vaccine.2017.07.100.

65. Doherty T, Di Pasquale A. Precision medicine and vaccination of older adults: from reactive to proactive (a mini-review). Gerontology. 2020;66:238–48. https://doi.org/10.1159/000503141.

66. Teresa Aguado M, Barratt J. Report on WHO meeting on immunization in older adults: Geneva, Switzerland. Vaccine. 2017;37(7):921–31. https://doi.org/10.1016/j.vaccine.2017.12.029.

67. H3Africa Consortium, R. C. Research. capacity. Enabling the genomic revolution in Africa. Science. 2014;344(6190):1346–8. https://doi.org/10.1126/science.1251546.

68. Cani P. Human gut microbiome: hopes, threats and promises. Gut. 2018;67:1716–25. https://doi.org/10.1136/gutjnl-2018-316723.

69. Banerjee A, Kulcsar K, Misra V, Frieman M, Mossman K. Bats and coronaviruses. Viruses. 2019;11(1):41.

70. Carlos WG, Dela Cruz CS, Cao B, Pasnick S, Jamil S. Novel Wuhan (2019-nCoV) Coronavirus. Am J Respir Crit Care Med. 2020;201(4):P7–8.

71. Peiris J, Lai S, Poon L, Guan Y, Yam L, Lim W, et al. Coronavirus as a possible cause of severe acute respiratory syndrome. Lancet. 2003;361(9366):1319–25.

72. Ksiazek TG, Erdman D, Goldsmith CS, Zaki SR, Peret T, Emery S, et al. A novel coronavirus associated with severe acute respiratory syndrome. N Engl J Med. 2003;348(20):1953–66.

73. Schoeman D, Fielding BC. Coronavirus envelope protein: current knowledge. Virol J. 2019;16(1):69.

74. Wu F, Zhao S, Yu B, et al. A new coronavirus associated with human respiratory disease in China. Nature. 2020;579:265–69. https://doi.org/10.1038/s41586-020-2008-3.

75. Corman VM, Landt O, Kaiser M, Molenkamp R, Meijer A, Chu DK, et al. Detection of 2019 novel coronavirus (2019-nCoV) by real-time RT-PCR. Eurosurveillance. 2020;25(3):2000045.

76. Phelan AL, Katz R, Gostin LO. The novel coronavirus originating in Wuhan, China: challenges for global health governance. JAMA. 2020;323(8):709–10.

77. Gorbalenya AE. Severe acute respiratory syndrome-related coronavirus–The species and its viruses, a statement of the Coronavirus Study Group. BioRxiv. 2020.

78. Cui J, Li F, Shi Z-L. Origin and evolution of pathogenic coronaviruses. Nat Rev Microbiol. 2019;17(3):181–92.

79. Donnelly CA, Malik MR, Elkholy A, Cauchemez S, Van Kerkhove MD. Worldwide reduction in MERS cases and deaths since 2016. Emerg Infect Dis. 2019;25(9):1758.

80. Poon L, Guan Y, Nicholls J, Yuen K, Peiris J. The aetiology, origins, and diagnosis of severe acute respiratory syndrome. Lancet Infect Dis. 2004;4(11):663–71.

81. World. Estimating mortality from COVID-19. https://www.whoint/news-room/commentaries/detail/estimating-mortality-from-covid-19.

82. Chen J. Pathogenicity and transmissibility of 2019-nCoV—a quick overview and comparison with other emerging viruses. Microbes Infect. 2020; https://doi.org/10.1016/j.micinf.2020.01.004.

83. Apicella M, Campopiano MC, Mantuano M, Mazoni L, Coppelli A, Del Prato S. COVID-19 in people with diabetes: understanding the reasons for worse outcomes. Lancet Diabetes Endocrinol. 2020; https://doi.org/10.1016/S2213-8587(20)30238-2.

84. Zhou F, Yu T, Du R, Fan G, Liu Y, Liu Z, et al. Clinical course and risk factors for mortality of adult inpatients with COVID-19 in Wuhan, China: a retrospective cohort study. Lancet. 2020; https://doi.org/10.1016/S0140-6736(20)30566-3.

85. Joseph T, Ashkan M. E-book "International Pulmonologist's consensus on COVID-19" 2nd edition, April 2020, India, p 92 https://www.saudedafamilia.org/coronavirus/artigos/international_pulmonologists_consensus.pdf.

86. World. Coronavirus disease (COVID-19) pandemic. https://www.whoint/emergencies/diseases/novel-coronavirus-2019

87. Lescure F-X, Bouadma L, Nguyen D, Parisey M, Wicky P-H, Behillil S, et al. Clinical and virological data of the first cases of COVID-19 in Europe: a case series. Lancet Infect Dis. 2020; https://doi.org/10.1016/S1473-3099(20)30200-0.

88. Thompson R. Pandemic potential of 2019-nCoV. Lancet Infect Dis. 2020;20(3):280.

89. Weiss SR, Leibowitz JL. Coronavirus pathogenesis. In: Advances in virus research, vol. 81. Amsterdam: Elsevier; 2011. p. 85–164.

90. Li F. Structure, function, and evolution of coronavirus spike proteins. Annu Rev Virol. 2016;3:237–61.

91. Knoops K, Kikkert M, van den Worm SH, Zevenhoven-Dobbe JC, van der Meer Y, Koster AJ, et al. SARS-coronavirus replication is supported by a reticulovesicular network of modified endoplasmic reticulum. PLoS Biol. 2008;6(9):e226.

92. van Boheemen S, de Graaf M, Lauber C, Bestebroer TM, Raj VS, Zaki AM, et al. Genomic characterization of a newly discovered coronavirus associated with acute respiratory distress syndrome in humans. MBio. 2012;3(6):e00473–12.

93. Wu A, Peng Y, Huang B, Ding X, Wang X, Niu P, et al. Genome composition and divergence of the novel coronavirus (2019-nCoV) originating in China. Cell Host Microbe. 2020; https://doi.org/10.1016/j.chom.2020.02.001.

94. Hu D, Zhu C, Ai L, He T, Wang Y, Ye F, et al. Genomic characterization and infectivity of a novel SARS-like coronavirus in Chinese bats. Emerg Microbes Infect. 2018;7(1):1–10.

95. Elbe S, Buckland-Merrett G. Data, disease and diplomacy: GISAID's innovative contribution to global health. Global Chall. 2017;1(1):33–46.

96. Shu Y, McCauley J. GISAID: Global initiative on sharing all influenza data–from vision to reality. Eurosurveillance. 2017;22(13):30494.

97. Lu R, Zhao X, Li J, Niu P, Yang B, Wu H, et al. Genomic characterisation and epidemiology of 2019 novel coronavirus: implications for virus origins and receptor binding. Lancet. 2020;395(10224):565–74.

98. Paraskevis D, Kostaki EG, Magiorkinis G, Panayiotakopoulos G, Sourvinos G, Tsiodras S. Full-genome evolutionary analysis of the novel corona virus (2019-nCoV) rejects the hypothesis of emergence as a result of a recent recombination event. Infect Genet Evol. 2020;79:104212.

99. Malik YS, Sircar S, Bhat S, Sharun K, Dhama K, Dadar M, et al. Emerging novel Coronavirus (2019-nCoV)-Current scenario, evolutionary perspective based on genome analysis and recent developments. Veterinary quarterly 2020;40(1):68–76.

100. Ceraolo C, Giorgi FM. Genomic variance of the 2019-nCoV coronavirus. J Med Virol. 2020; https://doi.org/10.1002/jmv.25700.

101. Cao Y, Li L, Feng Z, Wan S, Huang P, Sun X, et al. Comparative genetic analysis of the novel coronavirus (2019-nCoV/SARS-CoV-2) receptor ACE2 in different populations. Cell Discov. 2020;6(1):1–4.

102. Zhao W, Song S, Chen M, Zou D, Ma L, Ma Y, et al. The 2019 novel coronavirus resource. Yi Chuan. 2020;42(2):212.

103. China. 2019 Novel Coronavirus Resource (2019nCoVR). https://bigdbigaccn/ncov/?lang=en.

104. Song S, Ma L, Zou D, Tian D, Li C, Zhu J, et al. The global landscape of SARS-CoV-2 genomes, variants, and haplotypes in 2019nCoVR. BioRxiv. 2020.

105. Dowd JB, Andriano L, Brazel DM, Rotondi V, Block P, Ding X, et al. Demographic science aids in understanding the spread and fatality rates of COVID-19. Proc Natl Acad Sci. 2020;117(18):9696–8.

106. COVID-19 Host Genetics Initiative. The COVID-19 Host Genetics Initiative, a global initiative to elucidate the role of host genetic factors in susceptibility and severity of the SARS-CoV-2 virus pandemic. Eur J Hum Genet. 2020;28(6):715–8.

107. Li Q, Wu J, Nie J, Zhang L, Hao H, Liu S, et al. The impact of mutations in SARS-CoV-2 spike on viral infectivity and antigenicity. Cell. 2020;182(5):1284–94.e9.

108. Forster P, Forster L, Renfrew C, Forster M. Phylogenetic network analysis of SARS-CoV-2 genomes. Proc Natl Acad Sci U S A. 2020;117(17):9241–3.

109. Guan Q, Sadykov M, Mfarrej S, Hala S, Naeem R, Nugmanova R, et al. A genetic barcode of SARS-CoV-2 for monitoring global distribution of different clades during the COVID-19 pandemic. Int J infect Dis. 2020;100:216–23.

110. Eden JS, Rockett R, Carter I, Rahman H, de Ligt J, Hadfield J, et al. An emergent clade of SARS-CoV-2 linked to returned travellers from Iran. Virus Evol. 2020;6(1):veaa 027.

111. Mercatelli D, Giorgi FM. Geographic and genomic distribution of SARS-CoV-2 mutations. Front Microbiol. 2020;11:1800.

112. Korber B, Fischer W, Gnanakaran SG, Yoon H, Theiler J, Abfalterer W, et al. Spike mutation pipe-

line reveals the emergence of a more transmissible form of SARS-CoV-2. bioRxiv. 2020.

113. Elena SF, Sanjuán R. Adaptive value of high mutation rates of RNA viruses: separating causes from consequences. J Virol. 2005;79(18):11555–8.

114. Tao K, Tzou PL, Nouhin J, Gupta RK, de Oliveira T, Kosakovsky Pond SL, et al. The biological and clinical significance of emerging SARS-CoV-2 variants. Nat Rev Genet. 2021;22(12):757–73.

115. who. Tracking SARS-CoV-2 variants. https://www.who.int/en/activities/tracking-SARS-CoV-2-variants/.

116. Initiative C-HG. Mapping the human genetic architecture of COVID-19. Nature. 2021;600(7889):472–7.

117. Velavan TP, Pallerla SR, Ruter J, Augustin Y, Kremsner PG, Krishna S, et al. Host genetic factors determining COVID-19 susceptibility and severity. EBioMedicine. 2021;72:103629.

118. Hoffmann M, Kleine-Weber H, Schroeder S, Krüger N, Herrler T, Erichsen S, et al. SARS-CoV-2 cell entry depends on ACE2 and TMPRSS2 and is blocked by a clinically proven protease inhibitor. Cell. 2020; https://doi.org/10.1016/j.cell.2020.02.052.

119. Lu H. Drug treatment options for the 2019-new coronavirus (2019-nCoV). Biosci Trends. 2020; https://doi.org/10.5582/bst.2020.01020.

120. Hammond E. Genetic engineering and biological weapons. New technologies, desires and threats from biological research. EMBO Rep. 2003;4(Suppl 1):S57–60.

121. Xu H, Zhong L, Deng J, Peng J, Dan H, Zeng X, et al. High expression of ACE2 receptor of 2019-nCoV on the epithelial cells of oral mucosa. Int J Oral Sci. 2020;12(1):1–5.

122. Li W, Zhang C, Sui J, Kuhn JH, Moore MJ, Luo S, et al. Receptor and viral determinants of SARS-coronavirus adaptation to human ACE2. EMBO J. 2005;24(8):1634–43.

123. Hussain M, Jabeen N, Raza F, Shabbir S, Baig AA, Amanullah A, et al. Structural variations in human ACE2 may influence its binding with SARS-CoV-2 spike protein. J Med Virol. 2020; https://doi.org/10.1002/jmv.25832.

124. Ellinghaus D, Degenhardt F, Bujanda L, Buti M, Albillos A, Invernizzi P, et al. Genomewide association study of severe Covid-19 with respiratory failure. N Engl J Med. 2020; https://doi.org/10.1056/NEJMoa2020283.

125. Daniloski Z, Jordan TX, Wessels H-H, Hoagland DA, Kasela S, Legut M, et al. Identification of required host factors for SARS-CoV-2 infection in human cells. Cell. 2020; https://doi.org/10.1016/j.cell.2020.10.030.

126. Chen Y, Liu QY, Guo DY. Emerging coronaviruses: Genome structure, replication, and pathogenesis. J Med Virol. 2020;92(4):418–23.

127. Fehr AR, Perlman S. Coronaviruses: an overview of their replication and pathogenesis. Coronaviruses. Cham: Springer; 2015. p. 1–23.

128. Ye Q, Wang B, Mao J. The pathogenesis and treatment of the 'Cytokine Storm' in COVID-19. J Infect. 2020; https://doi.org/10.1016/j.jinf.2020.03.037.

129. Al-Bari MAA. Targeting endosomal acidification by chloroquine analogs as a promising strategy for the treatment of emerging viral diseases. Pharmacol Res Perspect. 2017;5(1):e00293.

130. Li G, Hu R, Zhang X. Antihypertensive treatment with ACEI/ARB of patients with COVID-19 complicated by hypertension. Hypertens Res. 2020;43(6):588-590. https://doi.org/10.1038/s41440-020-0433-1.

131. National Institutes of Health (US); Biological Sciences Curriculum Study. NIH Curriculum Supplement Series [Internet]. Bethesda (MD): National Institutes of Health (US); 2007. Available from: https://www.ncbinlmnihgov/books/NBK20364/.

132. Fierz W. Challenge of personalized health care: to what extent is medicine already individualized and what are the future trends? Med Sci Monit. 2004;10(5):Ra111–23.

133. National Institutes of Health (US); Biological Sciences Curriculum Study. NIH Curriculum Supplement Series [Internet]. Bethesda (MD): National Institutes of Health (US); 2007. Understanding Human Genetic Variation. Available from: https://www.ncbinlmnihgov/books/NBK20363/.

134. Mini E, Nobili S. Pharmacogenetics: implementing personalized medicine. Clin Cases Mine Bone Metab. 2009;6(1):17–24.

135. Agyeman AA, Ofori-Asenso R. Perspective: does personalized medicine hold the future for medicine? J Pharm Bioallied Sci. 2015;7(3):239–44.

136. Ramaswami R, Bayer R, Galea S. Precision medicine from a public health perspective. Annu Rev Public Health. 2018;39:153–68.

137. Channappanavar R, Perlman S, editors. Pathogenic human coronavirus infections: causes and consequences of cytokine storm and immunopathology. Seminars in immunopathology. Cham: Springer; 2017.

138. Prompetchara E, Ketloy C, Palaga T. Immune responses in COVID-19 and potential vaccines: Lessons learned from SARS and MERS epidemic. Asian Pacific J allergy Immunol. 2020;38(1):1–9. https://doi.org/10.12932/AP-200220-0772.

139. Cheng PK, Wong DA, Tong LK, Ip S-M, Lo AC, Lau C-S, et al. Viral shedding patterns of coronavirus in patients with probable severe acute respiratory syndrome. Lancet. 2004;363(9422):1699–700.

140. Valenzuela-Sánchez F, Valenzuela-Méndez B, Rodríguez-Gutiérrez J, Rello J. Personalized medicine in severe influenza. Eur J Clin Microbiol Infect Dis. 2016;35(6):893–7.

141. Meybodi HRA, Hasanzad M, Larijani B. Path to personalized medicine for type 2 diabetes mellitus: reality and hope. Acta Med Iran. 2017;55:166–74.

142. Hasanzad M, Sarhangi N, Meybodi HRA, Nikfar S, Khatami F, Larijani B. Precision medicine in non communicable diseases. Int J Mol Cell Med. 2019;8(Suppl1):1.

143. DeMerle K, Angus DC, Seymour CW. Precision medicine for COVID-19: phenotype anarchy or promise realized? JAMA. 2021;325(20):2041–2.

144. Lauring AS, Hodcroft EB. Genetic variants of SARS-CoV-2-what do they mean? JAMA. 2021;325(6):529–31.

145. Challen R, Brooks-Pollock E, Read JM, Dyson L, Tsaneva-Atanasova K, Danon L. Risk of mortality in patients infected with SARS-CoV-2 variant of concern 202012/1: matched cohort study. BMJ (Clinical research ed). 2021;372:n579.

146. Medzhitov R, Schneider DS, Soares MP. Disease tolerance as a defense strategy. Science (New York, NY). 2012;335(6071):936–41.

147. Yang X, Yu Y, Xu J, Shu H, Liu H, Wu Y, et al. Clinical course and outcomes of critically ill patients with SARS-CoV-2 pneumonia in Wuhan, China: a single-centered, retrospective, observational study. Lancet Respir Med. 2020; https://doi.org/10.1016/S2213-2600(20)30079-5.

148. Whirl-Carrillo M, McDonagh EM, Hebert J, Gong L, Sangkuhl K, Thorn C, et al. Pharmacogenomics knowledge for personalized medicine. Clin Pharmacol Ther. 2012;92(4):414–7.

149. Takahashi T, Luzum JA, Nicol MR, Jacobson PA. Pharmacogenomics of COVID-19 therapies. NPJ Genom Med. 2020;5(1):35.

150. Fricke-Galindo I, Falfán-Valencia R. Pharmacogenetics Approach for the Improvement of COVID-19 Treatment. Viruses. 2021;13(3):413.

151. Savarino A, Boelaert JR, Cassone A, Majori G, Cauda R. Effects of chloroquine on viral infections: an old drug against today's diseases. Lancet Infect Dis. 2003;3(11):722–7.

152. Devaux CA, Rolain J-M, Colson P, Raoult D. New insights on the antiviral effects of chloroquine against coronavirus: what to expect for COVID-19? Int J Antimicrob Agents. 2020;55:105938.

153. Sanders JM, Monogue ML, Jodlowski TZ, Cutrell JB. Pharmacologic treatments for coronavirus disease 2019 (COVID-19): a review. JAMA. 2020; https://doi.org/10.1001/jama.2020.6019.

154. Chan KS, Lai ST, Chu CM, Tsui E, Tam CY, Wong MM, et al. Treatment of severe acute respiratory syndrome with lopinavir/ritonavir: a multicentre retrospective matched cohort study. Hong Kong Med J. 2003;9(6):399–406.

155. Lee JY, Vinayagamoorthy N, Han K, Kwok SK, Ju JH, Park KS, et al. Association of polymorphisms of cytochrome P450 2D6 with blood hydroxychloroquine levels in patients with systemic lupus ery-

thematosus. Arthritis Rheumatol (Hoboken, NJ). 2016;68(1):184–90.

156. Wahie S, Daly AK, Cordell HJ, Goodfield MJ, Jones SK, Lovell CR, et al. Clinical and pharmacogenetic influences on response to hydroxychloroquine in discoid lupus erythematosus: a retrospective cohort study. J Invest Dermatol. 2011;131(10):1981–6.

157. Cappellini MD, Fiorelli G. Glucose-6-phosphate dehydrogenase deficiency. Lancet. 2008;371(9606):64–74.

158. Mason PJ, Bautista JM, Gilsanz F. G6PD deficiency: the genotype-phenotype association. Blood Rev. 2007;21(5):267–83.

159. Beutler E. G6PD deficiency. Blood. 1994;84(11):3613–36.

160. Beutler E, Vulliamy TJ. Hematologically important mutations: glucose-6-phosphate dehydrogenase. Blood Cell Mol Dis. 2002;28(2):93–103.

161. Vulliamy T, Beutler E, Luzzatto L. Variants of glucose-6-phosphate dehydrogenase are due to missense mutations spread throughout the coding region of the gene. Hum Mutat. 1993;2(3):159–67.

162. Paterson AD. HbA1c for type 2 diabetes diagnosis in Africans and African Americans: personalized medicine NOW! PLoS Med. 2017;14(9):e1002384.

163. Vulliamy TJ, D'Urso M, Battistuzzi G, Estrada M, Foulkes NS, Martini G, et al. Diverse point mutations in the human glucose-6-phosphate dehydrogenase gene cause enzyme deficiency and mild or severe hemolytic anemia. Proc Natl Acad Sci U S A. 1988;85(14):5171–5.

164. Mitcheson JS. hERG potassium channels and the structural basis of drug-induced arrhythmias. Chem Res Toxicol. 2008;21(5):1005–10.

165. Kannankeril PJ, Roden DM. Drug-induced long QT and torsade de pointes: recent advances. Curr Opin Cardiol. 2007;22(1):39–43.

166. Sanguinetti MC, Tristani-Firouzi M. hERG potassium channels and cardiac arrhythmia. Nature. 2006;440(7083):463–9.

167. Pearlstein R, Vaz R, Rampe D. Understanding the structure-activity relationship of the human ether-a-go-go-related gene cardiac K+ channel. A model for bad behavior. J Med Chem. 2003;46(11):2017–22.

168. Irwin. Some aspects of the history of biometric method in the twentieth century with a special reference to Prof. Linder's work. Experientia Suppl. 1976;22:11–20.

169. de Wit E, Feldmann F, Cronin J, Jordan R, Okumura A, Thomas T, et al. Prophylactic and therapeutic remdesivir (GS-5734) treatment in the rhesus macaque model of MERS-CoV infection. Proc Natl Acad Sci U S A. 2020;117(12):6771–6.

170. Arabi YM, Asiri AY, Assiri AM, Jokhdar HAA, Alothman A, Balkhy HH, et al. Treatment of Middle

East respiratory syndrome with a combination of lopinavir/ritonavir and interferon-β1b (MIRACLE trial): statistical analysis plan for a recursive two-stage group sequential randomized controlled trial. Trials. 2020;21(1):1–8.

171. Cao B, Wang Y, Wen D, Liu W, Wang J, Fan G, et al. A trial of lopinavir–ritonavir in adults hospitalized with severe Covid-19. N Engl J Med. 2020; https://doi.org/10.1056/NEJMoa2001282.

172. Ehmann F, Caneva L, Papaluca M. European Medicines Agency initiatives and perspectives on pharmacogenomics. Br J Clin Pharmacol. 2014;77(4):612–7.

173. Tarkiainen EK, Backman JT, Neuvonen M, Neuvonen PJ, Schwab M, Niemi M. Carboxylesterase 1 poly-morphism impairs oseltamivir bioactivation in humans. Clin Pharmacol Ther. 2012;92(1):68–71.

174. Whirl-Carrillo M, Huddart R, Gong L, Sangkuhl K, Thorn CF, Whaley R, et al. An evidence-based framework for evaluating pharmacogenomics knowledge for personalized medicine. Clin Pharmacol Ther. 2021;110(3):563–72.

175. Furuta Y, Komeno T, Nakamura T. Favipiravir (T-705), a broad spectrum inhibitor of viral RNA polymerase. Proc Jpn Acad Ser B. 2017;93(7):449–63.